The SCIENCE *of* PLANTS

INSIDE THEIR SECRET WORLD

DK

SMITHSONIAN

The SCIENCE *of* PLANTS

INSIDE THEIR SECRET WORLD

IN ASSOCIATION WITH Royal Botanic Gardens Kew

Penguin Random House

DK LONDON

Senior Editor Helen Fewster
US Editors Megan Douglass, Karyn Gerhard
Project Editors Nathan Joyce, Anna Limmerick, Simon Maughan, Steve Setford, David Summers, Angela Wilkes, Miezan van Zyl
Editorial Assistant Daniel Byrne
DK Media Archive Romaine Werblow
Production Editor Jacqueline Street-Elkayam
Managing Editor Angeles Gavira Guerrero
Associate Publishing Director Liz Wheeler
Publishing Director Jonathan Metcalf

Senior Art Editor Sharon Spencer
Project Art Editors Ray Bryant, Phil Gamble, Vanessa Hamilton
Jacket Designer Akiko Kato
Jacket Design Development Manager Sophia MTT
Senior Production Controller Meskerem Berhane
Managing Art Editor Michael Duffy
Art Director Karen Self
Design Director Phil Ormerod

Photographer Gary Ombler
Illustrators Dominic Clifford, Alex Lloyd

DK DELHI

Senior Picture Researcher Surya Sankash Sarangi
Senior Jackets Coordinator Priyanka Sharma-Saddi
Managing Jackets Editor Saloni Singh

Picture Research Manager Taiyaba Khatoon
Senior Jacket Designer Suhita Dharamjit
DTP Designer Rakesh Kumar
Senior DTP Designer Jagtar Singh
Production Manager Pankaj Sharma

Content previously published as *Flora* in 2018

This American Edition, 2022
First American Edition, 2018
Published in the United States by DK Publishing
1450 Broadway, Suite 801, New York, NY 10018

22 23 24 25 26 10 9 8 7 6 5 4 3 2 1
001–326348–May/2022

A catalog record for this book is available from the Library of Congress.
ISBN 978-0-7440-4843-8

Printed and bound in China

For the curious

www.dk.com

Contributors

Jamie Ambrose was an author, editor, and Fulbright scholar with a special interest in natural history. Her books include DK's *Wildlife of the World*.

Dr. Ross Bayton is a botanist, taxonomist, and gardener with a passion for the plant world. He has authored books, magazine features, and scientific papers encouraging readers to understand and appreciate the importance of plants.

Matt Candeias is the author and host of the *In Defense of Plants* blog and podcast (www.indefenseofplants.com). Trained as an ecologist, Matt focuses most of his research on plant conservation. He is also an avid gardener both indoors and out.

Dr. Sarah Jose is a professional science writer and language editor with a PhD in botany and a deep love of plants.

Andrew Mikolajski is the author of more than 30 books on plants and gardening. He has lectured in garden history at the English Gardening School at Chelsea Physic Garden and for the Historic Houses Association.

Esther Ripley is a former managing editor who writes on a range of cultural subjects, including art and literature.

David Summers is a writer and editor with training in natural history filmmaking. He has contributed to books on a range of subjects including natural history, geography, and science.

Smithsonian

Established in 1846, the Smithsonian—the world's largest museum and research complex—includes 19 museums and galleries and the National Zoological Park. The Smithsonian is a renowned research center, dedicated to public education, national service, and scholarship in the arts, sciences, and natural history. Smithsonian Gardens, an accredited museum and Public Garden, engages people with plants and gardens, informs on the roles both play in our cultural and natural worlds, and inspires appreciation and stewardship of living and archival collections and horticultural artifacts.

The Royal Botanic Gardens, Kew

The Royal Botanic Gardens, Kew is a world famous scientific organization, internationally respected for its outstanding collections as well as its scientific expertise in plant diversity, conservation, and sustainable development around the world. Kew's 320 acres (132 hectares) of landscaped gardens, and Wakehurst—Kew's Wild Botanic Garden in Sussex—attract over 2.5 million visits every year. Kew was made a UNESCO World Heritage Site in July 2003 and celebrated its 260th anniversary in 2019. Wakehurst is home to Kew's Millennium Seed Bank, the largest wild plant seed bank in the world.

Half-title page Bird of paradise (*Strelitzia reginae*)
Title page Northern sea oat (*Chasmanthium latifolium*)
Above Autumn trees changing color
Contents page *Nigella papillosa* 'African Bride'

contents

the plant kingdom

roots

stems and branches

leaves

flowers

seeds and fruits

plant families

BLUE LOTUS (*NYMPHAEA CAERULEA*)

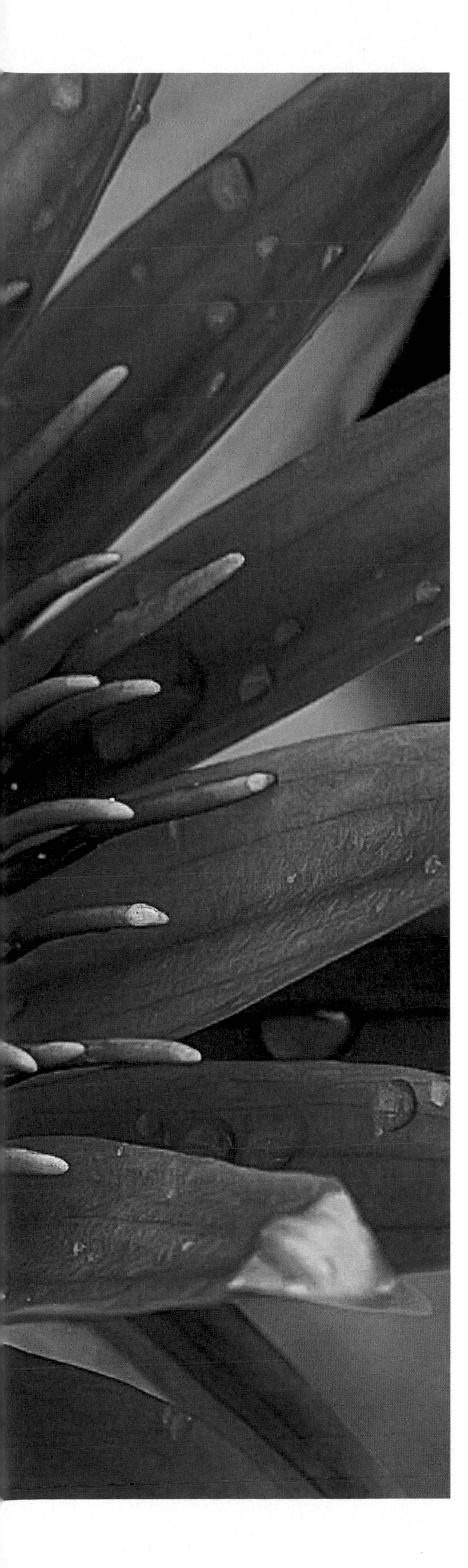

foreword

Plants are essential for life. They make our planet's atmosphere breathable, their decaying bodies create the soil under our feet, and they transform light energy into consumable nutrients that keep us alive. They also provide inspiration for artists: think of O'Keefe's poppies, Monet's water lilies, and van Gogh's sunflowers. When was the last time you examined the stunning intricacies of these essential life forms up close?

The Science of Plants brings together art and science to reveal botanical details in astounding and artful photographs. While previewing this book I was compelled to share the amazing images with coworkers. The otherworldly beauty of the magnified subjects made me feel like I had landed on Lilliput and happened upon old friends who had been supersized!

It also reminded me why I began my career in horticulture. At its most basic level, horticulture is the science and art of growing plants. This amalgamation of disciplines is what first attracted me to the profession and has held my interest for over forty years. This volume brought me back to Horticulture Studies 101, with each chapter examining a different plant part and how plants interact with the world around them. Thankfully, the distinctive photography that reveals fabulous details ranging from nectar bubbles to plant hairs and nettle stings far surpasses the old projected transparency sheets used by my professors.

Similar to this book, Smithsonian Gardens combines art and science in its many diverse garden and landscape exhibits. While their beauty is often what initially attracts visitors, the science they embody showcases our living collections and provides depths of engagement with our audiences. Our gardens are delightful examples of performative art. They change every season, indeed every day, of their existence. It is obvious that Smithsonian Gardens' staff horticulturists and gardeners are extraordinarily knowledgeable when it comes to plant science, yet so much of their work evolves into genuine artistry thanks to their natural skill in combining living collections.

Perhaps the sublime melding of art and science in *The Science of Plants* will launch or revitalize your own journey into the mesmerizing world of plants.

BARBARA W. FAUST
FORMER DIRECTOR, SMITHSONIAN GARDENS

the plant kingdom

plant. a living organism, usually containing chlorophyll—including trees, shrubs, herbs, grasses, ferns, and mosses—that typically grows in a permanent position, absorbing water and inorganic substances through its roots, and synthesizing nutrients in its leaves by photosynthesis.

WHAT IS NOT A PLANT?

Although they may look like plants, fungi are actually more closely related to animals. They cannot produce their own food—as plants do—and rely on the carbohydrates formed by normal plants. Many modern plants, notably forest trees and orchids, maintain some degree of dependence on fungi (see p.34). Algae is a general term used to denote a diverse range of organisms, including seaweeds, that do not have true roots, stems, or leaves, though they may appear green and leafy. Most of them live in water, and they are dominant in the sea. Lichens are composite organisms formed from algae (or certain bacteria) and fungi that have a mutually beneficial (symbiotic) relationship.

HAMMERED SHIELD LICHEN (*PARMELIA SULCATA*)

Orange fruiting body

JACK-O'-LANTERN MUSHROOM (*OMPHALOTUS ILLUDENS*)

SERRATED WRACK ALGAE (*FUCUS SERRATUS*)

what is a plant?

Plants are living organisms found virtually all over the planet, other than in permanently frozen or totally arid places. They range in size from majestic trees to tiny plants no larger than a grain of rice. Originally, all plants were aquatic and had roots simply to anchor them in place. Once they moved to land, many formed an association with fungi, which helped their roots obtain water and minerals. Plants differ from other organisms because they can make their own food by photosynthesis. Using chlorophyll—a green pigment in its cells—a plant absorbs energy from sunlight and uses carbon dioxide in the atmosphere to produce sugars. Unlike animals, which often stop growing once mature, plants may continue growing and producing new material every year, either to increase their size or to replace lost or damaged material.

Bright blooms
There are more than 350,000 species of flowering plants. Flowers are not just for show—they are the reproductive structures of a plant, and their shape and color are there to attract pollinators. These exotic orchid flowers are from a *Vanda* hybrid, many of which are native to tropical regions of Asia.

Opening flower bud
Flowering stems sometimes branch

MARCHANTIA POLYMORPHA

POLYTRICHUM SP.

ANTHOCEROS SP.

DIPHASIASTRUM DIGITATUM

types of plants

The smallest, simplest plants are known collectively as bryophytes and include liverworts, mosses, and hornworts. They often grow in places that are permanently damp, such as bogs or on the shady sides of rocks and tree trunks. Ferns are an ancient and diverse group of plants that have adapted to a wide range of habitats. They reproduce by means of spores. Gymnosperms produce cones, the females of which bear naked (not enclosed) seeds. Flowering plants are the most diverse and complex type of plants. They bear flowers, and the seeds that they produce are enclosed within fruits.

Evolving complexity
The first plants to grow on land were the ancestors of simple mosses, liverworts, and hornworts. Over the millennia, evolution has led to greater complexity, and angiosperms (flowering plants) now dominate the plant kingdom. From the fossil record it is clear that gymnosperms were once a much larger and more diverse group than they are today.

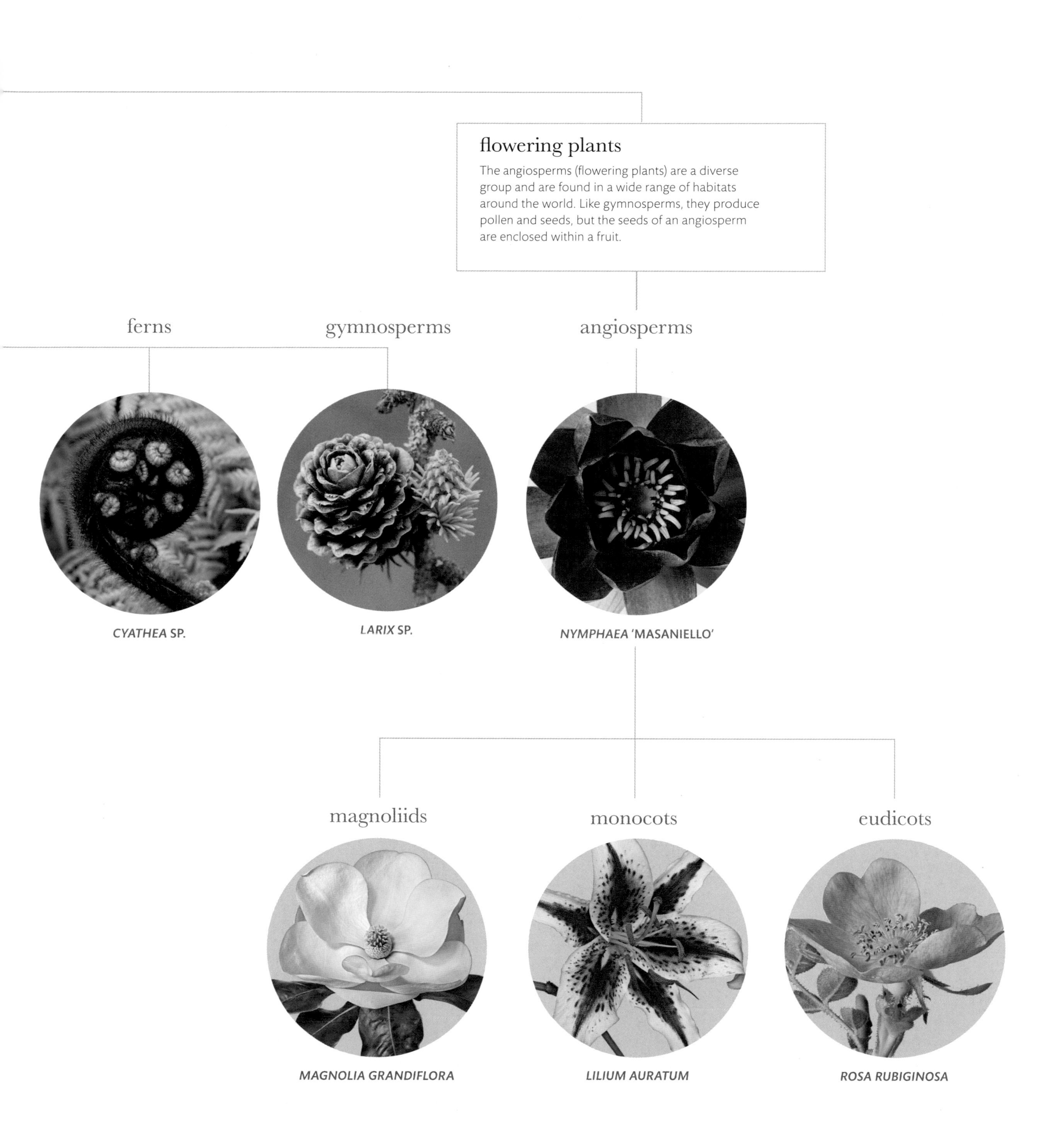
flowering plants
The angiosperms (flowering plants) are a diverse group and are found in a wide range of habitats around the world. Like gymnosperms, they produce pollen and seeds, but the seeds of an angiosperm are enclosed within a fruit.
ferns
gymnosperms
angiosperms
CYATHEA SP.
LARIX SP.
NYMPHAEA 'MASANIELLO'
magnoliids
monocots
eudicots
MAGNOLIA GRANDIFLORA
LILIUM AURATUM
ROSA RUBIGINOSA

classifying plants

Individual plants are formally named according to the system devised by Swedish botanist Carl Linnaeus (1707–1778), which used two-word names. In Latin and italics, a plant's name is made up of the genus to which it belongs, followed by the name of its particular species. Plants are grouped together according to the characteristics that they share. Historically, classification was determined by a plant's physical characteristics (notably flower structure) and biochemistry (the chemical compounds within a plant), and was often speculative or subjective. Nowadays, genetic evidence provides a more reliable means of understanding the relationships between plants.

A system of classification
This page of illustrations by Georg Dionysius Ehret, published in 1736, illustrates the Linnaean classification system, which differentiated species by looking at the sexual parts of the plants, notably the different numbers of male and female parts.

LOTUS FLOWER (*NELUMBO* SP.)

WATER LILY FLOWER (*NYMPHAEA* SP.)

PLANE TREE FLOWER (*PLATANUS* SP.)

Surprising relationships
DNA profiling has led to some unexpected discoveries. People often mistake the water lily (*Nymphaea*) for the lotus (*Nelumbo*), because they look alike. Genetic profiling has now shown, however, that the lotus is in fact more closely related to the plane tree (*Platanus*).

HIERARCHY OF TERMS

Botanists use the following ranks to arrange plants. Plants are organized into divisions, then classes, and so on, into increasingly specific groups according to the structure of their flowers, fruits, and other parts, as well as evidence from fossils and DNA profiling.

division
separates plants according to key features, e.g. angiosperms and gymnosperms

class
divides plants according to fundamental differences such as monocots and eudicots

order
groups families with a common ancestor

family
contains plants that are clearly related, e.g. the rose family

genus (genera)
a group of closely related species with similar features

species
plants with shared characteristics, which can sometimes interbreed

subspecies, forms, varieties
have features that differ from those typical of a species, or that are geographically distinct

cultivar
a cultivated variety of a plant that was produced from natural species or hybrids

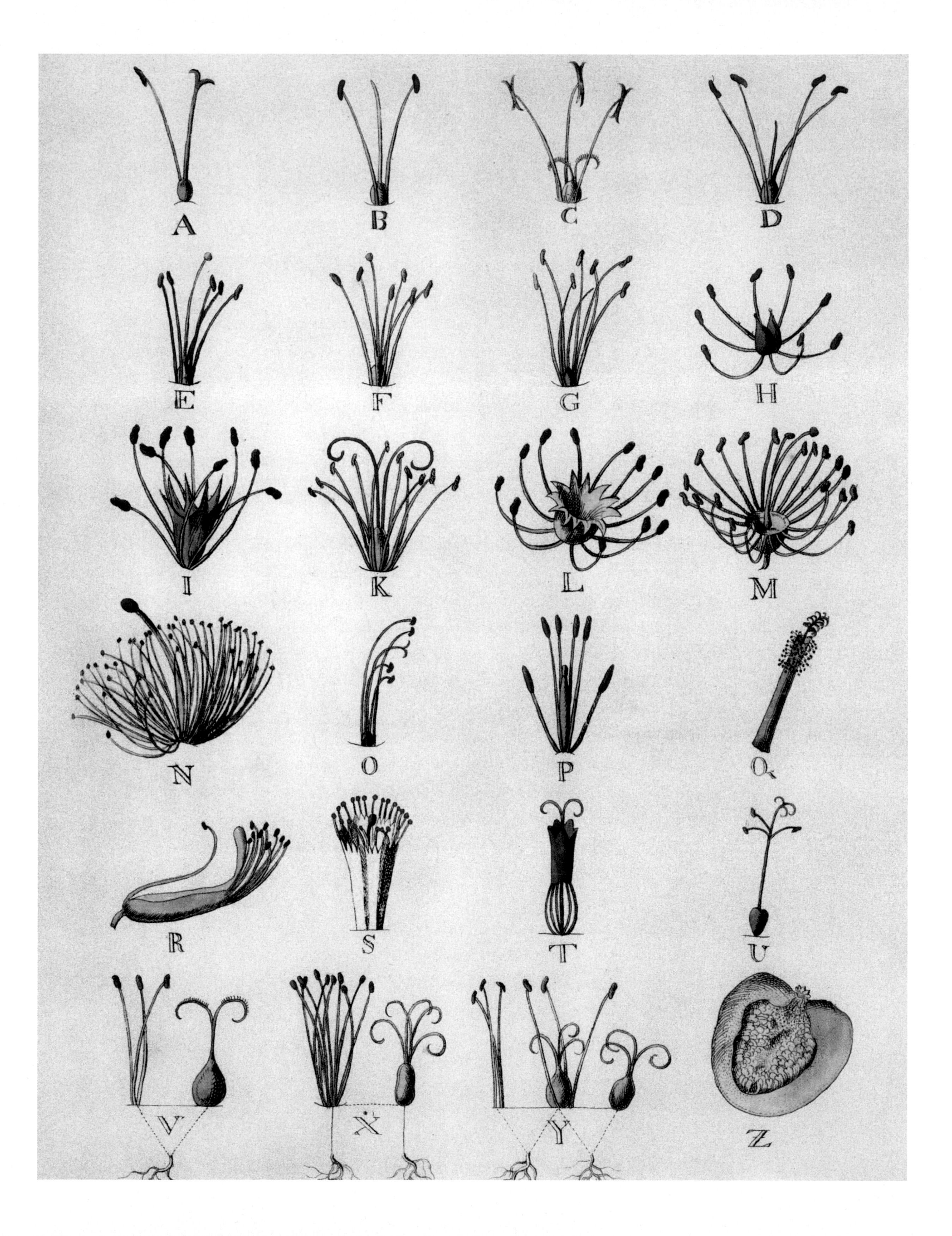
A
B
C
D
E
F
G
H
I
K
L
M
N
O
P
Q
R
S
T
U
V
X
Y
Z

roots

root. the part of a plant, usually underground, that anchors it in the earth and transports water and nutrients to the rest of the plant.

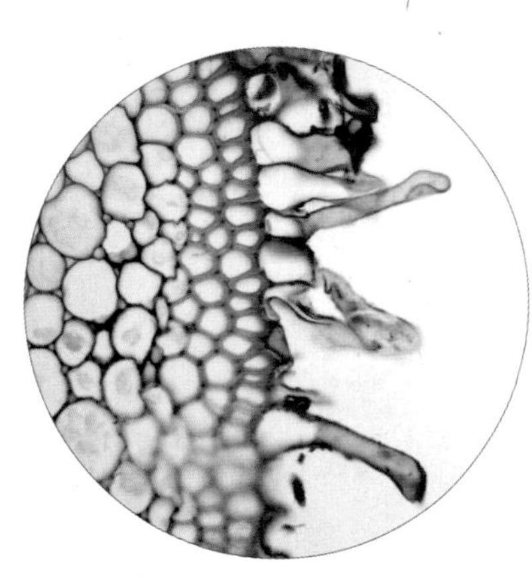

Root hairs
Fine root hairs, often only one cell wide, grow behind the tip of the root and move out into the soil to gather water and food for the plant. After a few days they fall off, but new ones grow out as the root extends.

fibrous roots

The root systems of flowering plants can be divided into two main types—fibrous or taprooted. Fibrous roots branch extensively, creating a vast, delicate network that spreads through the soil. This anchors a plant firmly in place and helps it find water and essential minerals from a wide area of ground.

Delicate, sprawling roots help prevent erosion by anchoring soil particles

Roots may penetrate deeply in search of moisture in dry soil

The underground network
Fibrous root systems connect plants to the vital resources spread throughout the soil. They are found in all ferns, and in many grasses and other flowering plants. Some trees produce a deep, thick taproot at first, then develop a fibrous root system as they age.

HOW ROOT HAIRS WORK

Root hairs wend their way between soil particles in order to absorb water and the dissolved minerals that it contains. Profuse, these hairs greatly increase the surface area of the roots and therefore the amount of water and nutrients available to the plant. Water and the minerals dissolved in it are absorbed into the cells by osmosis and transported via the cortex into the vascular system.

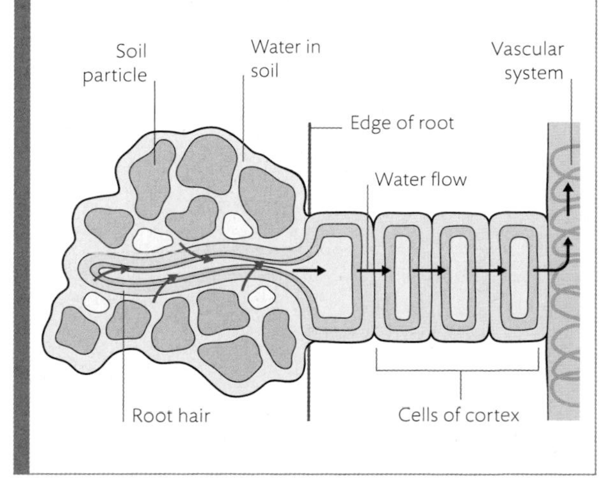

taproots

In contrast to fibrous root systems, taprooted plants normally develop a single dominant root with much smaller side roots. Some trees can develop either type—a taproot in loose soil, but a fibrous network in heavy soil. Most seedlings in the eudicot group of flowering plants (see p.15) start life taprooted, with the seed root—or radicle—dominating the root system. If the plant is not genuinely taprooted, this dies off, causing the roots to branch into a fibrous arrangement.

The taproot swells as it grows, storing food in the form of carbohydrates

Biennial taproots, like beets, grow in year one, then die in year two when the plant flowers and sets seed

The end of the taproot is long and slender so that it can penetrate deep into the soil

The deep red color comes from betalain pigments, which can be used as dyes or food coloring

Taproot food crops

Some plants, such as beets (*Beta vulgaris*), store carbohydrates as sugars within their taproots. These energy reserves are used to produce flowers and seeds (see pp.28–29). Taproot crops must be harvested while the roots are still rich in sugars, before the plant is able to bloom; after flowering, the roots become woody and unpalatable.

Fueled by sugars, the taproot can produce new leaves and stems to replace any harvested as salad crops

Thin side roots take up most of the water through minute root hairs

DEEP ROOTS

Taproots bring several advantages to their owners. Often growing deep into the soil, they can access water and nutrients beyond the reach of shallow-rooted plants. For weeds such as dandelions (*Taraxacum officinale*), a taproot also makes it difficult to remove the plant from the soil. More often than not, the leaves break off, leaving the intact root in place and ready to grow back. Many taproots can regrow from fragments left in the soil; even the smallest piece of dandelion root readily resprouts.

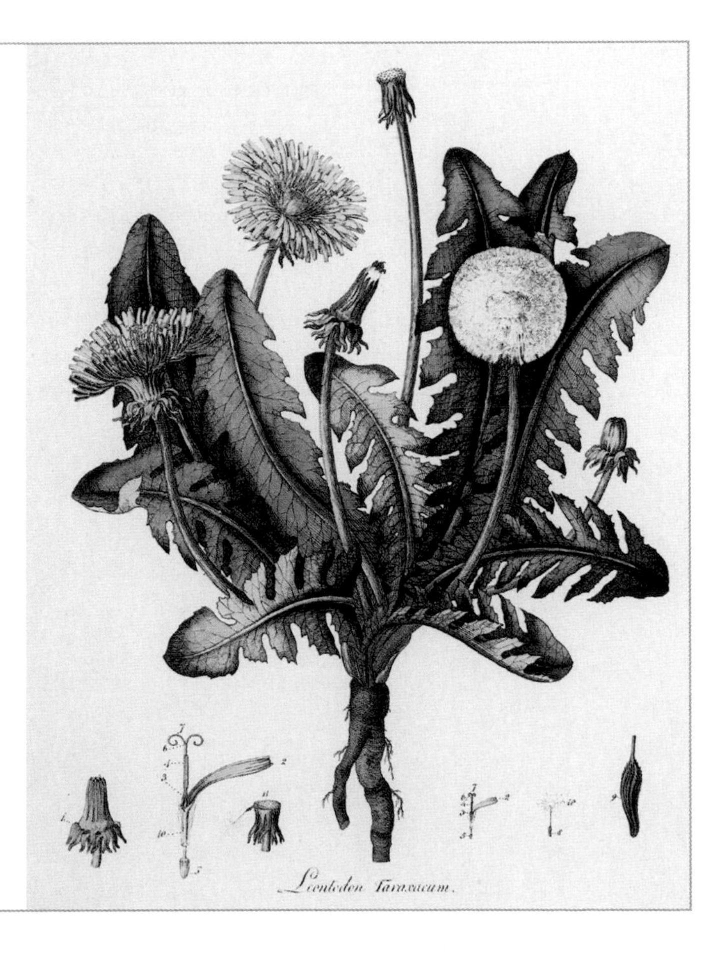

TARAXACUM OFFICINALE

plant supports

Anchorage is one of the major functions of any root system. For most plants, this is accomplished entirely underground, but where the soil is shallow—as in many rainforests—some species produce elaborate aboveground support systems. Coastal mangroves, growing in loose and unstable soil, do the same (see pp.54–55). Buttress, prop, and stilt roots all help to hold up the largest and heaviest of canopies, assuring a solid foundation.

SUPPORTIVE ROOT SYSTEMS

Buttress roots develop in shallow soil and are part of the primary root system. In contrast, prop and stilt roots develop from the main stem and branches above. Prop roots brace slender trunks and often appear in tiers. Stilt roots drop down from side branches.

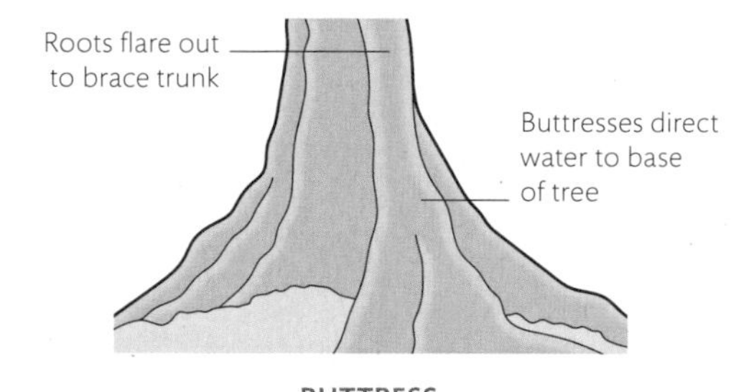

BUTTRESS

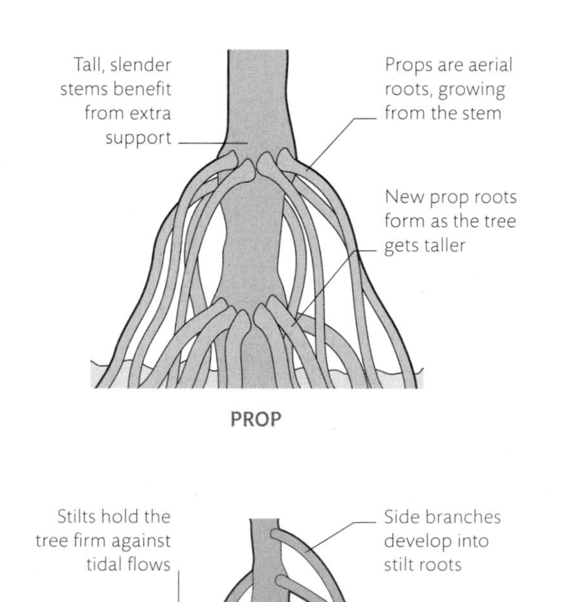

PROP

Stilts hold the tree firm against tidal flows

Side branches develop into stilt roots

STILT

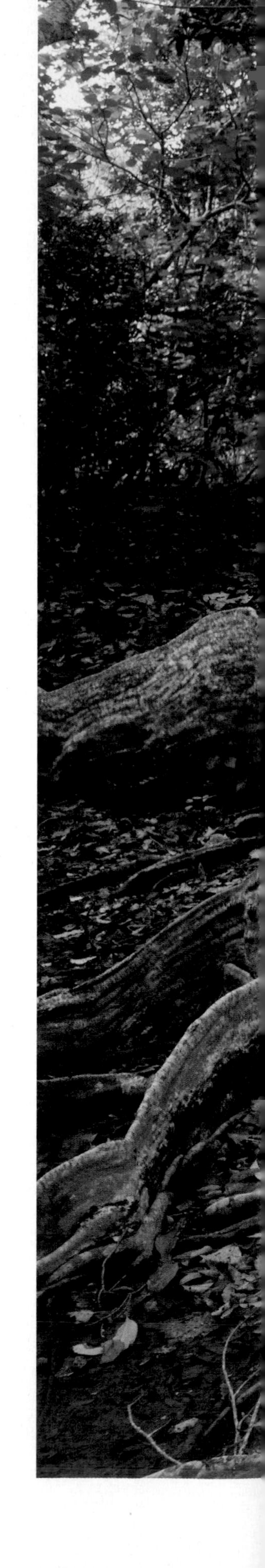

Buttress roots
Coastal trees such as the endangered looking-glass mangrove (*Heritiera littoralis*) need strong supports to protect against tidal flooding. Buttress roots interweave for greater strength and, if the tree's crown is larger on one side, more buttresses develop on the opposite side for stability.

Hydrangeas in acidic soil
Most hydrangeas have white flowers, but the color of mophead hydrangeas (*Hydrangea macrophylla*) is determined by the acidity of the soil. In alkaline soil above pH7, they are normally red to pink. In acidic soil below pH7, aluminum becomes soluble, and is absorbed by hydrangea roots. The aluminum ions bind to the red pigments in the blooms and turn the flowers blue.

Pink petals occur on hydrangeas in alkaline soil, where aluminum is unavailable

Healthy green leaves depend on the roots absorbing sufficient nutrients, such as magnesium and iron

A blue mophead develops when the hydrangea grows in acidic soil, where aluminum can be absorbed

absorbing nutrients

When roots absorb moisture from the soil, they also absorb vital minerals dissolved in the water that a plant needs in order to grow and thrive. The chemical structure of soil varies widely from one place to another, and deficiencies in the key elements that a plant needs can lead to stunted growth or discolored leaves. Leaves are, however, able to store surplus nutrients for later use, when there might be a shortage.

Iron deficiency
Plants need iron to manufacture important enzymes and pigments, including chlorophyll, the light-sensitive green pigment that they need for photosynthesis. If iron is in short supply, a plant produces less chlorophyll and its leaves turn yellow, as in this hydrangea. Iron can only be absorbed by the roots when dissolved in water, so in alkaline soil, where iron is less soluble, plants are more likely to suffer from an iron deficiency.

FOOD DISTRIBUTION

Nutrients absorbed from the soil are carried up from the roots to every part of the plant—stems, branches, leaves, and flowers—through bundles of miniscule pipes (xylem). The three most important minerals for a plant are nitrogen, phosphorus, and potassium. These are also the main components of many garden fertilizers.

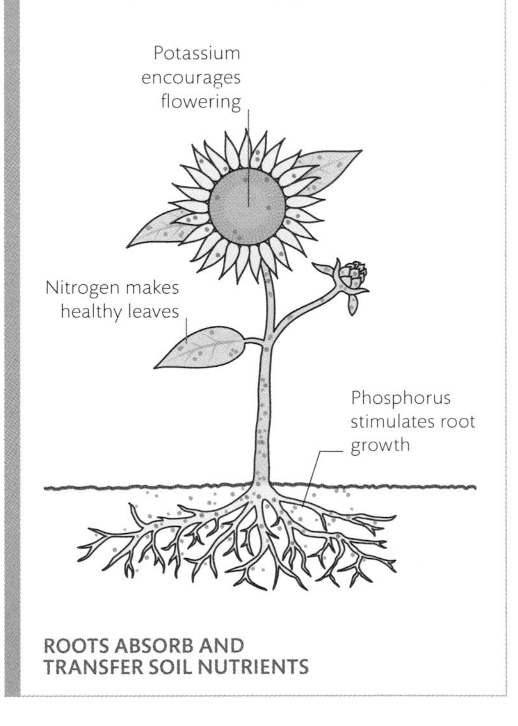

ROOTS ABSORB AND TRANSFER SOIL NUTRIENTS

Individual flowers and fruit are arranged in umbrellalike clusters (umbels), which curl inward when mature

Each individual flower produces two seeds

Carrot store

Most of the sugars stored within a taproot are used to produce flowers. Once a carrot plant has flowered, its taproot becomes bitter and inedible.

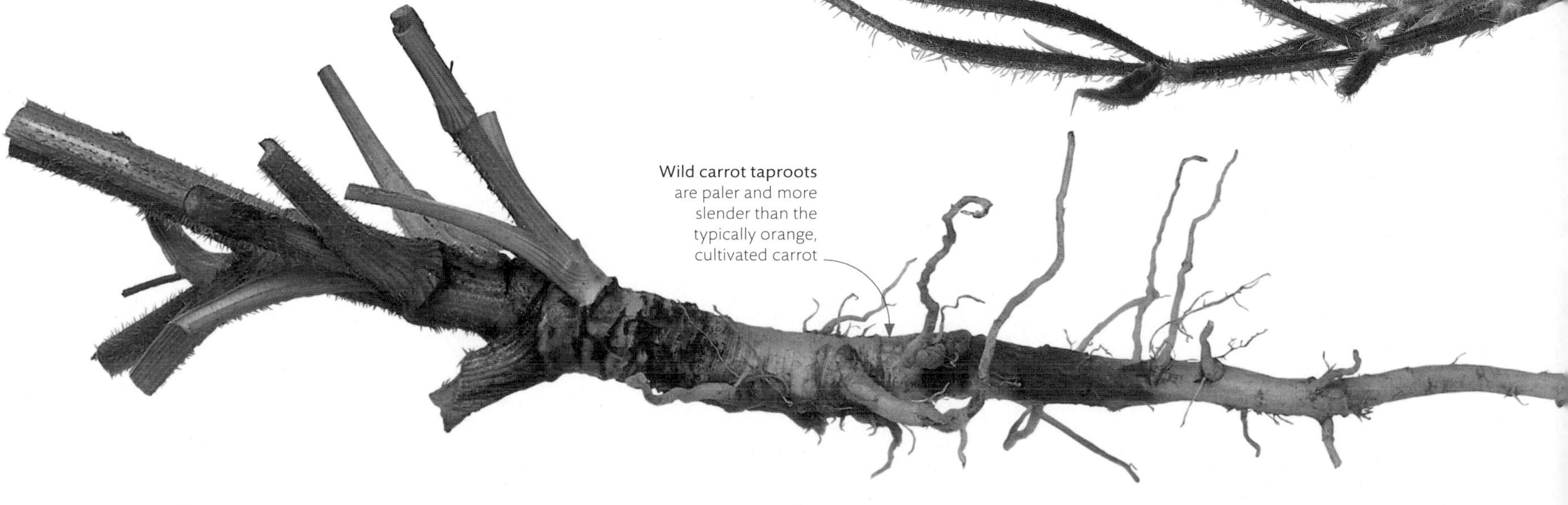

Wild carrot taproots are paler and more slender than the typically orange, cultivated carrot

Branching, hairy bracts extend under the carrot flower heads and may help to distribute seeds in the wind

UNDERGROUND ENERGY STORAGE

Carrots (*Daucus carota*) originate from Europe and Southwest Asia. The wild plants have slender, white roots, but years of cultivation by humans has resulted in the domesticated carrot, which has a much larger and more colorful root. Carrots live for two years. In the first year they develop a crown of leaves and a swelling, carbohydrate-packed root. These food reserves are used to make flowers in the second year.

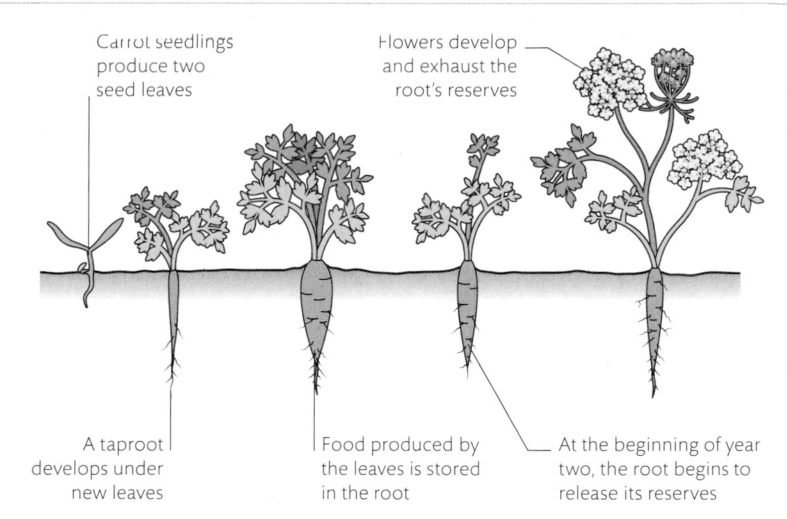

CARROT LIFE CYCLE

storage systems

Reproduction takes a lot of energy. Producing flowers, sugary nectar, and seeds requires a substantial investment from a plant. Biennials—plants that live for just two years—allow themselves a whole year in which to build up a reserve of carbohydrates before flowering in their second year of life. Once seeds are set, the exhausted plants die. Many biennial root crops store their reserves in taproots, but farmers interfere and harvest the nutrient-rich roots before the plants can use their resources to bloom.

The thickened flower stem transports carbohydrates away from the taproot

Food for seeds
Once the taproot has released the energy reserves it has stored and the carrot has flowered, seeds develop within the curled inflorescence. This can detach, scattering the seeds as it blows around in the wind. At this point, the taproot is shriveled and exhausted from the effort of producing hundreds of seeds.

Carrot taproots are thick, long, and usually unbranched

impressions of nature

Reacting against the formal constraints and rules of academic painting, the 19th-century French Impressionist painters immersed themselves in the natural world and worked at easels outdoors, so they could capture their fleeting impressions of the changing light on the landscape in a fresh and spontaneous way. The Postimpressionists who followed them took this approach a step further. Inspired by the geometric forms and dazzling colors they saw in the natural world, they created expressive, vibrant, semiabstract works bursting with energy.

Inspired by the Impressionists' fresh, new approach to art, Postimpressionists such as van Gogh and Cézanne adopted a simplified approach to painting, with little attempt at realism, and developed a visual language of their own. They emphasized the bold lines, geometric shapes, and luminous colors of the southern French landscape, paving the way for the 20th-century abstract artists to follow.

Vincent van Gogh painted from nature, but he used intense color and powerful brushstrokes to express his emotional response to it. In his search for a modern approach to art, he was greatly influenced by Japanese prints, with their bold, delineated shapes and flat expanses of color. His letters from the South of France simmer with joy. Believing he had found "his Japan," he wrote ecstatically about color and shape. "A meadow full of very yellow buttercups, a ditch with iris plants with green leaves … a strip of blue sky." He was preoccupied by the cypress trees and astonished that no one had painted them as he saw them.

***Tree Roots*, 1890**
At first look, this oil on canvas painting by Vincent van Gogh appears to be a jumble of bright colors and abstract forms. It is in fact a cropped study of gnarled tree roots, trunks, and boughs growing on a quarry slope. The colors are patently unreal. It was painted the morning before van Gogh died and is not finished, but it is astonishing for its vigor: it is full of sun and life.

> "I'll always love nature here, it's something like Japanese art, once you love that you don't have second thoughts about it."
>
> VINCENT VAN GOGH, *LETTER TO THEO VAN GOGH*, 1888

Japanese influence
Cypress Trees, a polychrome and gold-leaf screen, was painted c. 1590 by Kanō Eitoku, a major artist of the Kanō school of Japanese painting. The cropped composition; the swirling, expressive use of line; and the flat expanses of strong color are typical of the Japanese style of art that influenced van Gogh.

Clover's three-leaflet **foliage** is reflected in its Latin name, *Trifolium*

Clover stems and **foliage** provide useful protein for grazing animals

Enriching the soil
Red clover (*Trifolium pratense*) is grown by farmers as a cover crop, in order to protect bare soil from erosion and nutrient depletion during the winter. Because it is able to harness nitrogen, clover grows vigorously. In spring, it is plowed back into the soil, enriching the soil with nitrogen. These benefical properties make it ideal in crop-rotation systems.

Clover flowers are an important source of nectar for bees and other insects

Nitrogen-fixing legumes
Many different plants can "fix" atmospheric nitrogen into a form they can absorb, including sea buckthorn (*Hippophae*), California lilac (*Ceanothus*), and alder (*Alnus*). It is most common, however, in the legume family, which includes peas, beans, and trefoils, such as this *Trifolium subterraneum* (right). Legume nodules host several different sorts of bacteria, collectively known as rhizobia, that are essential for fixing nitrogen.

Fresh green leaves have a ready supply of nitrogen; lack of nitrogen can cause yellowing

nitrogen fixing

As a major component of proteins—one of the building blocks of life—nitrogen is an essential element for plants. While it is abundant in the air, it is largely unreactive and unusable, so plants make use of nitrogen-bearing compounds in the soil and absorb them through their roots. Some plants, however, are able to absorb atmospheric nitrogen in cooperation with bacteria and convert it into usable compounds. They are known as nitrogen-fixers.

ROOT NODULES

Plants are only able to fix nitrogen because of the symbiotic relationship between bacteria and the host plant. This takes place in specifically modified structures on the roots called nodules, which form when bacteria invade the cortex (outer layer) of growing root hairs. Within the nodules, bacteria produce an enzyme called nitrogenase. This converts gaseous nitrogen into soluble ammonia, a form in which the nitrogen can be used by the plant. In return, the bacteria are provided with food in the form of sugars.

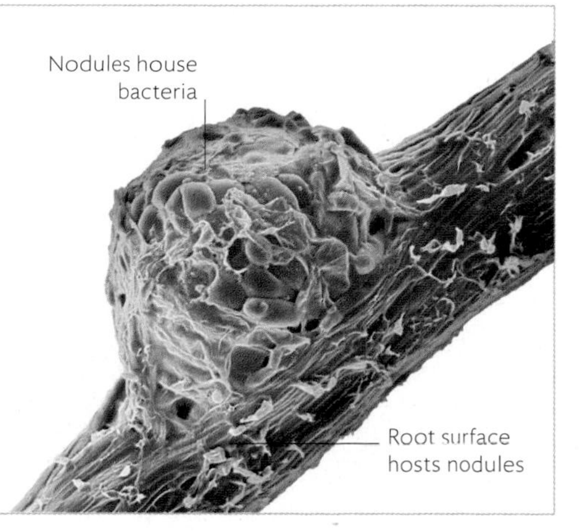

ROOT NODULE ON A PEA PLANT

Amanita muscaria

the fly agaric

It is impossible to talk about plants without mentioning fungi. Most plants have evolved a symbiotic relationship with certain types of fungi, known as mycorrhizae, which include the fly agaric. This relationship gives plants access to extra water and nutrients in exchange for carbohydrates.

The fly agaric fungus is native to much of the northern hemisphere and is spreading throughout the southern hemisphere as well. All that most people ever see of this fungus is the colorful fruiting body, or mushroom. Its umbrellalike cap is bright red to orange-yellow and usually studded with white warts. After releasing its spores, the mushroom rots away.

Through its main underground body, the mycelium, the fly agaric plays a vital role in the ecology of coniferous and deciduous woodlands. This sprawling mass of hairlike structures (hyphae), spread throughout the soil, can form symbiotic (beneficial) relationships with the roots of a huge variety of trees, including pines, spruces, cedars, and birches. Whereas some fungi penetrate tree root cells, the fly agaric forms a sheath that covers the roots. The sheath not only protects the roots from infectious microorganisms but also helps to transfer both nutrients and water to the roots. In return, the tree supports the fungus by supplying it with sugars produced by photosynthesis. The best places to look for fly agarics, therefore, are near the base of the trees with which it associates.

Because this fungus can thrive with so many different trees, it is now appearing in places where it is not native, possibly by hitching a ride on the roots of tree seedlings that are destined for plantations. Some experts fear that it may compete with and drive out important local species of mycorrhizal fungi.

Toxic toadstool
The fly agaric brews an impressive cocktail of chemicals that both help it break down soil nutrients and protect it from being eaten.

The tiny white warts are remnants of a veil of tissue that protected the spore-bearing cap as it emerged from underground

The skirtlike ring of tissue around the stem, together with the cap, identify the fly agaric

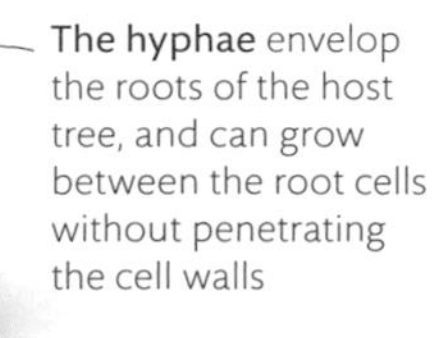

The hyphae envelop the roots of the host tree, and can grow between the root cells without penetrating the cell walls

Tip of the iceberg
A mushroom, like the one pictured here, is just a reproductive structure. The rest of the fungus lives underground and consists of countless hairlike structures called hyphae.

Heavy flower spikes rely on deep roots for support

Digging deep for flowers
Contractile roots anchor the bulb firmly within the soil, which helps to support the mass of flowers when they emerge above ground.

Bulbs also multiply underground, producing new plants that are clones of the parent

contractile roots

Roots provide anchorage for plants, keeping them in place through whatever the weather throws at them. But some plant species can actually shift their position in the earth using specifically adapted roots. Contractile roots pull a plant deeper into the soil by shrinking and extending, and they are commonly found in plants with bulbs, corms, or rhizomes (see p.87). Contractile roots also occur in a variety of other plants, including those with taproots. By plunging into the soil, contractile roots provide greater stability and ensure that maturing bulbs reach the right depth.

FINDING THE RIGHT DEPTH

Bulbous plants begin life as seedlings near the surface of the soil. However, if the developing bulb stays near the soil surface, then not only is it exposed to freezing temperatures and the drying rays of the sun, it is much more likely to be eaten by an animal. For its protection, contractile roots gradually pull the developing bulb down into the soil, where environmental conditions are more stable. Contractile roots widen before they extend, pushing the surrounding soil aside to form a channel through which the bulb is pulled.

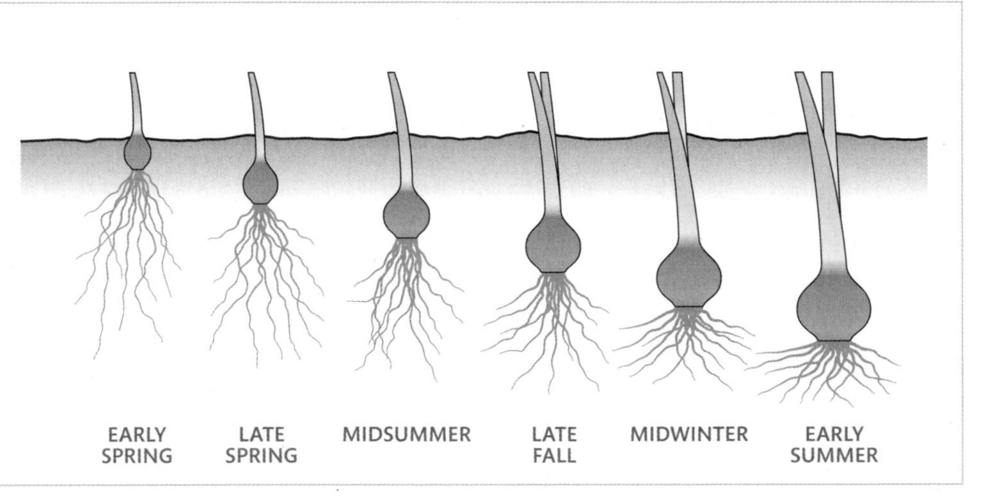

Surviving drought at depth
Hyacinths and many bulbous plants live in seasonally dry regions, where spring rainfall is followed by long summer droughts. Contractile roots pull bulbs deeper into the soil, where temperatures are cooler and they are less vulnerable to desiccation. Spring-flowering bulbs remain dry throughout the summer and their roots may shrivel entirely. When winter rains arrive, the roots regrow and the plant is ready to flower the following spring.

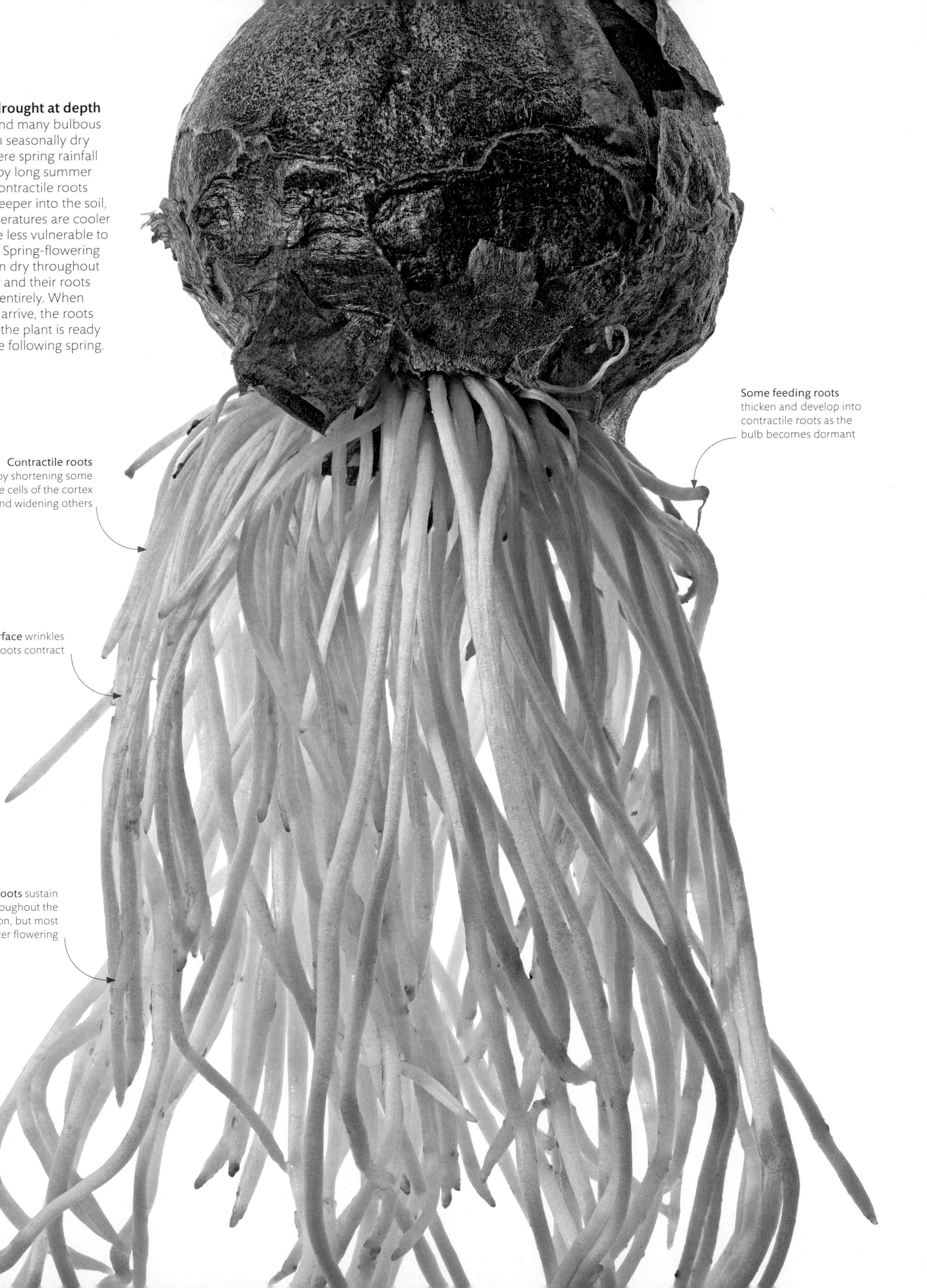

Some feeding roots thicken and develop into contractile roots as the bulb becomes dormant

Contractile roots work by shortening some of the cells of the cortex and widening others

The root surface wrinkles when the roots contract

Feeding roots sustain the plant throughout the growing season, but most die back after flowering

White berries attract birds, which distribute the seeds to other trees

aerial roots

Living high in the canopy, perched on the branches of trees, epiphytic plants need roots that hold them firmly in place. These aerial roots emerge along the stems and cling to any nearby surface. Unlike the roots of ground-dwelling terrestrial plants, which make direct contact with the soil, aerial roots can only absorb water from fog, mist, and rain. They are specially adapted to draw water from these sources. In some cases, aerial roots become green, and can manufacture food via photosynthesis (see p.129).

Leaves of epiphytes receive much more sunlight than those belonging to plants on the forest floor

Treetop traveler

The pearl laceleaf (*Anthurium scandens*) is an epiphytic plant that climbs along the branches of trees, sending out a profusion of aerial roots and greatly expanding its territory. The large mass of roots not only keeps the plant in place, but also increases the amount of water it can absorb.

Aerial roots appear white when dry, but turn green when moist

AERIAL ROOTS AND WATER

All roots are surrounded by a protective epidermis, but in some aerial roots, this is many layers thick. Known as velamen, it rapidly absorbs moisture and becomes transparent when wet, allowing any green cells under it to receive sunlight and photosynthesize. Velamen also protects these light-sensitive cells from harmful ultraviolet radiation.

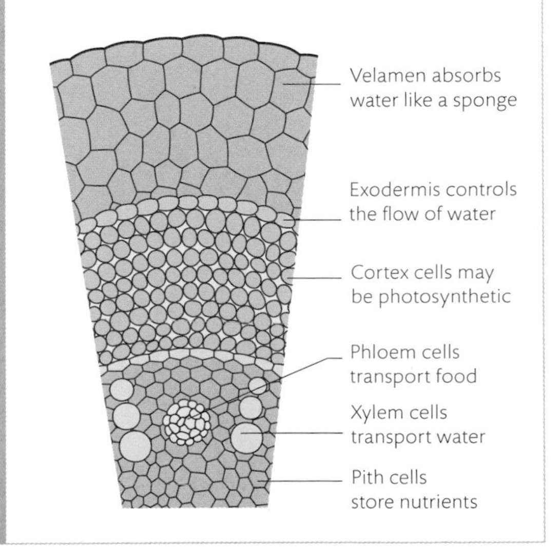

Ficus sp.

the strangler fig

Some fig trees have evolved a way of life that involves growing on other trees and eventually throttling them. Many different *Ficus* species exhibit the strangler lifestyle. As unsavory as this may sound, strangler figs are nonetheless vital components of tropical forests.

A strangler fig starts life as a tiny seed deposited on a tree branch by an animal. After germinating, the seedling's roots dig into whatever deposits have accumulated on the branch. Over time, those roots go in search of more nutrients, snaking their way down the trunk of their host. As soon as they touch the ground, the fig switches from being a harmless epiphyte into a lethal lodger. Like woody webbing, the roots of the fig grow larger and larger until they kill their host by strangulation.

Initially, strangler figs may give their host trees some protection via their roots, preventing them from being uprooted in tropical storms—but this benefit only lasts until the strangler throttles its host.

Hollow victory
When the dead tree host rots away, all that remains is the strangler fig, whose roots form a cast of the tree that once supported it. The hollow space inside is a safe habitat for birds, insects, and bats.

Strangler figs produce copious amounts of fruits. Each fruit is a syconium—an inverted inflorescence with its flowers on the inside. Tiny fig wasps are responsible for pollination. A female enters the syconium through a small hole and lays her eggs near the fig ovules. As she does so, she spreads pollen from another fig among the female flowers. The wasps are born, feed, and mate inside the fig. Impregnated females then collect pollen from male flowers and, leaving the males wasps behind, fly off to another fig, where the process starts again. Enough flowers are fertilized for the fruit to contain plenty of viable seeds. With luck, a tree-living animal will eat the fruit and deposit its seeds on the branch of another host tree, continuing the cycle.

This lifestyle enables the fig seeds to germinate close to the treetop canopy, which receives far more sunlight than the forest floor, where it would struggle to survive.

Sweet fruits
The energy-rich fruits of strangler figs are relished by a wide range of creatures. They carry the seeds far from the parent tree, passing them out in their droppings.

Fig fruits are packed with seeds, which remain viable after the fruits have been eaten and digested

living on air

Air plants, members of the pineapple family (Bromeliaceae), are named after their apparent ability to thrive on fresh air alone. Unlike most plants, which rely on roots to absorb water from the soil, they can absorb moisture from the air through scales on their leaves. Air plants belong to the genus *Tillandsia*, and although most species do have roots, these serve largely to secure them to a tree branch or rock.

Air plant leaves are covered with silver, shieldlike scales that absorb water from the hot, steamy air of the forests where they grow

TILLANDSIA TECTORUM

Life in the canopy

Epiphytes live perched on other plants. They are not parasites and do not feed from their hosts, but they benefit from living on branches high up in the rainforest, where they receive much more sunshine than they would on the forest floor. *Tillandsias* are not the only epiphytic plants—many ferns, orchids, and other bromeliads also live high in the treetops.

Colorful bracts help to attract pollinators when dark blue petals appear, indicating the flowers are rich in nectar and ready for pollination

The leaf scales become transparent when wet, turning the silvery leaves green

Wiry roots

Silky hairs

Air plant flowers produce numerous seeds equipped with fine, silky hairs. These enable the seeds to be blown away on the wind to new branches where they can lodge and grow.

TILLANDSIA TENUIFOLIA

AIR PLANT HOSTS

Tillandsia are usually found growing on branches or rocks. It is easiest for air plant seeds to wedge themselves into the crevices of rough surfaces, such as tree bark. However, smaller species, such as Spanish moss (*T. usneoides*), can adhere to the flimsiest of twigs, creating thick curtains of lichenlike gray stems.

TILLANDSIA IONANTHA

Holoparasites

Parasitic plants that have no leaves and are totally reliant on their host for food and water are known as holoparasites. Purple toothwort (*Lathraea clandestina*), shown right in a 19th-century chromolithograph, is an example. It often lives on the roots of poplar, willow, or other trees and only appears aboveground when it flowers.

parasitic plants

While most plants produce their own food via photosynthesis, there are some that cheat. Parasitic plants prey upon other plant species, stealing water and carbohydrates from them by penetrating their host's tissues using modified roots known as haustoria. Some plants, such as American mistletoe (*Phoradendron leucarpum*), attach themselves to stems and branches, but others live on the root systems of their hosts. Some parasites can only live when they are connected to a host, but others are able to survive independently.

Red bartsia (*Odontites vulgaris*) parasitizes a range of different species

Hemiparasites

Although parasites like yellow rattle and red bartsia have green leaves and can make their own food, they steal water from their hosts. These "hemiparasites" may also pilfer carbohydrates to supplement their own supply.

Yellow rattle (*Rhinanthus minor*) is a hemiparasite and can survive without a host plant

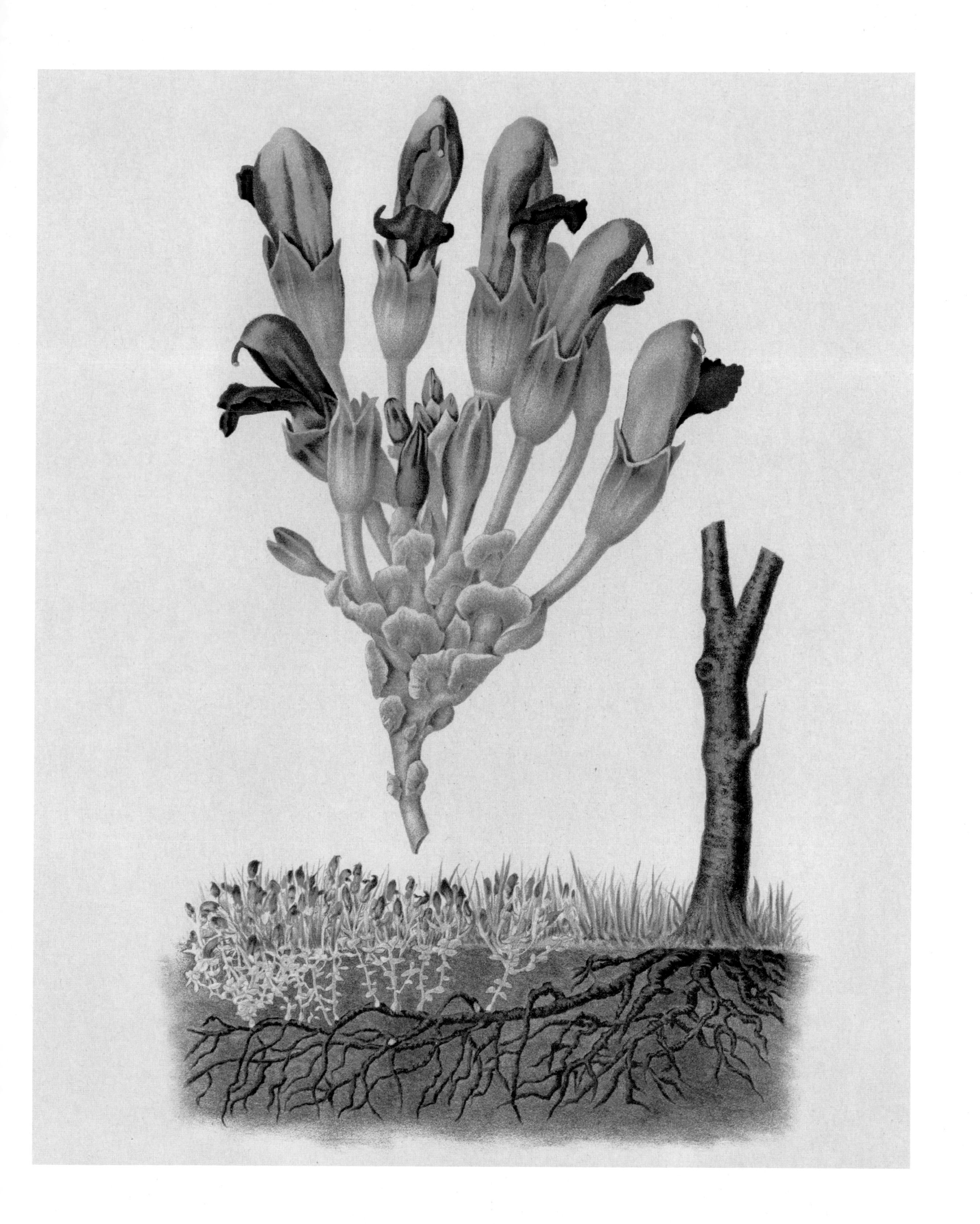

Waxy, whitish berries only form on female plants; they are beloved by birds, but are toxic to humans

Treetop parasite
Although its green leaves are fully capable of photosynthesis, mistletoe still steals water and nutrients from its host.

Elliptical, leathery leaves form in pairs

Viscum album

mistletoe

Ingrained in myth and folklore, mistletoe has a fascinating biology. This parasitic plant, and the similar species American mistletoe (*Phoradendron leucarpum*), live on various deciduous trees. It may cause deformities in its host, but it rarely kills it; in fact, if the host tree dies, so does the mistletoe.

Instead of growing its own roots, mistletoe develops specialized structures called haustoria that penetrate the vascular tissues of its host tree to absorb water and nutrients. Mistletoe is slow-growing, so healthy host trees can tolerate a few mistletoe plants without serious ill effects. Trees with heavy infestations, however, may be weakened and are less likely to survive additional stresses such as disease, drought, or temperature extremes.

Birds are crucial to the spread of mistletoe. After mistletoe's small flowers are pollinated, many white-to-yellow berries are produced. Birds love to feed on the berries, but they can only digest the soft pulp, so they either excrete or squeeze out the poisonous seed while eating the fruit. The seeds then stick to the bird's face, and are wiped off on to a branch. There, the sticky coating hardens, and glues the seeds to the branch. The seeds then send out their haustoria into the host, to complete the plant's life cycle.

Although parasitic, mistletoes play an important role wherever they grow. There are many species of mistletoe, and each one is a vital food source for birds and insects. Those animals, in turn, attract more wildlife, and it is now clear that mistletoes help to increase biodiversity in their native habitats. Moreover, the fact that they prefer certain trees to others helps prevent the species that they colonize from becoming dominant to the detriment of other types of trees.

Mistletoe in winter
Evergreen mistletoe is easiest to spot during the winter, when dense clusters up to 3 ft (1 m) wide can be seen dangling from naked trees. Each cluster is an individual mistletoe plant made up of many regularly forking branches.

ROOTS IN ANAEROBIC SOIL

Soil inundated with water becomes anaerobic (lacking in oxygen). Aquatic plant stems are hollow, allowing air to move down to their roots. Some air escapes and aerates the soil around the roots and rhizomes, in an area known as the rhizosphere.

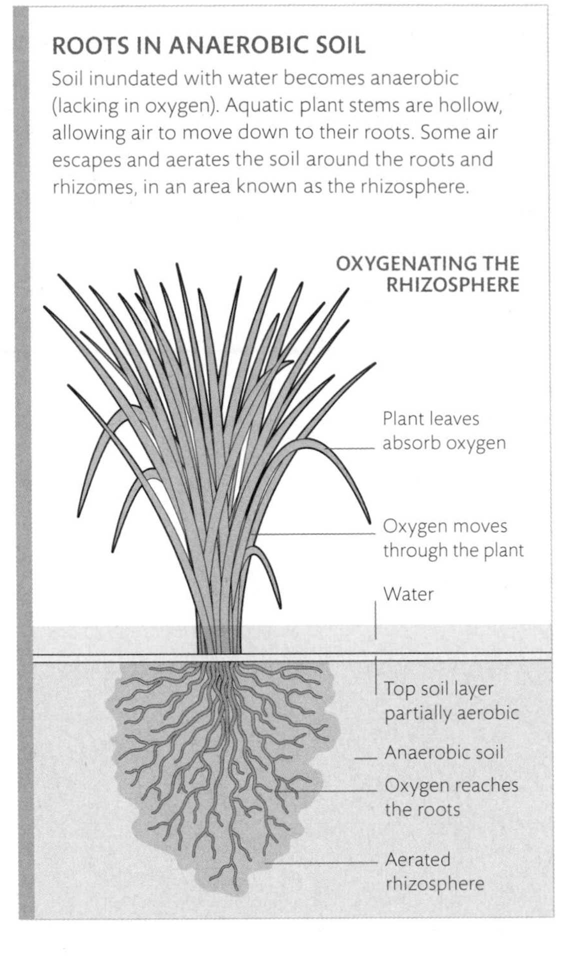

Horsetail stems are rich in silica, a mineral that makes them rough and unappetizing, which deters herbivores

Firm, vertical ridges give support to the hollow stems

Tiny leaves shrink and fuse to form toothed, collarlike sheaths at the nodes

Prehistoric plant
Commonly known as horsetails, *Equisetum* species have grown in damp areas, including those that are seasonally flooded, for more than 300 million years. Today's varieties are miniature versions of their ancestors, which could grow up to 3 ft (1 m) in diameter.

roots in water

All plant cells require a source of oxygen if they are to survive, so for aquatic plants, getting air down to their submerged roots is crucial. Hollow channels with a spongelike structure, called aerenchyma, run through the leaf, stem, and root tissues of aquatic plants. They allow air to flow through the cavities from above the water's surface down to the roots.

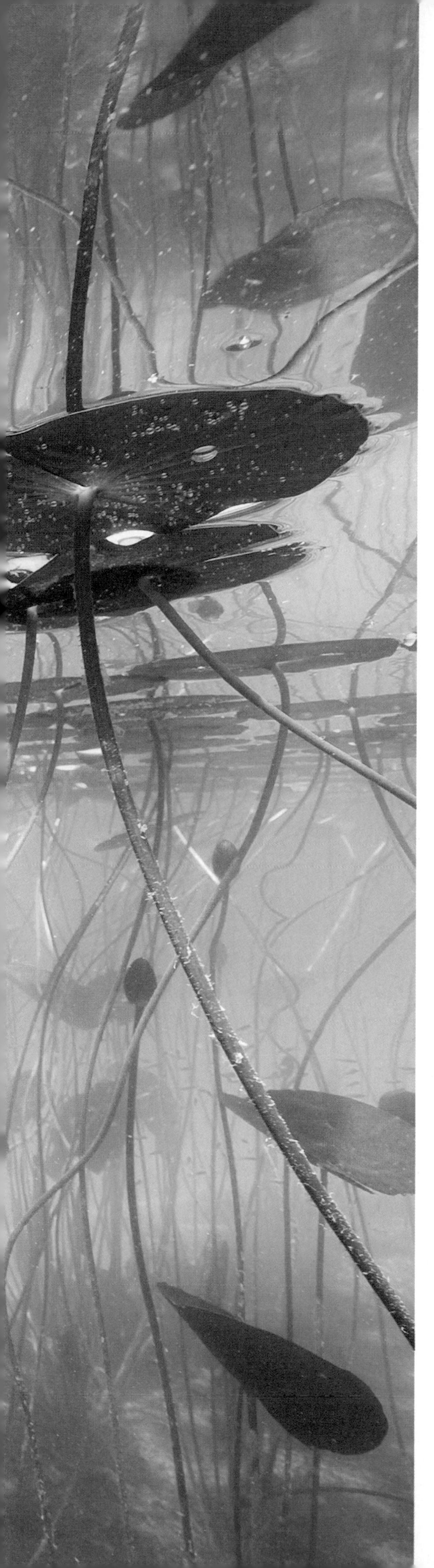

Nymphaea sp.

water lilies

The leaves of water lilies (*Nymphaea*) lie upon the water's surface with the utmost grace and their flowers punctuate the scene with subtle colors. DNA studies show that water lilies have one of the oldest lineages of any group of flowering plants.

There are more than 60 known species of *Nymphaea*, and they can be found growing in both tropical and temperate climates. Sisters to all other angiosperms except for the famed *Amborella*, water lilies are an evolutionary success story millions of years in the making. Look out over a lily pond and it is easy to mistake water lilies for floating plants. In fact, the leaves of *Nymphaea* plants are actually attached to long, slender stems that arise from a thick rhizome buried deep in the mud, and they are able to float with the help of large air sacs packed between their cells.

Aquatic elegance
Water lilies have become prized additions to water gardens the world over. Where they escape into the wild, they often crowd out the native plants, upsetting the delicate balance of many aquatic ecosystems.

When a water lily comes into bloom, the female parts of the flower (the stigma) mature first. The bowl-shaped stigma is filled with a thick, sticky fluid containing compounds that attract insects such as bees and beetles. As they crawl into the center in search of their sweet reward, pollen from other water lily blooms is washed off the insect and into the stigma, pollinating the flower. Some insects perish in the process and drown: it makes no difference to the water lily whether or not its pollinator survives.

After the first day, the flower stops producing fluid, and the stamens become active for a day or two, releasing pollen for insects to collect and deliver to other blooms, continuing the cycle. When the flower closes for the last time, its stem recoils, drawing the bud back underwater. This positions the developing seeds closer to the mud, where they can germinate.

The male and female parts of the flower mature at different times, reducing the chances of self-pollination

Opening bud
Water lily flowers only emerge above the water to be pollinated. Once this has happened, they retreat back below the surface, to give their seeds a better chance to germinate.

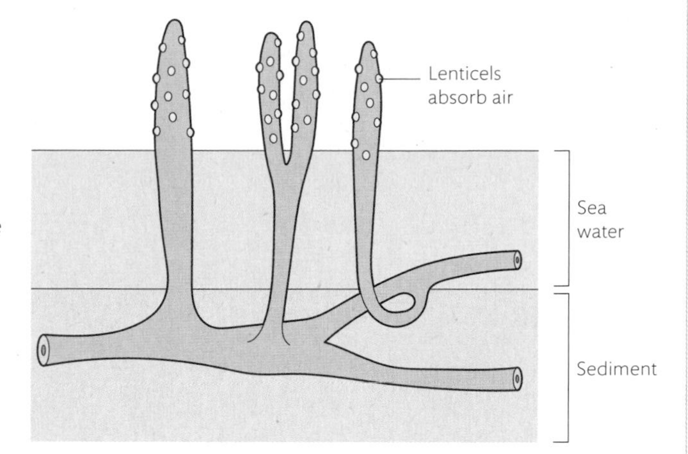

HOW ROOTS BREATHE
Tidal mangrove roots are submerged and exposed twice daily, and the soil they occupy is anaerobic, containing little or no oxygen. In order to breathe, some species develop upright extensions of the root system called pneumatophores, which act like snorkels. They absorb air through lenticels on their surface and transport it to the roots.

PNEUMATOPHORES

breathing roots

Shallow coastal waters represent one of the most challenging habitats for plants. Tidal flows and storm surges uproot them, especially as the soft sediment of the sea floor provides only minimal anchorage. Salty water also dries out plant tissues, and roots and stems are starved of oxygen because they are submerged. Mangroves are one of the few groups of trees and shrubs to thrive in such environments, and the forests they form protect coastal communities from storms and erosion.

Pneumatophores are root extensions that push their way through the swampy ground

Roots for anchorage
In order to resist the pull of the tide and survive tropical storms, some mangroves develop extensive networks of prop roots. These anchor the plant in the shallow soil and slow the flow of water so more sediment is retained around the roots.

Breathing air, excluding salt
Mangrove roots can breathe air during low tide through their pneumatophores. Some species can also exclude salt; the root membranes work like a filter, allowing water to enter while keeping out the harmful salt.

Rhizophora sp.

mangroves

Living in saline water is an immense challenge for plants, but one that mangroves have successfully overcome. Of all the trees and shrubs referred to as mangroves, relatively few are "true mangroves"—meaning those that live exclusively in salt water habitats, such as the various *Rhizophora* species.

The dehydrating effects of salt and the difficulties of obtaining fresh water make coastal conditions inhospitable. *Rhizophora* mangroves get around this problem by filtering out the salt. The trees sit on top of long, spindly prop roots that give them their characteristic appearance, and these prop roots are the key to the mangroves' success. Water entering the roots passes through a series of cellular filters that remove salts, giving the trees access to an endless supply of fresh water. By always keeping parts of their roots above water, the mangroves are also able to exchange carbon dioxide for oxygen gas, in order to avoid suffocation (see p.53).

Rhizophora mangroves play a vital role in sustaining coastal communities—both human and animal. Their roots hold on to sand, slow erosion, and reduce the intensity of waves, protecting and building up shorelines that would otherwise be washed away. Forests of mangroves also help to shield settlements and wildlife from hurricanes and tropical storms, and provide important feeding and nesting sites for many types of birds.

Rhizophora mangroves rely on the ebb and flow of the sea to colonize new sites. These trees are viviparous, which means that their seeds germinate on the branch before dispersal. The torpedo-shaped seedlings either plunge into the sand at the base of the parent tree or float away on the tide and, with luck, wash up on a faraway beach, ready to start a new forest.

Mangrove forest at high tide
Countless fish species spawn among the tangled mass of *Rhizophora* roots, and their offspring grow up in the shelter of these forests.

Low-tide mangrove
The prop roots of this red mangrove (*Rhizophora mangle*) curve down into the sand at a saltwater lagoon in the Bahamas. The uppermost part of the roots are never submerged so that gas exchange can take place continuously, enabling photosynthesis and respiration.

stems and branches

stem. the main body or stalk of a plant or shrub, usually rising above ground but occasionally subterranean.
branch. a limb that grows out from the trunk of a tree or from the main stem of a plant.

types of stems

Stems are the skeletons of plants, supporting and connecting roots, leaves, flowers, and fruits. They conceal a circulatory system that moves water and food around the entire plant. This is a huge variety of stem structures, from majestic trees and arching vines to spreading carpets and underground rhizomes, and they range in size from the wiry stems of miniature mosses to the massive trunks of forest redwoods.

Hard and soft stems
Secondary thickening is the process by which stems produce woody tissue, allowing them to become larger and stronger. Many plants, however, never become woody; their soft, herbaceous stems only last a single growing season.

Internodes contain a spongy pith and a sugar-rich liquid

Strong, upright sugarcane stems can reach up to 15 ft (5 m) tall

Tough stems are essential for this tall grass to support itself

Young ivy stems are soft and flexible to aid climbing; they become woody with age

Stems bearing flowers or flower heads (or later fruits) are called peduncles

ICELANDIC POPPY (*PAPAVER NUDICAULE*)

SUGARCANE (*SACCHARUM OFFICINARUM* 'KO-HAPAI')

COMMON IVY (*HEDERA HELIX*)

Bark protects woody stems from damage, water loss, and destructive insects

From stems to branches

As the layers of xylem (woody tissue) build up, the stems become much stronger and thicker. The tallest plants in the world are trees with woody stems.

Woody stems are durable enough to survive for many years

The contorted stems of the corkscrew hazel are a genetic abnormality, first discovered growing wild in a hedgerow

CORKSCREW HAZEL (*CORYLUS AVELLANA* 'CONTORTA')

SILVER BIRCH (*BETULA PENDULA*)

Monocot vascular bundles are scattered throughout the stem

inside stems

All stems have two key functions: support and transport. They hold the leaves up, allowing them to absorb sunlight, and then transport the carbohydrates that are manufactured in them around the plant. Water and minerals are also carried from the roots via the woody xylem tissue, made of dead cells, while food and other materials are transported around the stems through the living phloem cells.

Stems and vascular bundles
Inside the stem, xylem and phloem cells are packed together into vascular bundles. Among flowering plants, these are arranged differently within the stem depending on whether the plant is a monocot or a eudicot (see p.15). In monocots, bundles are scattered throughout the core of the stem, but in other flowering plants, and all eudicots, the bundles are arranged in a circle. This is clearly visible in trees: over time, most trees develop rings in their trunks, but monocot trees, such as palms, never do.

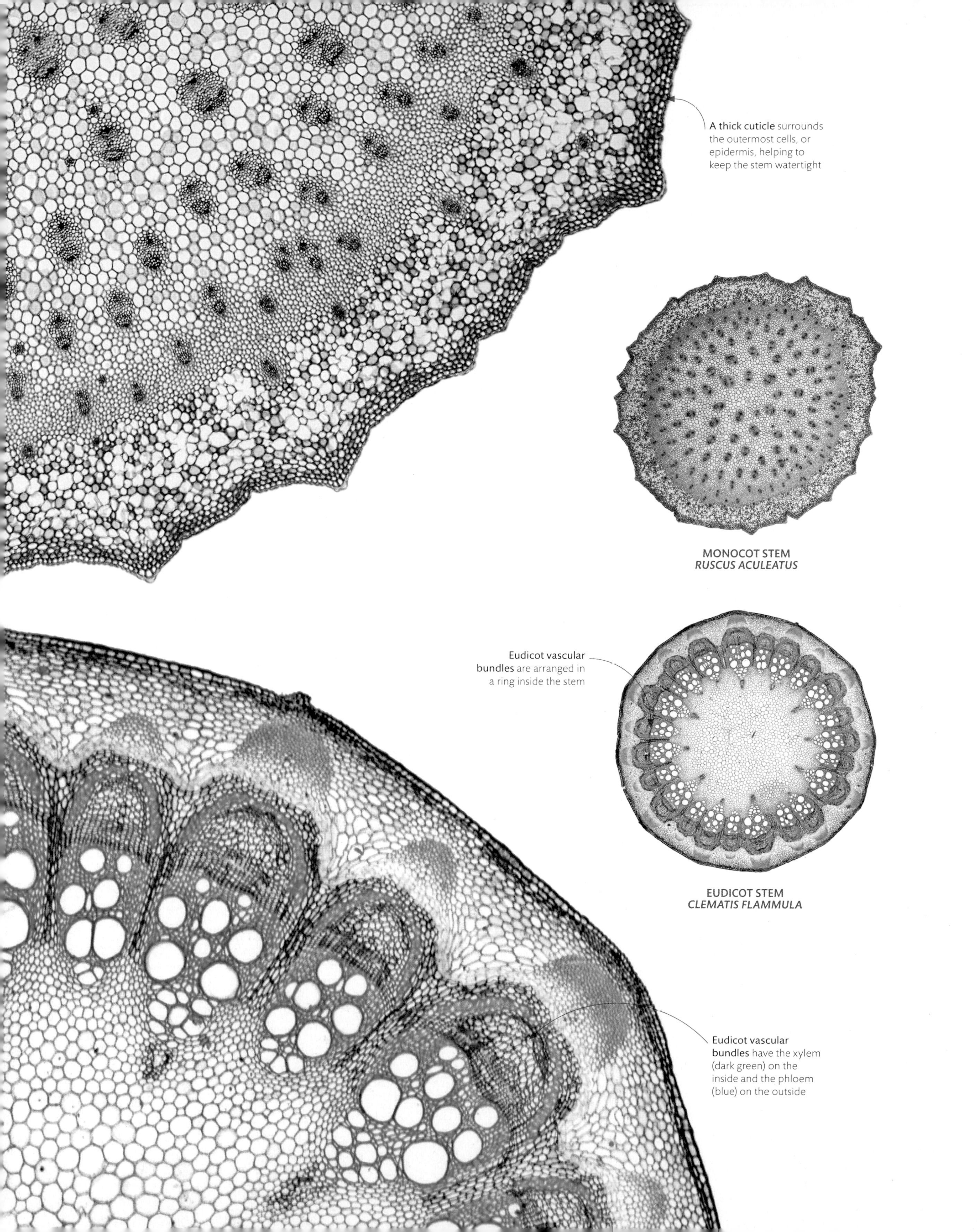

MONOCOT STEM
RUSCUS ACULEATUS

EUDICOT STEM
CLEMATIS FLAMMULA

***Great Piece of Turf*, 1503**
Albrecht Dürer's meticulous watercolor study of an everyday clump of grasses and weeds gains its power from perspective—the viewer is at ground level with the insects and small creatures that inhabit the turf. Artfully random and naturalistic against a plain background, the painting features a range of perfectly rendered plant species, including cocksfoot, creeping bent, smooth meadow grass, daisy, dandelion, germander speedwell, greater plantain, hound's-tongue, fool's watercress, and yarrow.

natural renaissance

Careful studies
Leonardo da Vinci's exquisite chalk and crayon sketches of plants and trees were often created as preparatory studies for his larger works, but they were also an important part of his investigations into botanical science.

Throughout the 200 years of the Renaissance, the scope of intellectual curiosity and human creativity seemed boundless. Major artists studied anatomy for sculpture and figure work, mathematics to solve problems of linear perspective, and the natural world in order to reproduce plant life and landscapes with complete accuracy. Their botanical sketches and watercolors are celebrated today for their naturalism.

Historically, plants were studied and illustrated in herbals (see pp.140–141) for identification, and in the Middle Ages artists used flowers, such as the lily (purity), to add symbolic meanings to religious paintings. These flowers featured in Italian artist Leonardo da Vinci's early works. However, in the late 15th century, the rediscovery of nature became the mainspring of Renaissance energy in Europe, inspiring the work of two giants, Leonardo in the south and German artist Albrecht Dürer in the north. Leonardo underpinned his great paintings with close studies of plant species and scientific investigations into botanical processes. He was an inspiration to Dürer, a master of oils, woodcuts, and engraving.

Dürer was renowned for his messianic self-portraits and visionary works on mythical and religious themes, but the private work he did was entirely different. His handful of quiet, perfectly observed watercolor studies of nature, which he probably intended to use to add realism to his religious paintings, include this uncultivated slice of summer meadow, a teeming microcosm of the natural world.

> “… I realized that it was much better to insist on the genuine forms of nature, for simplicity is the greatest adornment of art.”

ALBRECHT DÜRER, *LETTER TO REFORMATION LEADER PHILIPP MELANCHTHON*

tree trunks

A cross section of a tree trunk provides an unparalleled opportunity to peer into the past. Each year, the trunk—a woody stem—develops a new layer of tissue, the thickness of which is determined by environmental conditions. Good conditions lead to vigorous growth, which creates a wide ring. Stress caused by extreme temperatures or drought results in a thin ring. Studying these rings provides a glimpse into the weather conditions of the past.

Strong support
The columnlike shape of most tree trunks provides a physical support for the tree's framework of branches and thousands of leaves. Tree trunks can grow massively tall and are incredibly strong. They can hold themselves upright without needing a structure to climb around or wrap themselves over.

The bark splits as the trunk expands, but new layers form underneath

Phloem tissue lies just beneath the bark and transports food around the tree

Vascular cambium produces new xylem layers

Each pale ring is composed of early wood, which forms when the tree starts to grow in spring

Dark rings are late wood, which forms later in the year, just before the tree becomes dormant

STRUCTURE OF A TREE TRUNK

Within the woody stems of trees and some flowering plants are rings of xylem and phloem, the tissues that transport water and food around the plant. A thin layer of phloem sits just under the bark, while multiple layers of xylem form the growth rings visible when a tree is cut down. Every year, tissue known as vascular cambium produces new layers of xylem on top of the previous year's layer; the layers can be counted to estimate the age of the trunk. The younger, outer layers of xylem continue to transport water and are known as sapwood; the older layers on the inside gradually become blocked and form heartwood. On the outside of the woody layers, cork cambium produces new bark, to cover and protect the expanding tree trunk.

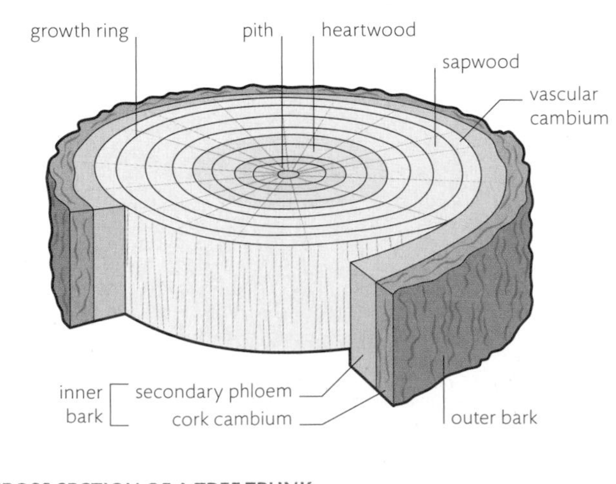

CROSS SECTION OF A TREE TRUNK

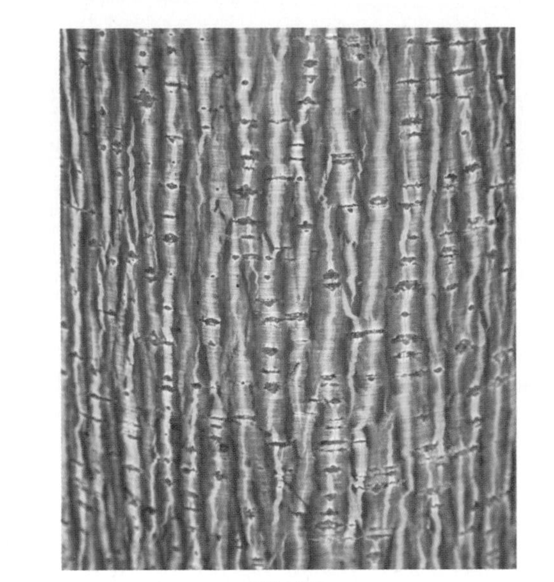

CORKY
Quercus suber

STRIATE
Acer pensylvanicum

RIDGED
Castanea sativa

LENTICELLATE
Prunus serrula

SCALY
Pinus sp.

SPINY
Ceiba speciosa

FLAKING
Platanus sp.

PEELING
Eucalyptus gunnii

PEELING IN STRIPS
Carya ovata

bark types

The protective "skin" of woody plants, bark keeps out invading insects, bacteria, and fungi, and retains precious moisture. It can also defend a tree from fire, and trees that shed layers of bark deter clinging vines and epiphytes from taking hold. Within the bark there are two important layers of dividing cells called cambium. They are relatively shallow, and damage to them can impede growth and, in extreme cases, kill the tree.

Coat of many colors
Only woody trees produce bark, so it is found in conifers and eudicots, but not in ferns and monocots. As it ages, bark splits, and it does so in many ways, creating a great diversity of patterns, textures, and colors.

SMOOTH
Betula populifolia

FISSURED
Liriodendron tulipifera

PAPERY
Acer griseum

Lenticels in the bark of a poplar tree (*Populus* sp.)

Strips of bark can feel rubbery because they contain a waxy, waterproof substance called suberin

The outer bark splits and may peel or flake as a trunk increases in girth

Populus tremuloides

the quaking aspen

With striking white trunks and leaves that tremble in the wind, few sights are as captivating as a grove of quaking aspen trees in the fall. This remarkable species has the widest distribution of any tree in North America, ranging from Canada all the way south into Mexico.

The quaking aspen lives for an extremely long time, although not in the traditional sense. Each tree is either male or female, making sexual reproduction possible, but it rarely reproduces by seed. Instead, once a tree is established, it grows multiple offshoots from its roots. Each offshoot is capable of growing into a new tree, so entire aspen groves can be made up of clones of a single tree. Over time, the trees eventually die, but the rootstock itself can continue to produce new trees for hundreds, or even thousands, of years. The largest known grove of aspen clones, the Pando grove in Utah, is 80,000 years old and covers 100 acres (40 hectares).

The stark white bark helps to protect the tree from overheating and reduces the risk of sunscald during winter, when the bark thaws and then freezes again. By reflecting most solar rays, the tree is able to maintain lower temperatures on sunny winter days. A close look at the bark reveals a greenish tinge: this is photosynthetic tissue, so even before its leaves emerge in spring, the tree is busy harvesting sunlight for photosynthesis.

The tree's habit of cloning itself enables stands to regrow even after forest fires. In fact, fire is essential for maintaining the quaking aspen's habitat. Without cleared land, aspens would eventually be shaded out by trees such as conifers.

Forest growth
The uniformity seen in this stand of quaking aspens suggests it might be a clonal group. Aspens respond rapidly when the ground is cleared; as more light reaches the soil, they quickly send up new, fast-growing shoots.

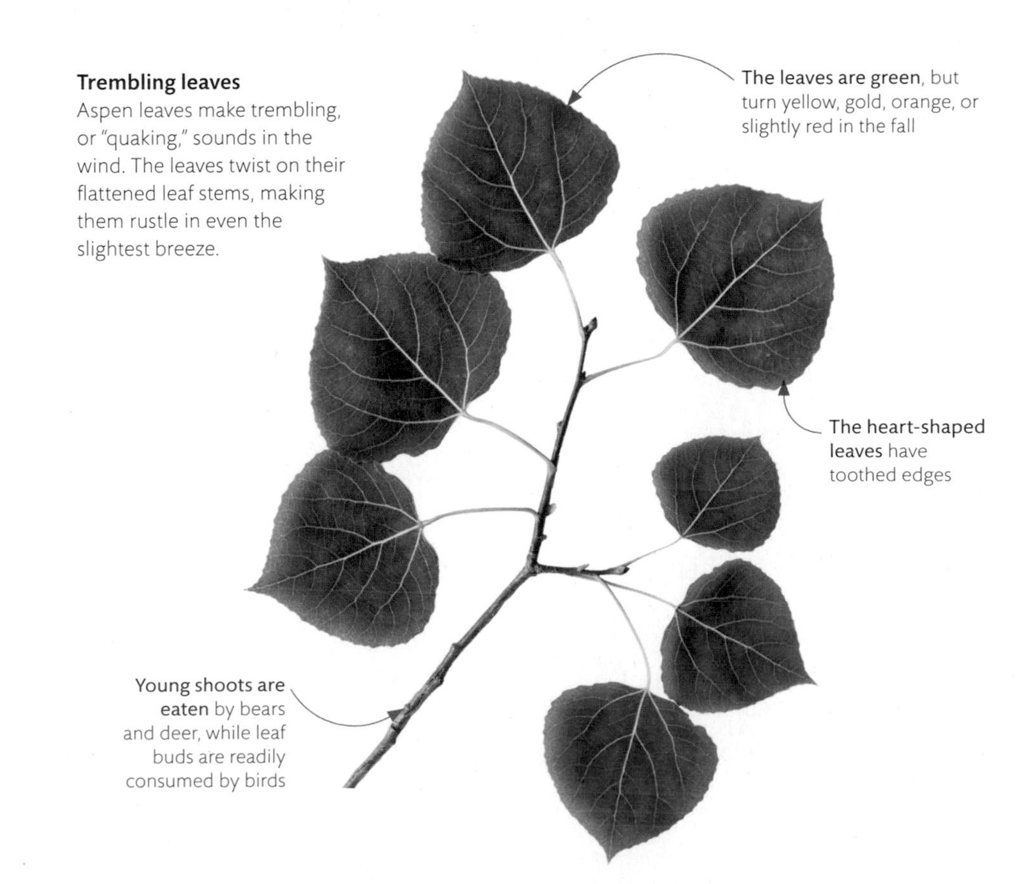

Trembling leaves
Aspen leaves make trembling, or "quaking," sounds in the wind. The leaves twist on their flattened leaf stems, making them rustle in even the slightest breeze.

The leaves are green, but turn yellow, gold, orange, or slightly red in the fall

The heart-shaped leaves have toothed edges

Young shoots are eaten by bears and deer, while leaf buds are readily consumed by birds

BRANCH POSITION AND SHAPE

The position of branches on a tree is governed by the arrangement of the new buds, or growing points. Where buds are arranged alternately along the stem, so are the resulting branches, and the tree develops a broad, rounded canopy. In many conifers, the buds are arranged in whorls, which means the branches also develop in whorls. As the lowermost branches continue to elongate, new ones develop above. As they are younger, these branches are also shorter, so the result is a triangular-shaped tree with the longest branches at the base.

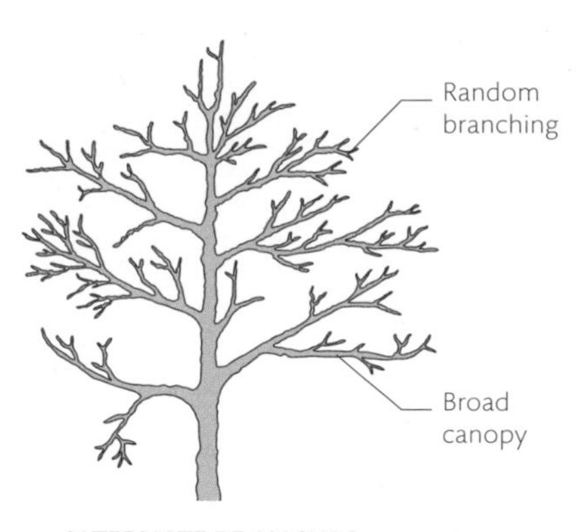

ALTERNATE BRANCHES

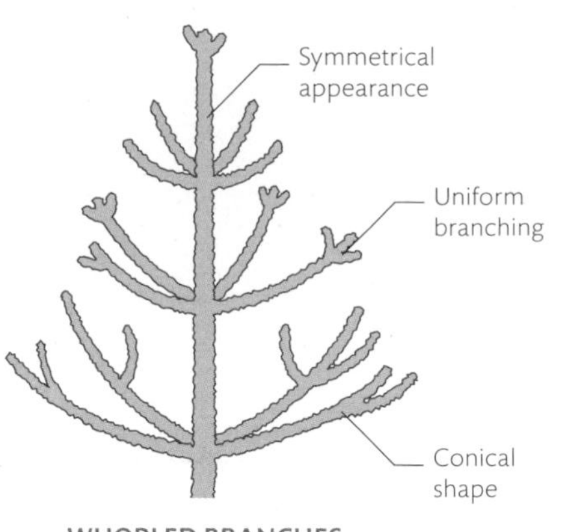

WHORLED BRANCHES

Curious conifer construction

The monkey puzzle tree (*Araucaria araucana*) gets its name from its dense covering of spine-tipped leaves, which even a monkey would struggle to scale. As a young plant, it develops a symmetrical, whorled branching habit. Mature monkey puzzles are generally less uniform; the lowermost branches are shed with age, and pests, diseases, storms, lightning strikes, or other factors can damage branches and spoil the perfect profile.

Sharp spines protect the leaves from herbivores

Pollen cones are found at the end of the branches of male monkey puzzle trees

Whorls of tough, leathery leaves are arranged along the branches to maximize light absorption

ARAUCARIA ARAUCANA

Pollen is dispersed by the wind, ready to pollinate seed cones on female trees

branch arrangement

The way in which trees arrange their branches determines the overall shape of the tree. Perhaps the two most familiar shapes are the conical form of most conifers and the wide, cloudlike canopy of broadleaved trees. Individual branches are arranged to maximize the amount of light that the leaves can receive.

Terminal buds form at top of twig or stem

Scaled buds may be covered by resin for extra protection in winter

Pseudoterminal buds form when branches die back to an axillary bud

Flower buds tend to be larger than leaf buds

Unscaled buds have no scales, but may still be protected by hairy bracts

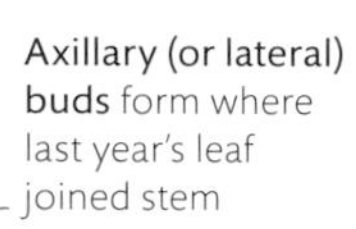

Axillary (or lateral) buds form where last year's leaf joined stem

Leaf scars are sometimes left by the old leaves when they fall

Juglans mandshurica var. *sieboldiana*

Aesculus hippocastanum

Quercus frainetto

Magnolia campbellii

TERMINAL BUD

AXILLARY BUD

PSEUDOTERMINAL BUD

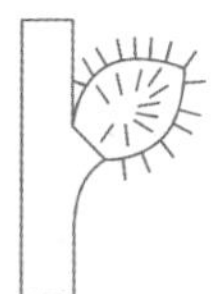

UNSCALED BUD

winter buds

Leaf buds vary enormously and are highly distinctive, both in shape and in the way in which they grow. Along with a tree's shape, they are a vital aid to identifying a tree during winter. Examining how buds grow along a stem is revealing. They can form opposite each other or on alternate sides of a stem, rising around it at intervals. Bud scales, which protect the developing leaves and flowers, also vary in shape, color, and number.

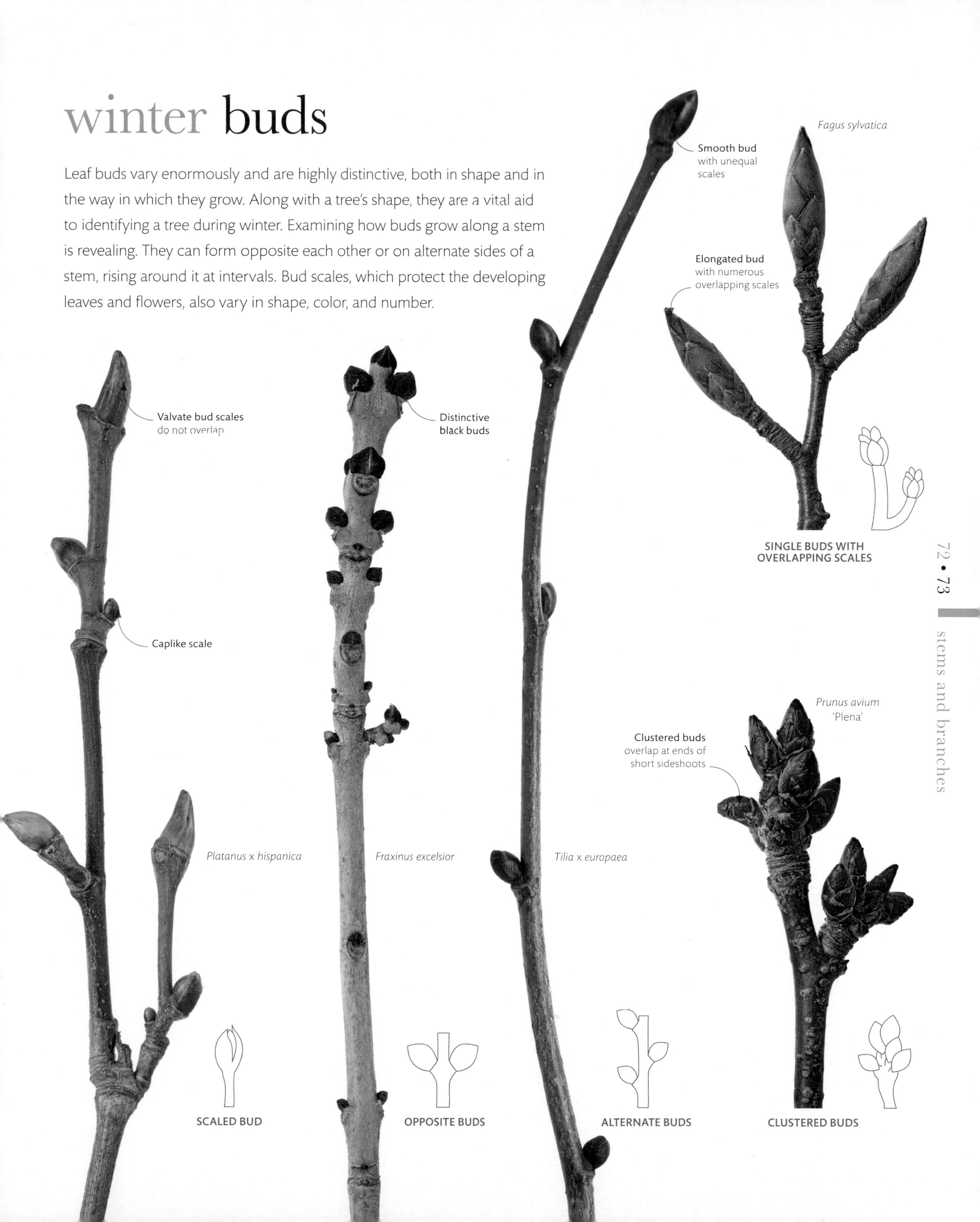

Magnolia petals and sepals are almost identical and are known collectively as tepals

The tough, leathery tepals are strong enough to prevent damage by large insect pollinators

Layers of hairy bracts enclose and protect the developing flower buds of magnolias

FLOWERS BEFORE FOLIAGE

When deciduous trees and shrubs such as forsythia, alder, witch hazel, and deciduous magnolias produce flowers before their leaves emerge, it is known as hysteranthy.

FORSYTHIA VIRIDISSIMA

Inside a flower bud

Magnolias were one of the earliest groups of flowering plants to appear on Earth, and their blooms have features that are not found in more recently evolved species. They do not have distinct sepals and petals, and their flower buds are protected by deciduous bracts rather than sepals.

Magnolia stamens are unlike the stamens of most other flowers because they lack distinct anthers and filaments

Scales and scars
Magnolia stems are distinctive, even when bare of leaves. The hairy bracts (sometimes referred to as bud scales) that protect the flower buds, and the circular scars beneath them, are easily recognized. The shield-shaped leaf scars are another characteristic.

Bracts either fall as the flower opens, or they are shed prior to flowering

Crusty lichens grow on the long-lived stems and branches

Silky bract hairs may be silver or light brown, or sometimes absent

Distinctive scars are left on the stems by fallen leaves

insulated buds

As well as leaf buds, the woody stems of trees and shrubs also bear the buds of next year's flowers. Deciduous magnolias form their flower buds in the late summer and fall, remaining dormant in the winter. It is usually possible to distinguish between flower and leaf buds by their shape or size.

Buds that develop into leaves are much smaller than flower buds

Trunk in bloom
Archidendron ramiflorum is a member of the legume family and is native to Queensland, Australia. Rather than having showy petals, the flowers attract pollinators with their flashy stamens. The spherical bloom clusters develop on woody stems in the shade of the rainforest canopy, where the bright white flowers stand out in the gloom.

cauliflorous plants

Flowers and fruit usually develop on new shoots, but in some trees and shrubs, the blooms erupt directly from the woody trunk and primary branches. Known as cauliflory, this strategy is more common in the tropics than in cooler regions. The reasons why some plants adopt cauliflory remain mysterious, but it may be an evolutionary adaptation to allow animals that live lower down in the forest canopy easy access to the flowers and fruits. Despite the name, cauliflowers are not cauliflorous, but produce congested flower clusters at their stem tips.

COCOA TREE FRUIT

Cocoa plants (*Theobroma cacao*) produce their flowers and fruit on woody stems shaded by the foliage. The flowers are pollinated by midges, which prefer a dappled environment. Other cauliflorous trees include breadfruit (*Artocarpus altilis*), papaya (*Carica papaya*), calabash (*Crescentia cujete*), and many tropical figs (*Ficus* sp.). Outside of the tropics, rare examples of cauliflory among temperate trees and shrubs include the eastern redbud (*Cercis canadensis*) and the Judas tree (*C. siliquastrum*), which bear pink flowers on mature branches before the new leaves appear in spring.

THEOBROMA CACAO

Flower buds emerge from growing points, or meristems, located at nodes along the woody stems
The bold tufts of white stamens provide pollen for potential pollinators
Spent flowers will develop into vivid red, coiled pods
The stamens can grow up to 2½ in (5 cm) long

stem defenses

Herbivores depend on plants for sustenance, but plants can defend themselves. By arming their stems with thorns, spines, or prickles, they can deter at least some of their adversaries. The names for these three defenses are often used interchangeably, yet each is thought to have developed from a particular part of the plant in evolutionary history.

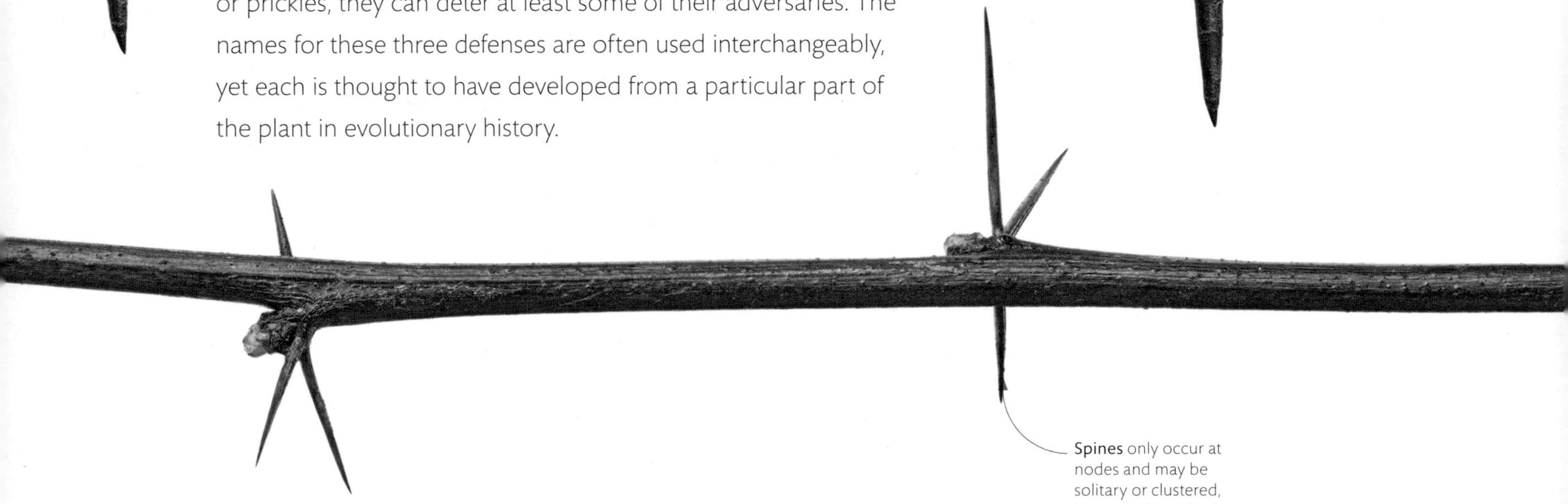

Thorns are solitary, although they sometimes form their own branches

Spines only occur at nodes and may be solitary or clustered, but never branched

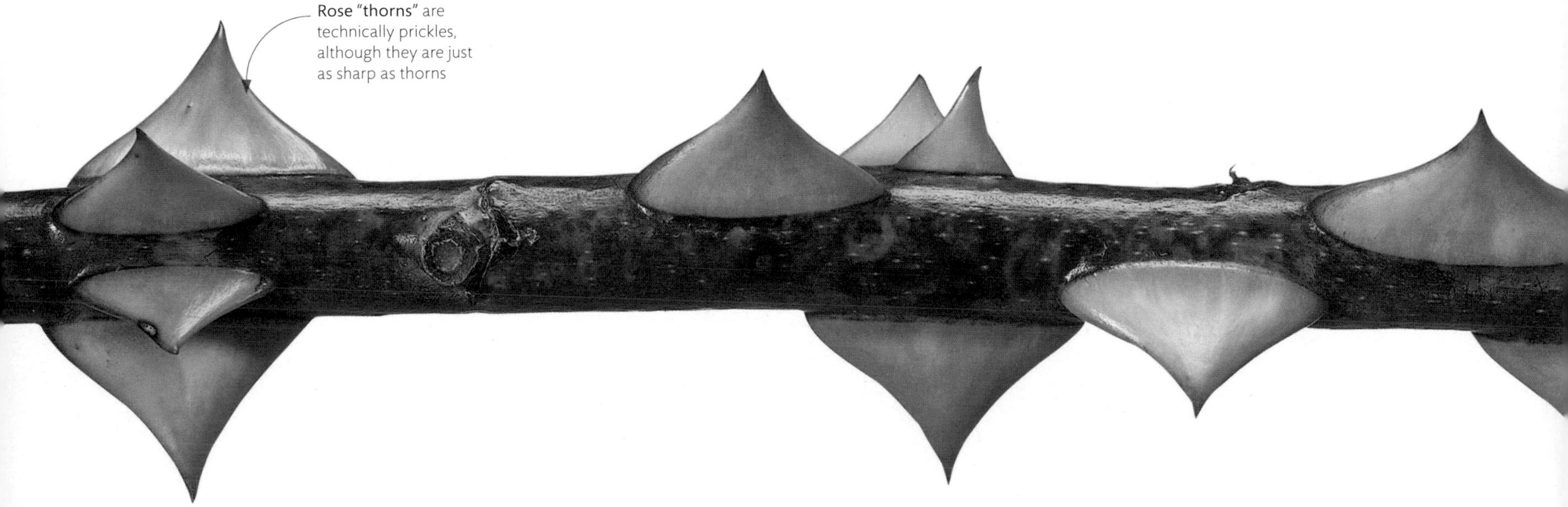

Rose "thorns" are technically prickles, although they are just as sharp as thorns

Thorns are positioned at nodes, as if they were side branches

Thorn

Derived from stems, thorns contain vascular tissue and can be branched; some even bear their own leaves. They are typically stiff and woody, as on this hawthorn (*Crataegus* sp.). Thorns also can be found on the branches of orange trees (*Citrus*) and firethorn (*Pyracantha*).

Spine

Like thorns, spines contain vascular tissue, but they are derived from leaves or leaf parts, such as stipules or petioles. They are never branched but often occur in clusters at the nodes, as on this barberry (*Berberis* sp.). Spines are found on most cacti and acacias.

Leaves on spiny stems develop above the spine

Prickle

Unlike thorns and spines, prickles, such as on this rose (*Rosa* sp.), develop as outgrowths of the cortex and epidermis of a plant, and so do not contain vascular tissue. They are also not restricted to stems and can be found on bark, leaves, and fruits.

Prickles can occur all along the stem and are not restricted to the nodes

Ceiba pentandra

the kapok tree

Many rainforest trees grow to great heights and display huge buttress roots. Few, though, are more impressive than the majestic kapok. Given the right conditions, this important canopy species can grow as tall as 230ft (70m), with buttresses that project up to 65ft (20m) from the trunk.

The deciduous kapok tree is found in the Americas, growing from southern Mexico to the southernmost limits of the Amazon rainforest. It is also found in parts of West Africa. Exactly how the tree came to be native to both regions has been the subject of scientific inquiry. By analyzing kapok DNA, experts now believe that the species made it to Africa after seeds were dispersed across the ocean from Brazil.

The kapok plays a vital role in the local ecology and culture wherever it grows. Its textured bark provides a habitat for bromeliads and other epiphytes, and also for reptiles, birds, and amphibians. The tree's ability to invade disturbed areas also makes it a key pioneer species, as it is one of the first trees to colonize open land after forest clearances.

The kapok's flowers open at night and emit a foul odor that attracts bats, its main pollinators. The tree is able to alter its pollination strategy according to the strength of the local bat population. In places where bats are plentiful, kapoks rely on them to spread pollen from tree to tree; where there are few bats however, the trees self-pollinate, guaranteeing at least some reproductive success each year.

After pollination, kapoks become laden with seedpods, each of which opens up to release about 200 seeds. The seeds are surrounded by cottonlike fibers that help them to disperse on the slightest breeze. Unopened seedpods float in water, so it is likely that the kapok initially traveled from the Americas to Africa by sea.

Prickly giant
A kapok's massive trunk can be up to 10ft (3m) in diameter. The huge prickles keep animals from gnawing on its bark. The prickles eventually drop off as the tree ages.

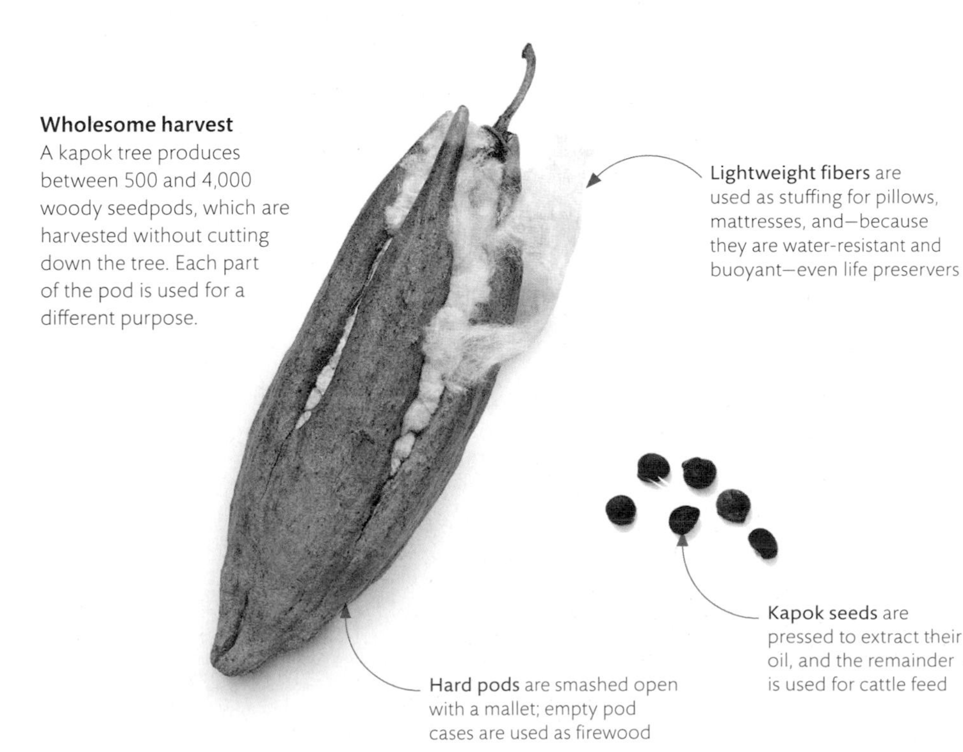

Wholesome harvest
A kapok tree produces between 500 and 4,000 woody seedpods, which are harvested without cutting down the tree. Each part of the pod is used for a different purpose.

Lightweight fibers are used as stuffing for pillows, mattresses, and—because they are water-resistant and buoyant—even life preservers

Hard pods are smashed open with a mallet; empty pod cases are used as firewood

Kapok seeds are pressed to extract their oil, and the remainder is used for cattle feed

stems and **resin**

Trees are constantly under attack from a huge variety of insects, birds, fungi, and bacteria, which attempt to break through the bark and feed on the tissues below, either directly or through existing wounds. Many trees produce sticky resin to heal breaks in the bark and ensnare pests. Some resins even contain chemicals that attract predatory insects that will feed on the tree's attackers. Resin gradually hardens, and fossil resin, known as amber, often contains the remains of ancient insects.

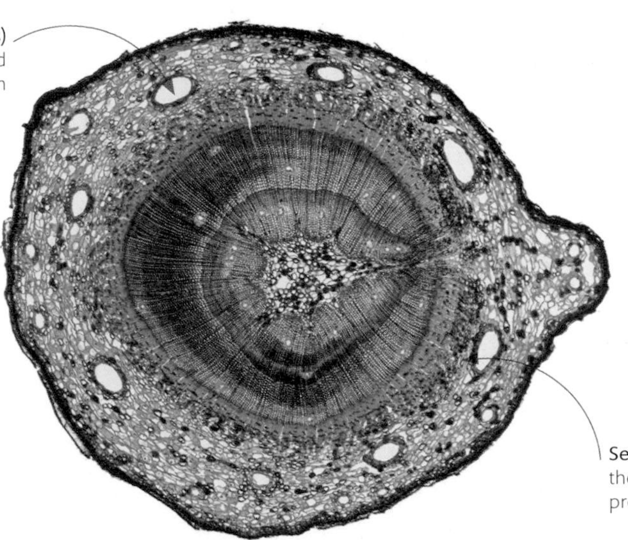

CROSS SECTION OF A PINE TRUNK

Releasing resin
Some trees only produce resin in response to damage, while others, such as this pine—stained to reveal the cell structure—have resin ducts in their wood as a part of their natural makeup.

Sticky protection
The dripping resins produced by trees such as this oak are designed to heal wounds, whether inflicted by pests, or caused by physical damage from bad weather or fire. Plant resins are mixtures of organic compounds, and are used for many things—from perfumes and varnishes to adhesives. They are also the source of valuable commodities such as frankincense, turpentine, myrrh, and tar.

Dracaena draco

the dragon tree

The dragon tree looks as though it belongs in a fantasy novel. This strange but beautiful monocot gets its name from the red resin, known as "dragon's blood," that it exudes when wounded. A member of the asparagus family, *Dracaena draco* has evolved a unique treelike growth habit.

The dragon tree is endemic to parts of North Africa, the Canary Islands, Cape Verde, and Madeira. For the first few years of its life, the dragon tree is a single stem ending in a tuft of long, slender leaves. After 10 to 15 years' growth, the tree flowers for the first time. Long spikes full of fragrant white flowers erupt from the leaves, followed by bright red berries. A cluster of new buds then emerges from the top of the plant, and the buds develop into miniature versions of the original. They continue to grow for another 10 to 15 years before the branching process begins again. Over time, this repeated branching gives the tree its unusual, umbrellalike shape. The lifespan of this species is thought to be about 300 years, but accurate aging is difficult, as the trees do not produce annual growth rings.

The dragon tree's branches produce aerial roots that gradually snake their way down the trunk until they reach the soil. The roots emerge from wounds. If enough damage has been done to the tree, the roots can function as a new trunk, developing into a clone of the parent tree.

The dragon tree's blood-red resin was once highly valued as a medicine and an embalming fluid and is now used to stain and varnish wood. The resin is obtained by making cuts in the bark, but repeated wounding puts the tree at risk of infection. Dragon tree numbers in the wild are in decline, due to aggressive resin harvesting in the past and habitat loss today.

Dragon grove
In the wild, the dragon tree grows in nutrient-poor soils. The thick trunk branches into upright arms ending in rosettes of lance-shaped, blue green leaves up to 2ft (60cm) long. The species' status is currently listed as vulnerable.

Dragon's blood oozes from wounds as a viscous liquid that dries and hardens

The deep-red resin has been used in dyes and traditional medicines since ancient times

Hardened resin
The tree bleeds a crimson resin called "dragon's blood," hence the species' common name. In nature, resin serves as a form of defense, deterring herbivores and keeping out pathogens.

storing food

Some plants have modified stems, roots, or leaf bases that live permanently underground. Swollen and densely packed with nutrients, these subterranean bulbs, corms, tubers, and rhizomes are dormant for part of the year, then sprout new shoots when the growing conditions are right. They are hidden from herbivores and can often spread underground to expand a plant's territory.

From bulb to bloom
Bulbous plants—which include familiar plants such as onions—have a short, squat stem called a basal plate, to which fleshy leaves (scales) are attached. The scales store the nutrients and water that the plant needs in order to produce flowers. These hyacinth bulbs flower in spring. Afterward, the leaves photosynthesize to create more food, which is stored for the next year's flowering.

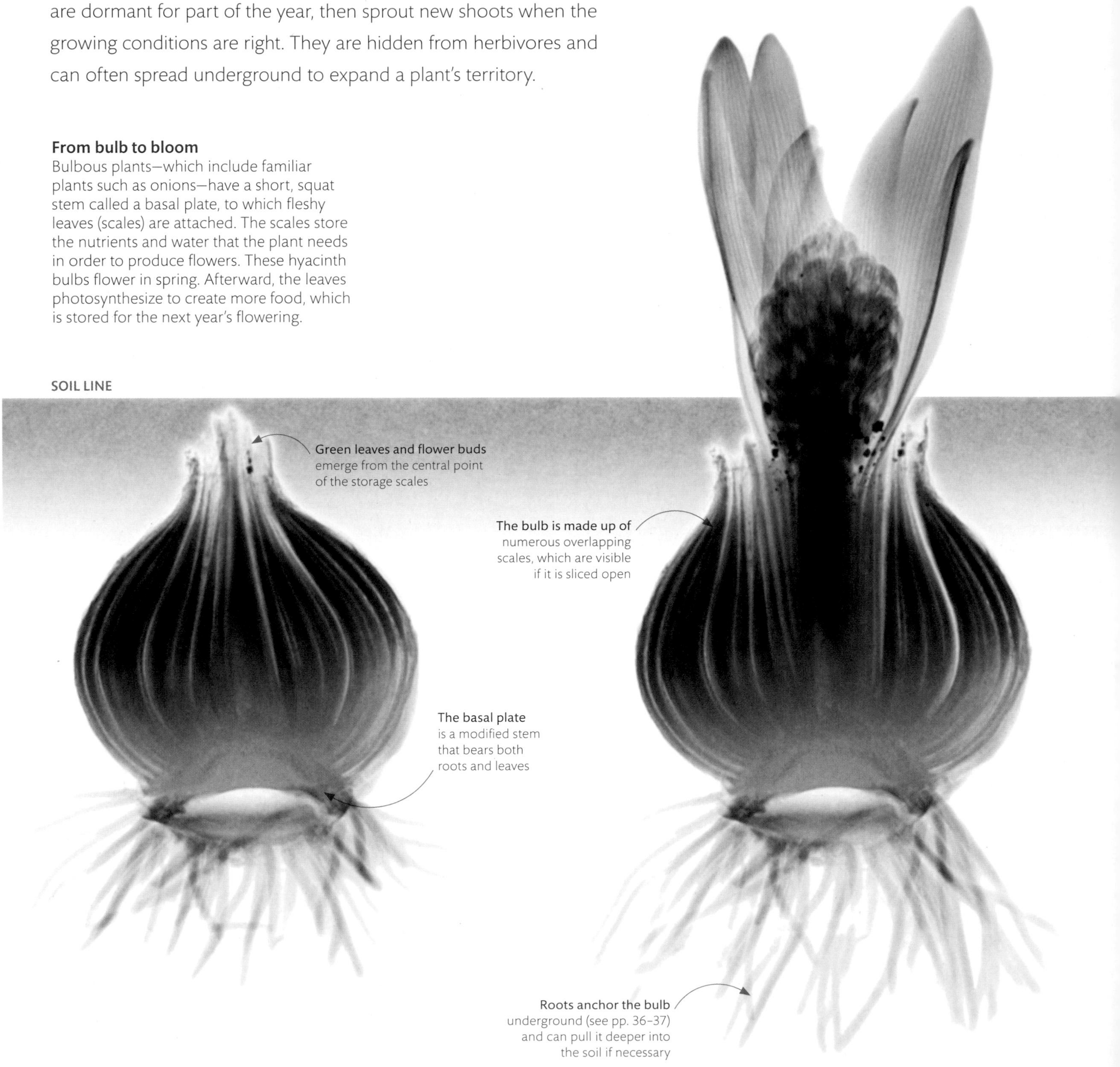

SOIL LINE

Green leaves and flower buds emerge from the central point of the storage scales

The bulb is made up of numerous overlapping scales, which are visible if it is sliced open

The basal plate is a modified stem that bears both roots and leaves

Roots anchor the bulb underground (see pp. 36–37) and can pull it deeper into the soil if necessary

X-RAY OF A DORMANT HYACINTH BULB

EMERGING BULB WITH A DEVELOPING FLOWER HEAD

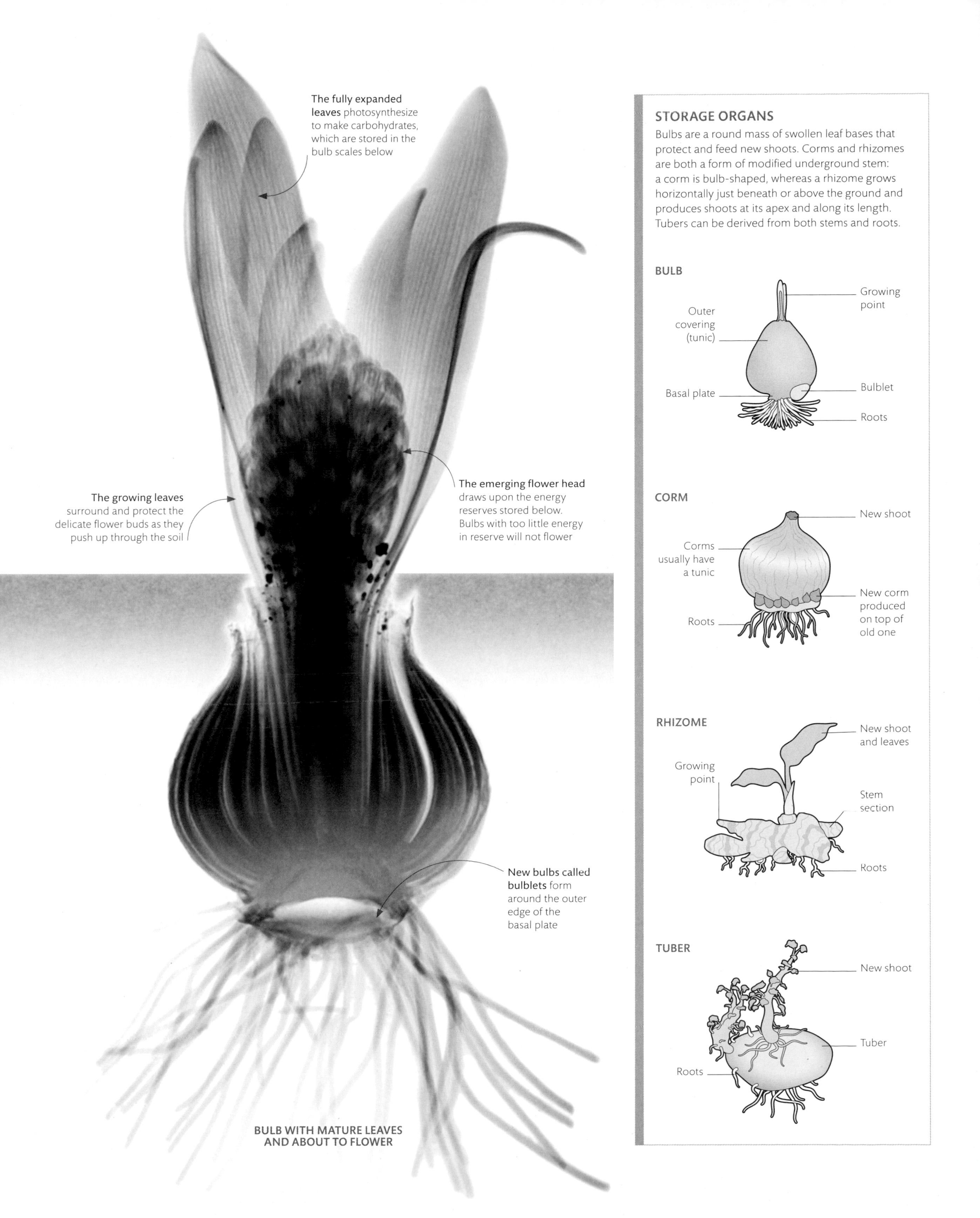

BULB WITH MATURE LEAVES AND ABOUT TO FLOWER

HOMES FOR ANTS

The living quarters that ant plants provide for ants are known as domatia and can develop in different parts of the plant. *Dischidia* vines house ants in swollen leaves, whereas *Lecanopteris* ferns shelter them within their rhizomes. *Myrmecophila* orchids accommodate ants in hollow, bulblike swellings of the stem, and some acacias host them within hollow spines. *Myrmecodia* and *Hydnophytum* both have swollen stem tubers with intricate internal structures, giving ants chambers for various uses.

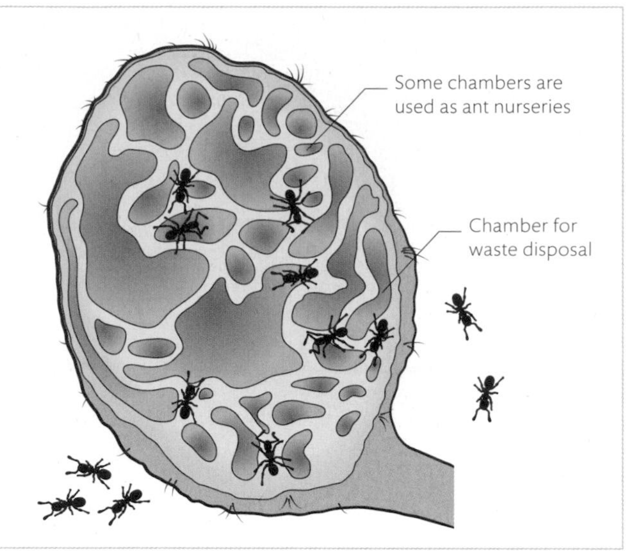

DOMATIA IN A *HYDNOPHYTUM* TUBER

Good tenants
The elaborate stem tubers of the ant-house plant (*Myrmecodia beccarii*) include rough-walled cavities where ants deposit their waste, as well as remains of their prey and corpses. Nodules on the walls of the cavities absorb nutrients from this slurry, providing this tree-dwelling plant with vital elements that it would otherwise find hard to access. The plant shown right, from Australia and grown at Kew in 1888, is depicted in an illustration from *Curtis's Botanical Magazine*.

beneficial relationships

Insects can be a benefit or a burden for plants. Although some provide their services as pollinators, others are ruthless leaf-eaters that weaken the plant to satisfy their appetites. A handful of plants, however, have struck up a mutually beneficial (symbiotic) relationship with ants. So-called "ant plants" provide a secure home for the ant colony and, in return, the ants protect the plant by attacking anything that comes near it. As many ant plants are epiphytes, unconnected to the soil and its nutrients, the rich fertilizer formed from the ants' waste is also a crucial food supply.

Partners in crime
Some ant plants often grow together. Here, the brown stem tuber of *Hydnophytum* is nestled within the yellow leaves of a *Dischidia* species. Both plants provide a home for ants.

fibrous trunks

Not all trees are true trees. Well-known species such as pines and firs (conifers), or oaks and maples (deciduous) have woody trunks with characteristic growth rings and an outer layer of bark. Tree ferns and banana plants, however, have quite a different stem structure and do not produce wood or bark. Their upright stems are supported by densely packed fibers; roots; or tightly packed, overlapping sheaths of leaves.

STEM STRUCTURES

A tree fern trunk is really an upright rhizome supported by a dense, encircling mass of roots and fibers. Banana stems are not really stems at all, but overlapping layers of leaf sheaths; the true "stem" is a rhizome (see pp.86–87) that is concealed underground.

Roots and fibers make up the outer surface

A vertical rhizome forms the central cylinder

TREE FERN (*DICKSONIA ANTARCTICA*) CROSS SECTION

Hollow chambers strengthen the sheaths

The leaves are arranged in spirals

Overlapping sheaths provide support

BANANA PLANT (*MUSA* SP.) CROSS SECTION

Tree ferns
Rhizomes are a type of swollen stem that usually grow horizontally and act a food store for the plant, but in tree ferns they grow upward. Masses of roots and fibers emerge from the rhizome and grow around it, forming a thick, protective mantle that supports it in an upright position.

Plants of the Coast of Coromandel
William Roxburgh's *Plants of the Coast of Coromandel* (1795), which included this illustration of the toddy palm (*Borassus flabellifer*), was published under the direction of Sir Joseph Banks, the longtime president of the Royal Society.

Company style
Painted by an unknown artist, this watercolor illustration of a fan palm (*Livistona mauritiana*) is attributed to the Company School, a group of Indian artists who worked under the patronage of the East India Company, using the distinctive Indo-European Company style.

plants in art

west meets east

During the 18th and 19th centuries, with the expansion of British influence into India, scientists and natural historians engaged by the East India Company began to explore and document the richness and diversity of the country's flora. Highlights of the resulting body of work were striking illustrations that display a unique fusion of Western science and Eastern art.

William Roxburgh (1751–1815), often referred to as the father of Indian botany, was the director of the Calcutta Botanic Garden when he began to commission local artists to create the botanical illustrations used in his important books. More than 2,500 life-size paintings would be featured in his landmark *Flora Indica*.

Influenced by the miniature painters of the 16th- and 17th-century Mughal Empire, the Indian artists created a style that combined the precise detail of Western botanical illustration with the decorative approach and immediacy of their own art. This hybrid style of artwork was well suited to botanical illustration. Working with the naturalist Johan König, who had been a pupil of the great Swedish taxonomist Carl Linnaeus, Roxburgh commissioned many artworks that would become "type" drawings identifying particular wild species.

> "The artworks embody essential qualities of their diverse patronage."
>
> PHYLIS I. EDWARDS, *INDIAN BOTANICAL PAINTINGS*, 1980

Characteristic arrangement
Painted in watercolor on paper, this hand-painted copy of an illustration commissioned by William Roxburgh depicts the Indian redwood (*Caesalpinia sappan*). The way that the plant has been cropped, so that parts of it are not visible, is characteristic of this style of painting. The illustration is inscribed "Received from Rodney, 9th June 1791." This refers to the East India Company merchant ship that delivered artworks from India to London.

HOW CLIMBERS FIND SUPPORTS

Plants cannot see, so they must find another way to locate a support. Some vines detect shade and grow toward it, as that is likely to lead them to the foot of a tree. Others can follow trails of chemicals that lead to a suitable host, and avoid those that lead to other vines. Young stems revolve as they grow, which can help them snag a neighboring branch. Once in position, twining stems or tendrils hook on to the support.

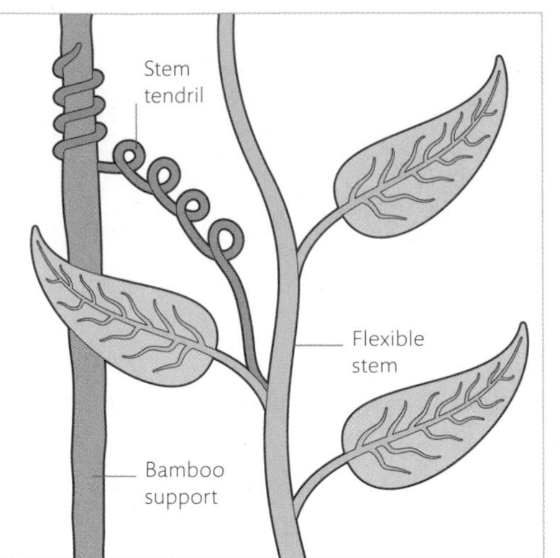

Emerging flowers have the potential of releasing many seeds, further expanding the plant's territory

Long leaf stalks allow the leaves to reach away from the support and face the sun

Bindweed stems can quickly choke out neighboring plants, reaching over 10 ft (3 m) in length each year

To expose their leaves to light, vines climb up other plants and escape the shade

Takeover by twining

Bemoaned by many gardeners, the twining stems of hedge bindweed (*Calystegia sepium*) are one of the secrets of its success. Growing through shrubs and perennials, it quickly smothers their foliage with its own as it successfully competes with them for sunlight. Underground, the plant is equally vigorous, spreading its white rhizomes in all directions.

By flowering up high, the blooms are in easy reach of bee, moth, and butterfly pollinators

Bindweed stems twine counterclockwise when viewed from the tip of the stem

Twining stems can become rigid and woody with age, as with this wisteria

twining stems

Vines and other climbing plants use a number of different methods when scrambling up a support. Tendrils, aerial roots, and hooked prickles all attach themselves to supports, but in some climbers, the stem itself can cling by twining. Some twining plants twist clockwise, while others twist counterclockwise; this distinction, which may have a genetic basis, can be used to tell some species of climbers apart. Beans and bindweed twine counterclockwise, while hops and honeysuckle twine clockwise.

Touch and feel
The ability of a stem to twine around another is due to thigmotropism. When climbing stems and tendrils detect the presence of a support, one side of the growing point begins to grow faster than the other, making the stem bend.

climbing techniques

For plants growing on the forest floor, sunlight is a limiting factor, but vines and other climbers have the ability to scramble up trees and shrubs toward the light. Climbers generally have elongated internodes (the length of stem between each leaf joint) to help cover the distance, but they employ other structures to get a grip, including tendrils, aerial roots, and twining stems.

Tendrils recognize their own stems and avoid twining around them

Tendrils coil when the cells on each side grow at different rates

Tendrils have the ability to feel, and may be more sensitive than the human sense of touch

Surface hairs detect foreign objects and stimulate twining

Springlike tendrils
Many members of the cucumber family, including this loofah (*Luffa cylindrica*), produce tendrils. Derived from modified leaves, they attach to a branch, then coil up, pulling the vine toward its support structure.

Upwardly mobile roots
When roots develop above ground, they are said to be "aerial." Like common ivy (*Hedera helix*) and many others, this *Rhaphidophora elliptifolia* uses aerial roots to cling to and climb along tree branches.

Vine leaves access more sunlight due to the climbing stems

Aerial roots attach to tree bark or other vertical structures

SELF-ADHESIVE SUCKERS

Vines such as Boston ivy and Virginia creeper (*Parthenocissus* sp.) use tendrils to cling to surfaces, but their tendrils are unusual in that they have adhesive pads at the tips. These suckers can hold more than 250 times their own weight, firmly fastening the vine to its support.

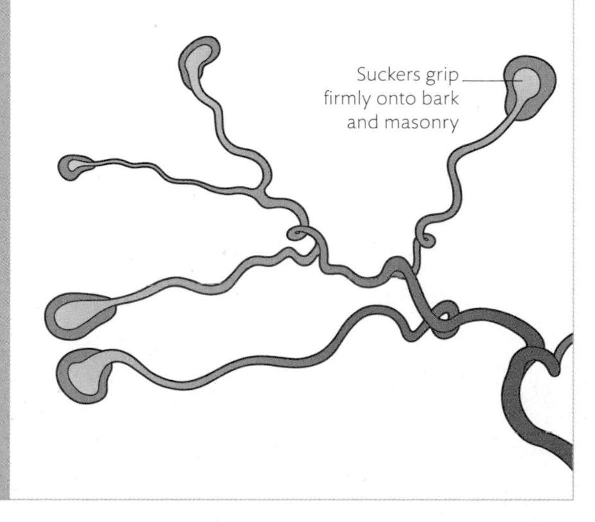

SPINES AND WATER CONSERVATION

Cactus spines are derived from leaves and protect their succulent stems from the attention of thirsty animals. Spines intercept water and direct it down to the ground; they also shade the plant from sunlight and slow the passage of air around the plant. Both strategies reduce water loss.

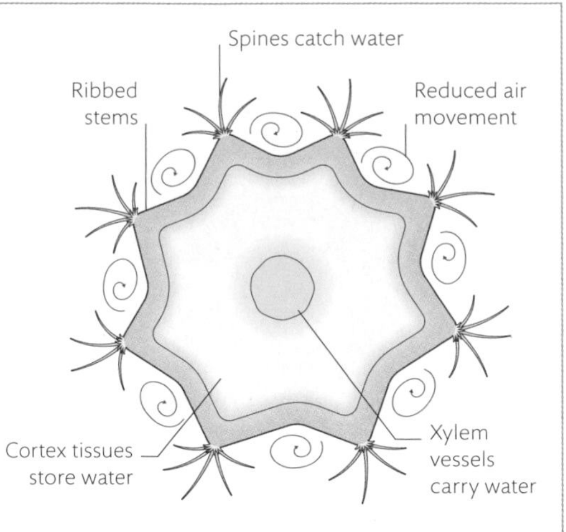

CROSS SECTION OF A TYPICAL CACTUS STEM

Water storage
The most famous members of the cactus family are giants like the saguaro cactus (*Carnegiea gigantea*, see p.100), but tiny treasures like this *Mammillaria infernillensis* are also well adapted to drought. Its thick skin has a waxy coating that greatly reduces water loss.

storage stems

Cacti are renowned for their ability to hoard water in their succulent stems. Most species live in arid areas where rainfall is scarce, so when the heavens open, they must take advantage of the deluge. To absorb as much water as possible, and quickly, many cacti have ribbed stems that expand like an accordion. This prevents them from splitting as they become rehydrated.

Lost leaves
Prickly pears such as *Opuntia phaeacantha* produce tiny leaves on new stem segments, but they are quickly shed to conserve water.

Individual stem segments can be dropped during extended droughts

Large spines are surrounded by tiny, hairlike spines called glochids, which detach and irritate animal skin

Tufts of white hair reduce water evaporation and reflect sunlight, cooling the cactus
Sharp spines take the place of leaves in order to protect the plant from predators

Carnegiea gigantea

the saguaro cactus

What Western film would be complete without a backdrop of giant saguaro cacti? These towering succulents are icons of the Sonoran Desert in Arizona and northwest Mexico, and a special community of bats, birds, and other animals has evolved around them.

By far the most impressive aspect of the saguaro cactus is its size. Individuals regularly reach 50 ft (15 m) or more in height and can weigh as much as 4,400 lb (2,000 kg). Much of that bulk consists of stored water—a precious commodity in the desert. On the rare occasions when it rains, the ribs that run the length of the saguaro's stem can expand to allow the plant to swell and absorb as much water as possible through its extensive, shallow root system. Once stored, the water needs protecting: the saguaro is covered with spines and bristles that not only deter herbivores from feeding on its succulent tissues, but also create shade and reduce airflow close to the saguaro's skin, minimizing water loss to the air.

In spring, saguaro cacti usually put on an impressive floral display. Dense clusters of bright white flowers appear from the apex of the main stem and arms. The flowers are pollinated by birds and insects by day, and by bats at night. The nectar in the saguaro's blooms contains compounds that help female lesser long-nosed bats produce enough milk to feed their young. After the flowers come the fruits, providing an energy-rich food for a wide range of desert animals.

The saguaro has a particularly close relationship with the Gila woodpecker. This bird excavates a hole in the cactus in which it can build its nest. Vacated woodpecker holes are subsequently used as shelter and nesting sites by many other birds, mammals, and reptiles.

Saguaro sentinels
Saguaro cacti stand like alert guards watching over the desert. Slow-growing but long-lived, saguaros can survive for 200 years or more. They only grow arms after they reach 50 to 100 years old.

Crested saguaro
Occasionally saguaro cacti take on a crested form. This fanlike appearance is due to changes in the growing tip (apical meristem). The cause is not known: it may result from a genetic mutation, or from physical damage by lightning or frost.

Fanlike crests are thought to develop on only about 1 in every 200,000 cacti

New arms can continue to sprout on the crest

leafless stems

Leaves produce food for plants, but moisture quickly evaporates through their large surface area. In the dry, harsh climate of the desert, some plants have adapted by not growing leaves. Instead, their green succulent stems take on photosynthesis. By taking in carbon dioxide at night and storing it in their stems, these plants can photosynthesize with pores closed during the hot sunlight hours.

Succulent stems only open their pores to allow gas exchange at night

Stems at work

Many succulents, such as this *Euphorbia woodii*, a South African member of the spurge family, show the same adaptation to their environment as cacti. They do not have spines, but their very reduced leaves mean that they rely on their succulent stems for photosynthesis to produce the carbohydrates that all plants need to grow.

A poisonous sap in the stems of this succulent *Euphorbia* deters herbivores

SPIKED STEMS

Cacti display a great variation in the size and shape of their succulent stems, but almost all of them are leafless. A few species still produce leaves, but in most, these have evolved into spines. Spines protect the plant from grazing animals, reduce air circulation, and help to add shade. Cacti native to humid rainforests also lack leaves, although their flattened stems may look leaflike.

The green pigments in the stem that make photosynthesis possible are found just below the outer layer (epidermis)

Strawberry runners
A number of species colonize new territory using stolons. They include the edible strawberry (*Fragaria* x *ananassa*), spider plant (*Chlorophytum comosum*), and mock strawberry (*Duchesnea indica*, syn. *Fragaria indica*)—shown here in a drawing made in India for Robert Wright in 1846. Their stems produce new plants at the nodes, and become independent once severed from the parent.

new plants from stems

Stems make it possible for a plant to expand its range in several ways. The ground-hugging or prostrate stems of some plants can root as they spread, while subterranean rhizomes do the same underground. Some plants that are otherwise upright produce slender, elongated horizontal stems that creep across the soil or just beneath it and form baby plants at the nodes; these stems are called stolons or runners.

STEMS UNDERGROUND

Many plants develop rhizomes, corms, or tubers, which are derived from stems. These underground "stems" grow on or just below the soil surface, and they not only allow plants to survive adverse conditions, but they also provide a means of propagation. Any pieces broken off the parent can root and form new plants. *Crocosmia* corms can also produce stolons that distribute new corms a short distance away from the parent.

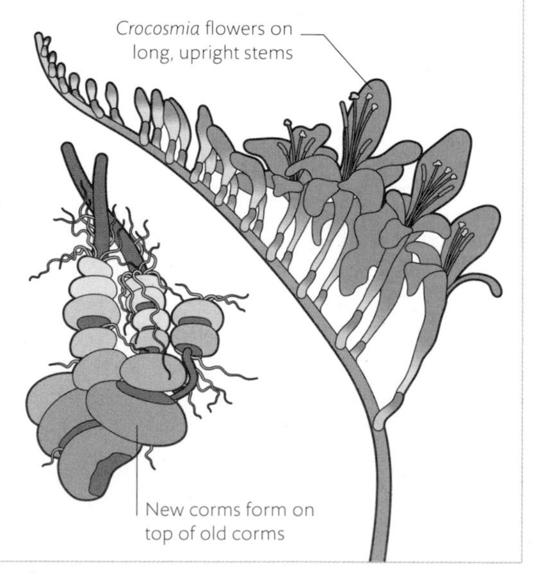

***CROCOSMIA* CORM OFFSETS**

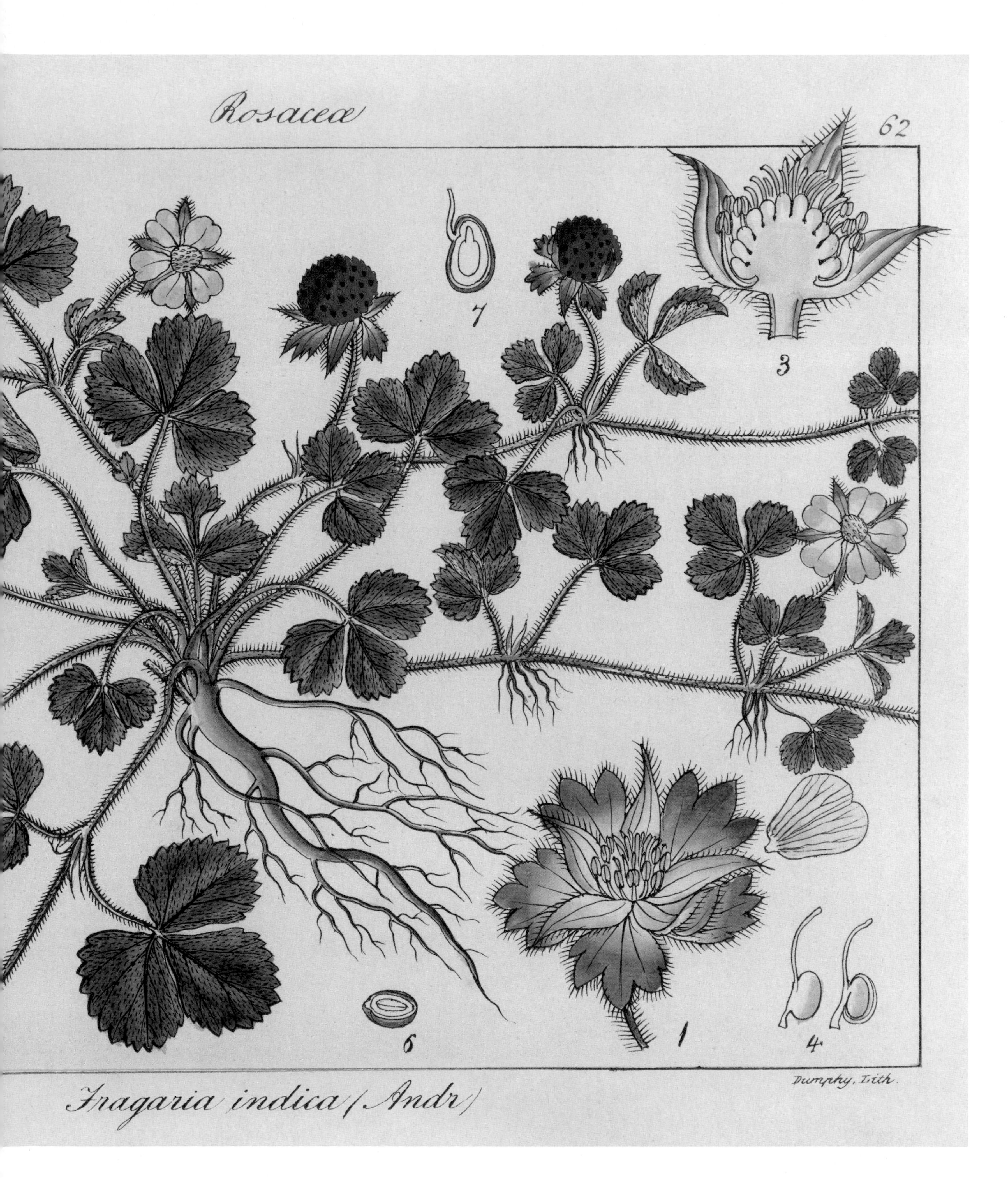
Rosaceæ
62
7
3
6
1
4
Dunphy, Lith.
Fragaria indica (Andr)

Phyllostachys edulis

moso bamboo

The canopy of a grove of moso bamboo can reach a level of up to 100 ft (30 m) above the ground, making it easy to think these giant bamboos are trees. But moso bamboo is a grass, albeit a very tall, woody one. Like other grasses, it is characterized by a jointed stem called a culm.

Although it is synonymous with Japanese culture, moso bamboo (*Phyllostachys edulis*) is actually native to mountain slopes in warm, temperate regions of China, and only naturalized to Japan. This species is of great economic importance throughout Asia, where it is used as a food source, a building material, and a fiber for making textiles and paper.

Moso bamboo has an astonishing rate of growth, and new stems are able to add more than 3 ft (1 m) to their height each day. Growth is also vigorous below the surface of the soil: dense mats of roots and rhizomes (underground stems) spread out relentlessly, throwing up new shoots to colonize territory. Such vegetative growth is the plant's main reproductive strategy. As a result, entire hillsides can be made up of clones of a single individual. While it can also reproduce sexually, this species flowers only once every 50 to 60 years. When it does flower, however, it produces many thousands of seeds that are quick to germinate.

Moso's aggressive growth habit is cause for concern when it is introduced outside its native range. Individuals can quickly escape gardens and invade surrounding areas. Producing impenetrable root mats, heavy leaf litter, and dense shade, they easily smother other plant species.

Young moso shoots are edible, but like many bamboos, moso protects itself with a potent chemical cocktail that includes oxalic acid and cyanide compounds. With adequate boiling, the compounds break down and make the shoots safe to eat.

Bamboo canopy, Japan
Covering 6 sq miles (16 sq km), the Sagano Bamboo Forest near Kyoto is treasured for the beauty and tranquility of its dense bamboo groves.

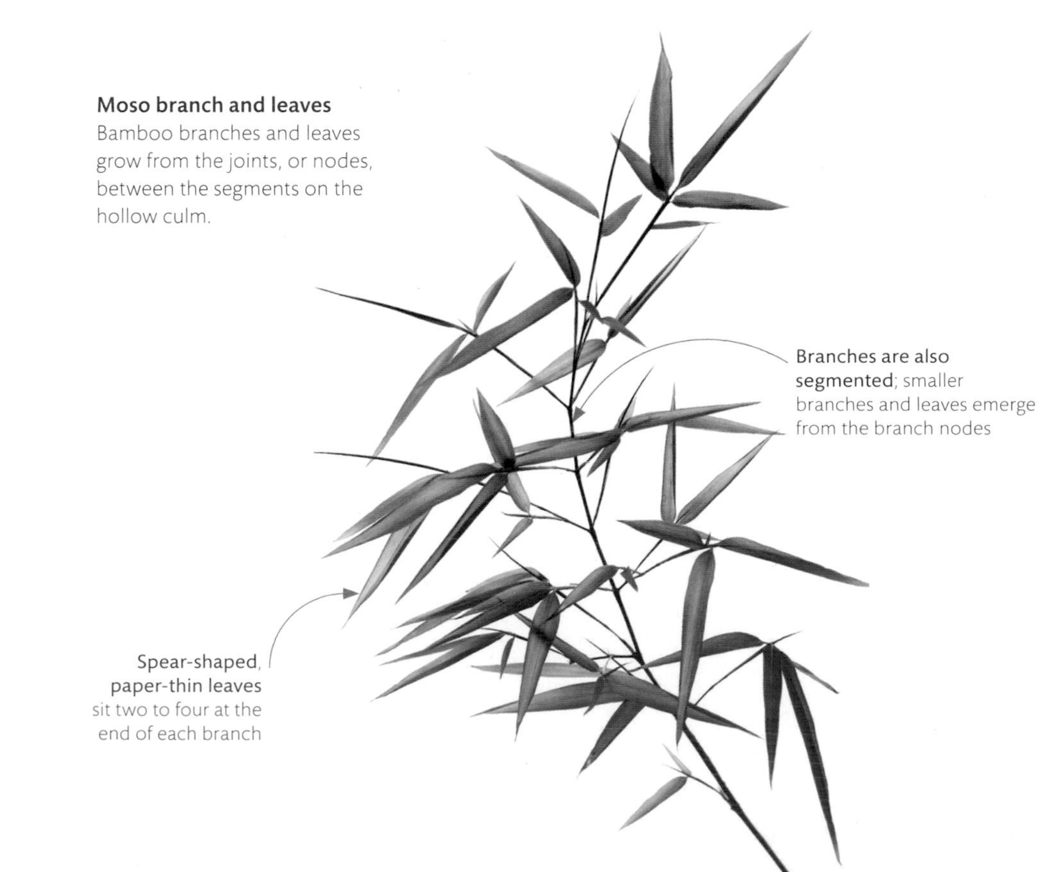

Moso branch and leaves
Bamboo branches and leaves grow from the joints, or nodes, between the segments on the hollow culm.

leaves

leaf. a flat, usually green, structure attached to a plant stem directly or by a stalk, in which photosynthesis and transpiration take place.

Deciduous broad-leaves
Many plants, such as this sugar maple (*Acer saccharum* subsp. *saccharum*), have large, flat leaves, maximizing the surface area available for photosynthesis.

leaf types

Leaves can be evergreen, which stay on the plant year round, or deciduous, which are discarded during certain times of year. Leaves are a costly investment for a plant and evergreen leaves may minimize the demands on its resources in the long term. However, if they are unlikely to survive year-round, having cheap, deciduous leaves may be more beneficial.

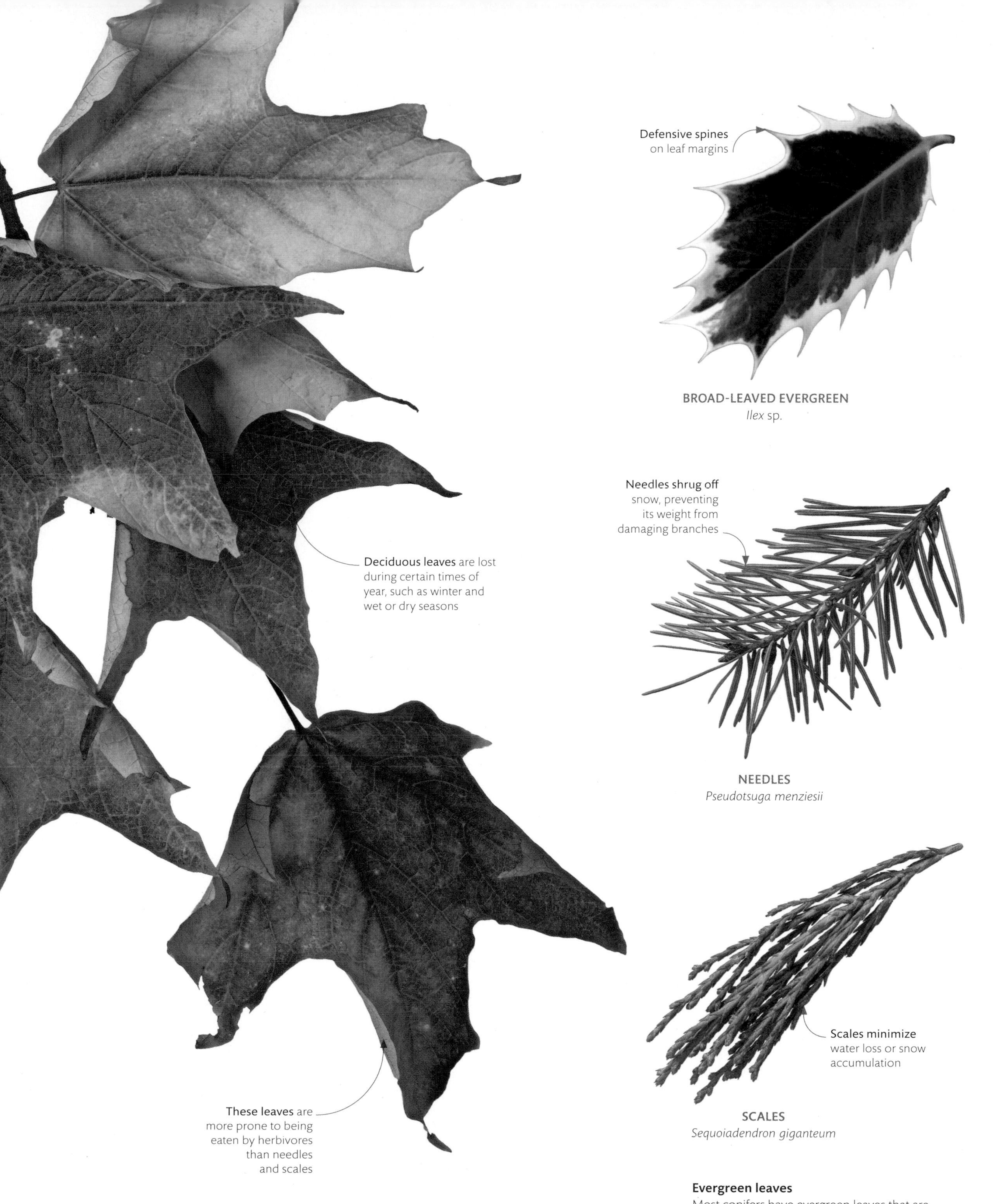

Evergreen leaves
Most conifers have evergreen leaves that are reduced to needles or scales. These have a smaller surface for photosynthesis, but they can photosynthesize for more of the year. They are also adapted to survive cold winters.

leaf structure

Most leaves are full of cells that harvest light for photosynthesis. These mesophyll cells are supplied with water and nutrients by a network of veins. The veins also carry carbohydrates made during photosynthesis to other parts of the plant. Pores on the leaf surface—stomata—open to take in carbon dioxide and close to prevent water loss. Evaporation from the rest of the leaf surface is halted by a waterproof, waxy coating called the cuticle.

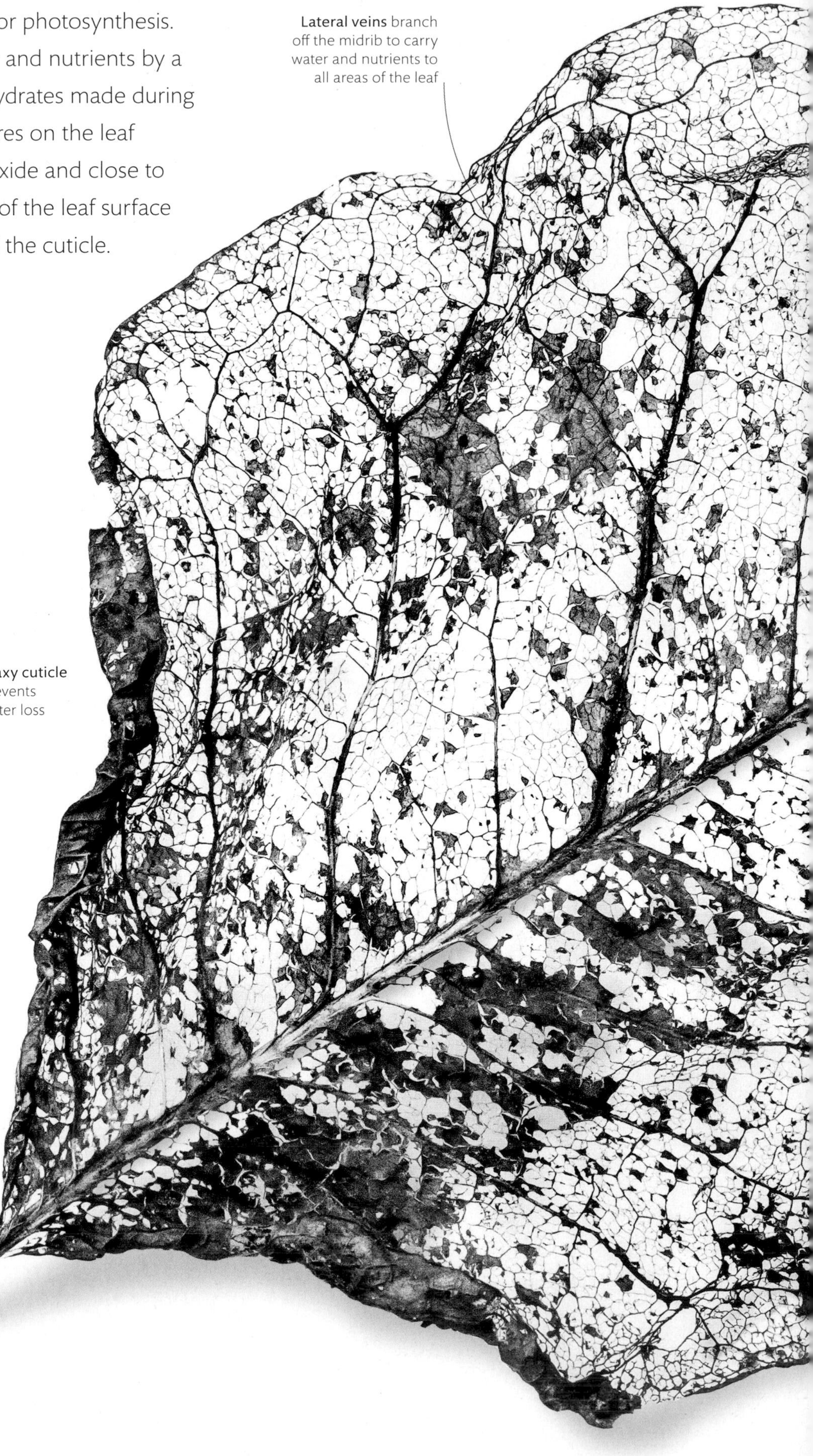

Lateral veins branch off the midrib to carry water and nutrients to all areas of the leaf

Inside foliage
Taro leaves (*Colocasia esculenta*) are covered by a single layer of cells called the epidermis (blue). The interior has palisade mesophyll (green) and spongy mesophyll (yellow) cells. The gray structure is a vein, or vascular bundle, made up of xylem and phloem vessels.

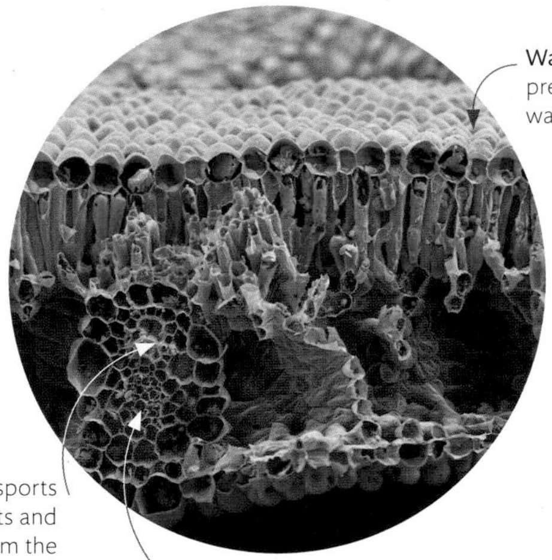

Waxy cuticle prevents water loss

Xylem transports nutrients and water from the root to the shoot

Phloem carries carbohydrates made during photosynthesis to other parts of the plant

The leaf stem extends along the center of the leaf, known as the midrib

Margins and edges may be smooth, bumpy, or serrated, giving leaves their characteristic shapes

The midrib—the central vein—is surrounded by thick, supportive tissues that give the leaf rigidity and structure

Leaf blades are usually flat in order to harvest the maximum light and to minimize how far nutrients and carbon dioxide need to travel

Skeletal framework

The skeleton of a leaf is formed by its veins, which carry water, nutrients, and food throughout the plant. Like our own bones, leaf veins also provide support. Monocot leaves have parallel veins, but the veins of most plants form a branched network, as in this *Magnolia*.

STRUCTURAL VARIATIONS

Although most flowering plant groups have similar leaf structures, the leaves of grasslike monocots have a single type of mesophyll, while those of other plants have two distinct layers: palisade mesophyll—the main photosynthesis site—and spongy mesophyll, through which carbon dioxide can move.

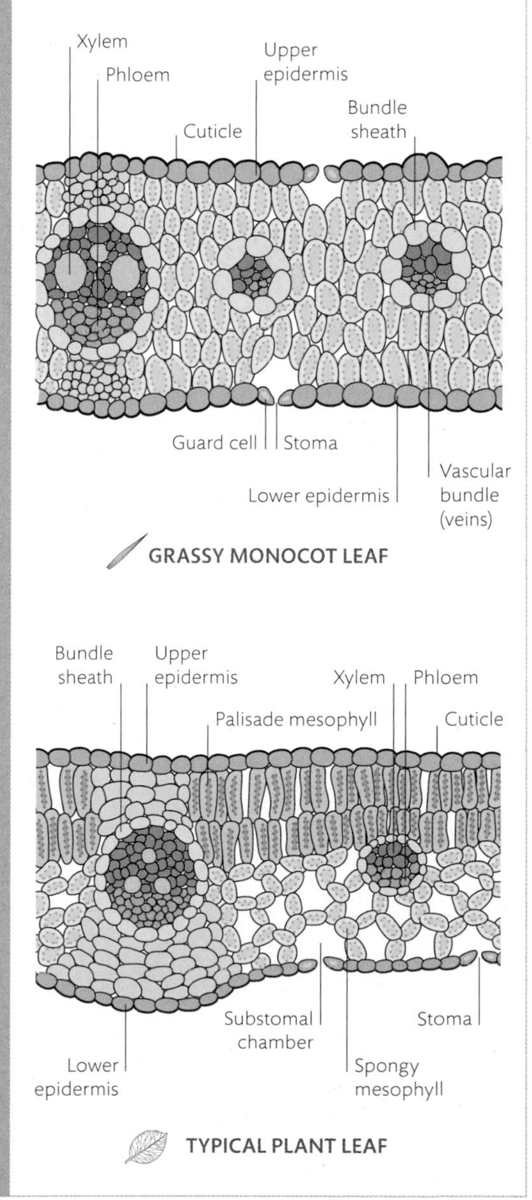

compound leaves

Compound leaves are divided into two or more leaflets, providing plants with more photosynthetic surface area for the same investment in resource-hungry support tissues, such as branches. When leaflets radiate from a single common point the leaves are called palmate. When leaflets originate from several locations along the central stalk, or rachis, the leaf is said to be pinnate.

TRIPINNATE (ALTERNATE)
Pteridium sp.

Pinnae are arranged in a staggered fashion along the rachis

Fern fronds
Most ferns, such as this *Pteridium* leaf, exhibit varying degrees of pinnation. The leaflets are referred to as pinnae, and these are often further subdivided into pinnules. Pinnules, in turn, are divided into segments.

Pairs of leaflets develop along the rachis
UNIPINNATE (PARIPINNATE)
Tamarindus indica
Terminal leaflet forms at the end of the rachis
UNIPINNATE (IMPARIPINNATE)
Carya ovata
Leaflets are themselves pinnately divided
BIPINNATE (OPPOSITE)
Leucaena leucocephala
Winged petiole (leaf stalk) resembles the true leaf
UNIFOLIATE
Citrus hystrix
Two leaflets grow off a single petiole
BIFOLIATE
Hymenaea courbaril
Three leaflets comprise a single palmate leaf
TRIFOLIATE
Trifolium sp.
Four palmately arranged leaflets on a single petiole
QUADRIFOLIATE
Marsilea crenata
Palmately compound leaves with five leaflets
5-PALMATE
Aesculus pavia
Variable numbers of leaflets form, depending on the growing conditions
MULTIFOLIATE
Cannabis sp.

Acanthus wallpaper design, 1875
Morris's wallpapers and fabrics featured stylized designs based on large, repeating patterns of flowers, leaves, or fruit. In this, his most expensively produced wallpaper, he used the deeply cut leaves of the acanthus, a plant that has appeared in architecture and art since ancient times. Acanthus was printed by the London firm Jeffrey & Co., using 15 natural dyes and 30 separate wood blocks for each full repeat.

> “... any decoration is futile ... when it does not remind you of something beyond itself.”
>
> WILLIAM MORRIS, *LECTURE ON PATTERN*, 1881

Art of glass
This decorative stained glass window design is by the Italian artist Giovanni Beltrami (1860–1926). The delicate leaf and floral motifs are typical of the Art Nouveau style.

plants in art

designed by nature

The Arts and Crafts movement of the late 19th century developed as a reaction to the effects of industrialization on the lives of ordinary people and to the shoddy quality and design of mass-produced goods. Nostalgia for the simple ways of the past, fine materials, and honest craftsmanship were the core tenets of the movement, and its leading craftsmen and designers were primarily inspired by the forms of the natural world.

Horse chestnut, 1901
Scottish artist Jeannie Foord's botanic paintings were composed with a designer's eye. Their homage to the simple, naturalistic beauty of everyday leaves and flowers typifies the values of the Arts and Crafts movement.

The driving force behind the Arts and Crafts movement was the British craftsman William Morris. Virtually all of his designs for wallpapers and textiles show twining tendrils, leaves, and flowers. They were named after the plants they featured, but Morris's designs were stylized evocations of their forms, rather than botanically accurate reproductions.

Morris's study of ancient herbals. medieval woodcuts, tapestries, and illuminated manuscripts informed his designs, and he revived traditional crafts such as woodblock printing and hand weaving. He urged design students to correct their "mannered work" with a diligent study of nature, study of the different ages of art, and imagination.

Influenced in part by the Arts and Crafts movement, Art Nouveau artists and designers regarded nature as the underlying force of life and developed their own distinctive vocabulary of motifs based on the organic shapes of swirling plant roots, tendrils, and flowers, often fused with images of sensuous women.

developing **leaves**

Like all parts of a plant, leaves develop from clusters of dividing cells. Many species of trees, notably conifers, produce leaves continually, but deciduous broad-leaved trees can only produce them at specific times of the year. In the fall, deciduous trees produce hardened, dormant buds that contain partially developed leaves. The buds are protected during the winter and establish new leaves quickly in the spring when the risk of frost is over.

The emerging leaves slowly expand into a flat shape

Red pigments protect young sycamore maple leaves from being damaged by light

Young leaf tissues are soft and easily damaged

Timing it right

New leaves have to emerge at just the right time; if the buds burst too early, the young leaves risk being damaged by frost; if they emerge too late, they lose valuable growing time. Trees such as this sycamore maple (*Acer pseudoplatanus*) monitor the number of cold days to estimate when winter is over, and wait for warmer temperatures before the buds start bursting.

WHY LEAVES ARE GREEN

Leaves contain chlorophyll, a pigment that harvests light energy for use in photosynthesis (see pp.128–129). Chlorophyll is stored on stacks of membranes within tiny particles in the leaf cells called chloroplasts. Sunlight contains every color, and chlorophyll absorbs them all except for green. The green light reflects out of the leaf, or passes through it, which makes the leaf look green.

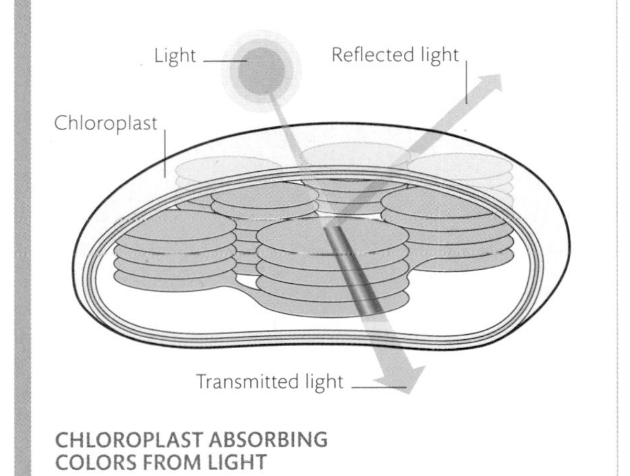

CHLOROPLAST ABSORBING COLORS FROM LIGHT

Expanded sycamore maple leaves have five palmate lobes

unfurling ferns

Developing fern fronds are tightly curled into structures called fiddleheads, which protect the delicate growing tips at their center. As they slowly unfurl, the lower parts of the leaf toughen and begin to photosynthesize, providing the energy that powers the development of the rest of the leaf. This uncoiling process, known as circinate vernation, is mainly found in ferns and palmlike cycads.

SPORE-BEARING FRONDS

Ferns do not flower; they produce spores (see p.338) in speckles on their leaf undersides. Each speckle, or sorus, has a protective cover—the indusium, shown as a semicircular structure on the drawing below—which shrivels to enable the fern to release its spores.

DRYOPTERIS FILIX-MAS

Why ferns are furled
Ferns produce relatively few large leaves, but each represents a major investment of resources. The furled leaves only have a limited ability to photosynthesize, but remain shielded from herbivores. The loss of photosynthesis is outweighed by the reduction in damage from grazing insects.

***Blechnum orientale* fiddlehead** is edible; it is used in traditional medicines

The soft tissues of the developing leaf toughen as the fiddlehead slowly unfurls

BLECHNUM ORIENTALE

Hairs protect the developing fiddleheads from insect herbivores

The lower fronds begin to photosynthesize as soon as they unfurl

DICKSONIA ANTARCTICA

Cibotium glaucum **fiddleheads** are the size of real violin heads, but they develop into leaves more than 9 ft (2.5 m) long

CIBOTIUM GLAUCUM

When viewed from the shoot tip, leaves often form spirals in precise mathematical ratios

The arrangement of the developing leaves protects the vulnerable shoot tip

Alternate leaves

Most plants, such as this porcelain berry (*Ampelopsis glandulosa*), produce leaves at alternate intervals along their stems. At the shoot tip, the plant growth hormone auxin flows into the developing leaves. The lack of auxin on the opposite side of the shoot tip makes a new leaf develop at the farthest point from the previous leaf.

TYPICAL ARRANGEMENTS

While alternate leaf patterns are the most common, some plant species develop pairs (opposite) or groups of leaves (whorls) at the same point along the stem. These paired and grouped leaves are able to collect sunlight from multiple directions at once. Whorls are relatively widely spaced along the stem to avoid shading the leaves below.

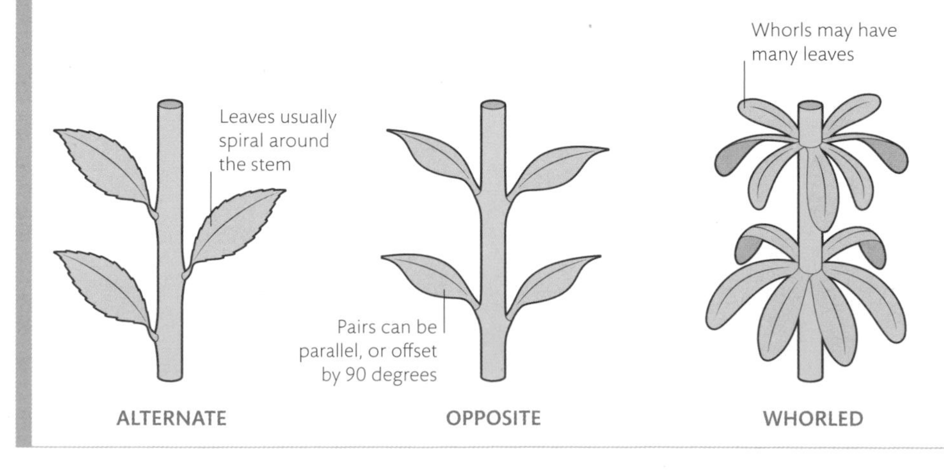

leaf arrangement

It is important to avoid being shaded by neighbors, but plants also have to avoid shading their own leaves. Leaf arrangement, or phyllotaxis, occurs in patterns that are specific to each species and prevents the upper leaves from blocking light to the lower branches, enabling the plant to absorb as much sunlight as possible.

Brightly colored berries are positioned in gaps between the leaves to attract birds and other seed dispersers

Complex composition
Laelia tenebrosa, Philodendron hybrid, Calathea ornata, Philodendron leichtlinii, Polypodiaceae (1989) by Pandora Sellars not only shows Sellars' exceptional ability to render plants accurately, but is also an example of her artist's sense of composition, capturing the way light plays across the leaves.

***Brittany 1979* (detail)**
This watercolor on vellum painting of a fallen leaf from a European pear tree (*Pyrus communis*), is a portrait of exquisite beauty. McEwen's focus is on nature's treasure lying underfoot in this painting from his series of leaves in fall colors or stages of decay. Recording the location and year of each painting, he captured the hues and imperfections of leaves with botanic integrity and an artist's sensibility.

plants in art

botanic reinvention

There was a revolution in botanic art when artists stopped seeking perfect specimens and chose to reveal the beauty in ordinary vegetables, fruits, and flowers, with all their minute imperfections, and leaves ravaged by beetles or in stages of decay. Rory McEwen, a 20th-century British artist, was a pioneer of this approach and is widely regarded as the first botanical painter to portray the natural world with the mind of a modern artist.

McEwen's radical reworking of botanical art emerged from the general ferment of change in the 1960s. Working on vellum, rather than paper, he found that his watercolors had an extraordinary translucency and intensity on the vellum's silky smooth and nonporous surface, just as illustrations did in medieval illuminated manuscripts.

McEwen painted with scientific precision, using tiny, fine brushes, and applied the same meticulous techniques to every subject, regardless of whether it was a heritage bloom, an onion, or a fallen leaf that he had picked up from the pavement. He took the time to paint his subjects in minute detail, highlighting their beauty of form and color regardless of so-called imperfections.

Another British artist who took botanical art to new heights in the 20th century was Pandora Sellars. Her work appeared in many botanical publications, and her artistic ability and sensibility brought worldwide recognition. Sellars started her career as a botanical artist when she found that a camera was incapable of capturing the color and form of the orchids in her husband's greenhouse.

***Indian Onion painted in Benares 1971* (detail)**
This onion, depicted in glowing shades of purple and pink with a papery brown skin, is almost tangible. Painted in translucent watercolor, it seems to be poised in space. McEwen's series of paintings of onions were among his most influential and original works.

> “A dying leaf should be able to carry the weight of the world.”
>
> RORY McEWEN, *LETTERS*

Waxy surfaces and drip tips help rainforest leaves shed rainwater rapidly

leaves and the water cycle

Plants use less than five percent of the water they take from the soil; the rest evaporates from the surface of the leaves into the surrounding air. This seemingly wasteful process, known as transpiration, can be a problem in dry climates, but it is vital in many ways. Transpiration makes it possible for water to move up against the pull of gravity through even the tallest trees, carrying with it nutrients from the soil needed for growth. In hot climates, the evaporating water cools the leaves, just as sweating cools human skin.

The red lower leaf surfaces maximize light absorption in shady conditions

TRANSPIRATION

When the stomata are opened to take in carbon dioxide, water continuously evaporates from the leaves. This creates negative pressure, drawing water up the stem from the roots through the plant's vascular system—bundles of fine tubes known as xylem.

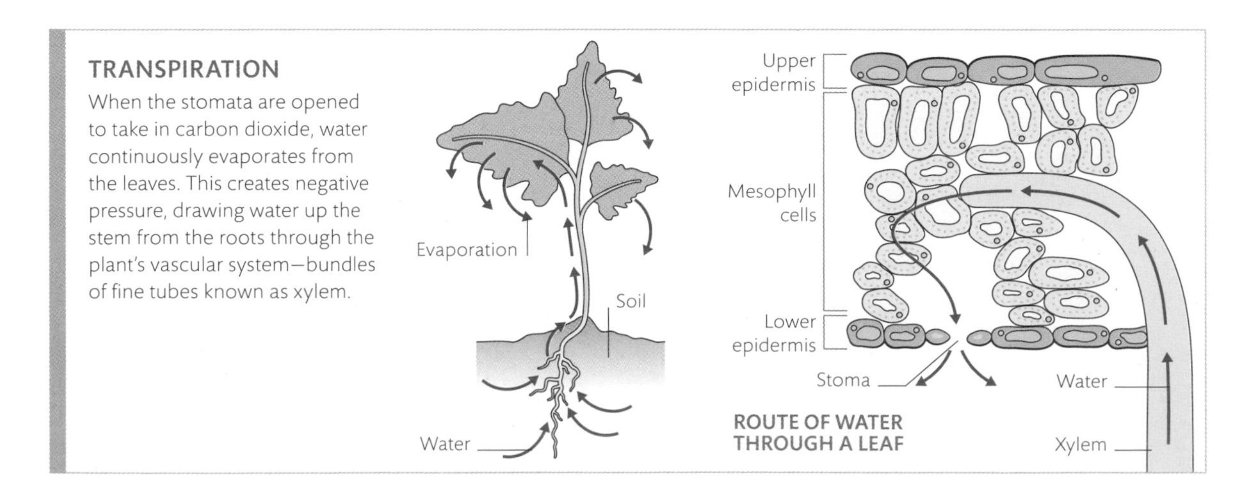

Wasting water?
Costus guanaiensis grows in tropical South America. Water is so abundant in the rainforest that plants can produce large leaves to maximize light absorption without the risk of dehydrating due to their high rate of transpiration. As a result, approximately 30 percent of all the rain that falls over land each year passes through the leaves of rainforest plants.

Rainforest leaves have many stomata to maximize the uptake of carbon dioxide

The long leaves of *Costus guanaiensis* are typically up to 2 ft (60 cm) in length

leaves and **light**

A plant's leaves gather sunlight and convert its energy into food by means of a complex process called photosynthesis. Plants use light and a light-sensitive green pigment called chlorophyll in their leaves to transform carbon dioxide from the air and water in the soil into sugars to feed themselves. As a by-product, they also produce the oxygen that supports almost all life on earth.

The green color of leaves comes from chlorophyll, a pigment that absorbs light energy

Broad surface area
Philodendron ornatum leaves are large because the plant grows in the shade and needs to survive with little light. As with most plants, a network of veins takes water absorbed by the roots to the leaf; it also transports sugars produced during photosynthesis to the rest of the plant.

The long leaf stem makes it possible for the plant to tilt the leaf toward the sun

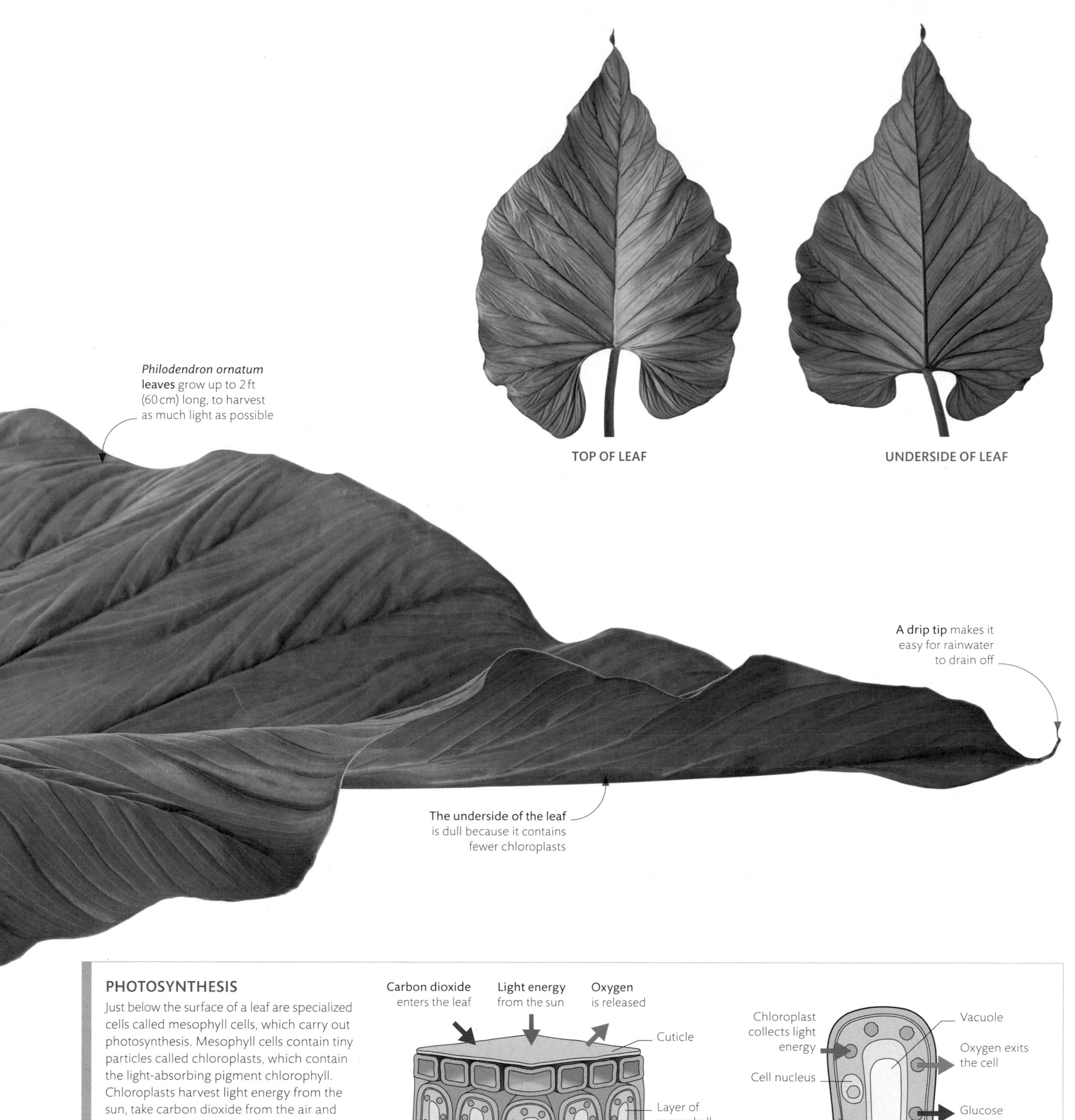

PHOTOSYNTHESIS

Just below the surface of a leaf are specialized cells called mesophyll cells, which carry out photosynthesis. Mesophyll cells contain tiny particles called chloroplasts, which contain the light-absorbing pigment chlorophyll. Chloroplasts harvest light energy from the sun, take carbon dioxide from the air and water (absorbed from the soil by roots and delivered to the leaf by the plant's vascular system), and convert it all into glucose. This is then packaged into sucrose, a form of sugar that the plant uses as food. During photosynthesis, oxygen is also released into the air through the leaf's pores (stomata).

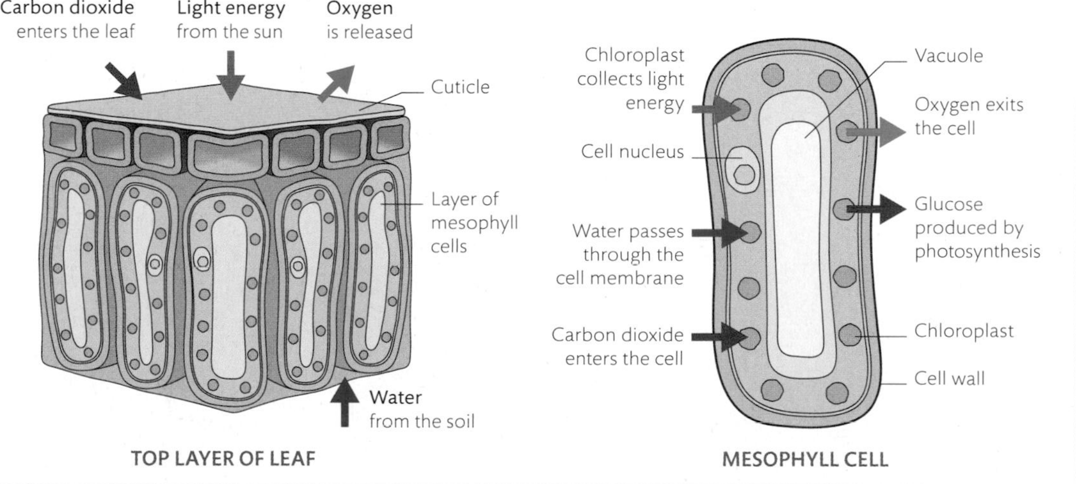

Thick tissues and stiff veins help a large leaf to maintain its shape

SMALL LEAF
Eucryphia sp.

Tiny surface area reduces heat loss in cold climates

TINY WIDTH
Pinus sp.

Large leaves enable understory plants to harvest enough light

LARGE LEAF
Colocasia esculenta

GIANT LEAF
Gunnera manicata

Colossal canopy
The enormous leaves of the giant rhubarb (*Gunnera manicata*) can reach up to 10 ft (3 m) in diameter. This plant originates from the warm, wet mountains of Brazil, where its oversized leaves allow it to dominate other plants in the battle for sunshine.

leaf size

Leaves range in size from less than 1/25 in (1 mm) to more than 82 ft (25 m) in length in some raffia palm trees. Large leaves have a greater surface area for photosynthesis but also transpire more water, which cools down plants in wet tropical regions. Plants in cold alpine regions have small leaves that lose less heat, minimizing the chance of frost damage, while desert species form tiny leaves (or none at all), to reduce the amount of water evaporation.

Midsize leaves maximize photosynthesis and help prevent excessive water loss in temperate climates

MIDSIZE LEAF
Acer japonicum

leaf shapes

Leaves grow in a wide array of shapes and sizes, each enabling a plant to thrive in its natural habitat. Leaf shapes balance the plant's need to take in light with that of preventing water loss or resisting damage from wind and rain. Simple leaves grow in one piece; compound leaves have several parts.

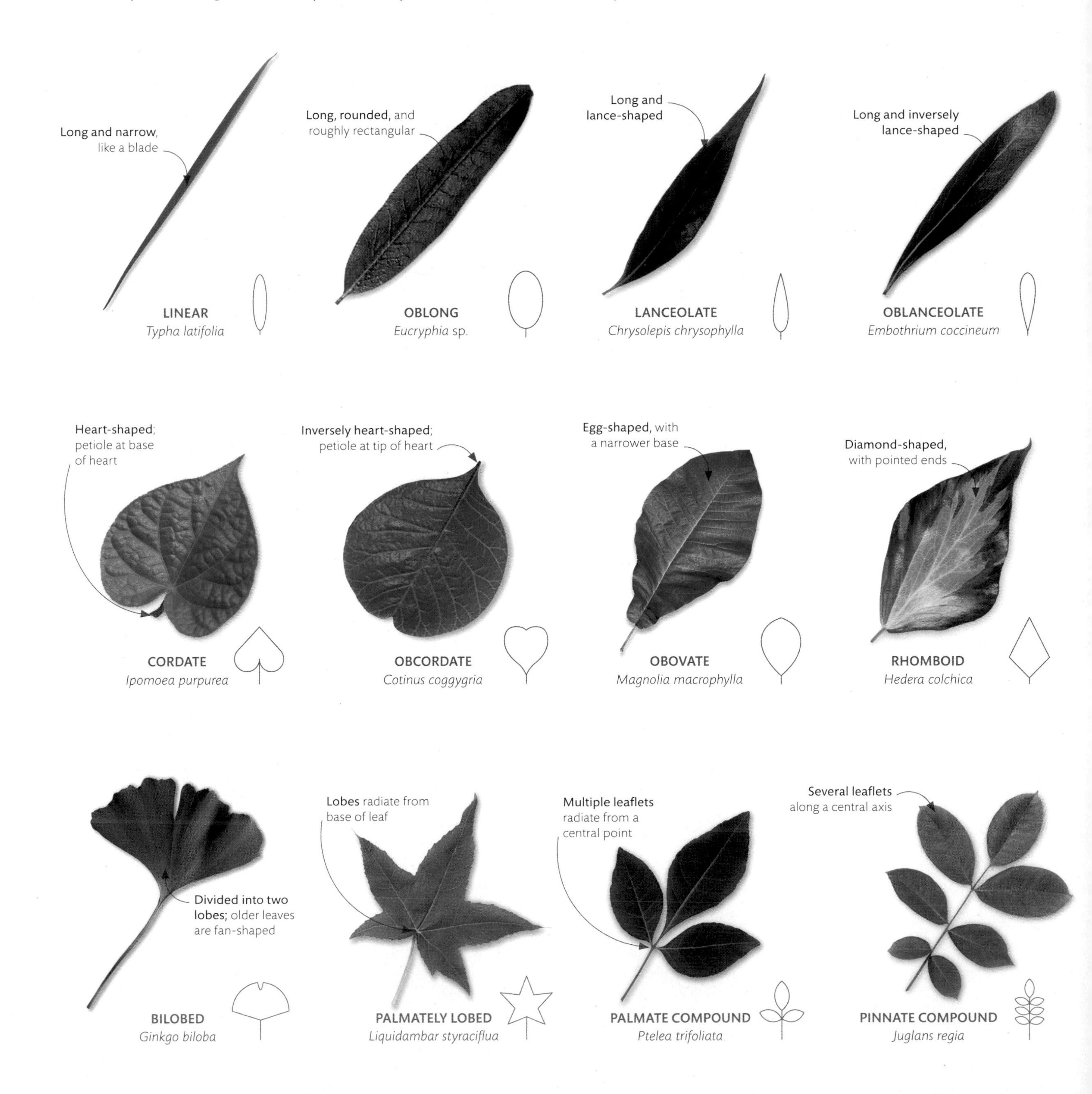

Evolving differences

Why do species growing in the same environment have different-shaped leaves? Evolution is a factor. Over time, a plant's DNA changes as individual plants with leaf shapes better suited to their particular needs—and their environment—are the ones that survive and reproduce. Plants with less optimal leaf shapes die. Evolution does not always create a perfect leaf shape, but selects the best option available.

DESIGNED FOR DRAINAGE

Drip tips are usually aligned with major veins, which channel water from the leaf. Each lobe of more complex leaves, such as those shown here, or the segments (pinnae) of fern fronds, forms its own drip tip. Combined with a leaf's water-repelling, waxy outer layer (cuticle), drip tips enable rainforest plants to cope with extremely heavy rainfall.

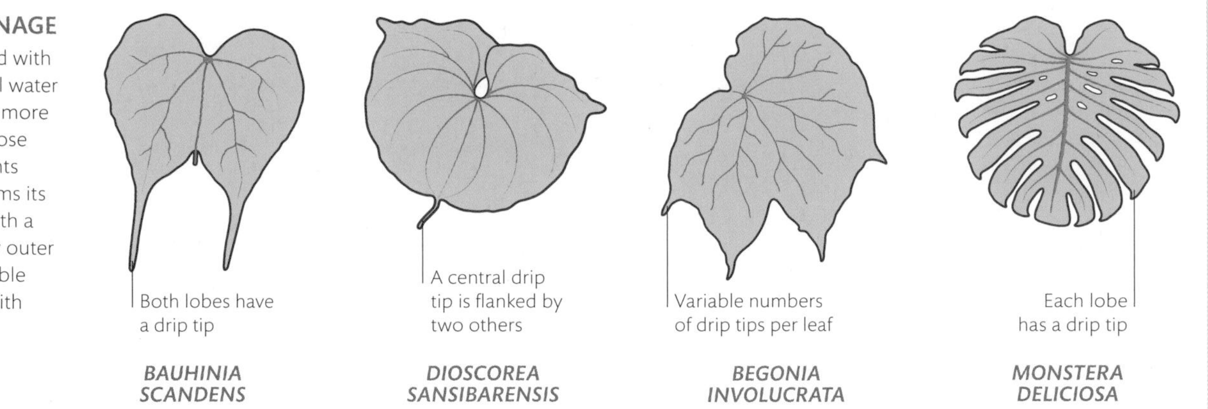

Queen anthurium

Native to South America, the queen anthurium (*Anthurium warocqueanum*) has leaves that are more than 3 ft (1 m) long. This large surface area receives a lot of rain, but it runs off rapidly with the help of a drip tip. Drip tips are more common in understory plants such as anthurium than in leaves that grow at the top of the rainforest canopy, which dry rapidly in the sun.

The waxy cuticle helps water run off

Debris on the leaf surface, which could shade leaves, is carried away by rain flowing down the leaf

The midrib of the leaf acts like a gutter, draining water toward the drip tip

The weight of water drops makes the leaf tilt downward, sending the water to the drip tip

The drip tip is also known as an acumen

The long leaf can turn toward the brightest source of light

A collective vein circles the leaf near its edge, channeling water into the leaf's central gutter. All *Anthurium* species have a collective vein

The petiole holds the leaf at an angle, directing water downward

Leaves on Asian figs have drip tips to cope with heavy rainfall

ASIAN FIG LEAF

drip tips

As an adaptation to heavy rainfall, many tropical rainforest leaves form drip tips—elongated leaf tips that allow water to drain off rapidly. The exact benefit of this is unclear. Some researchers think that water left standing on a leaf might encourage the growth of harmful fungi, algae, or bacteria, while others believe that removing the water may help the leaf to regulate its temperature or prevent the droplets from reflecting sunlight, which would hinder photosynthesis.

leaf **margins**

The edge of a leaf, known as its margin, is a distinguishing feature that can be used to identify plant species. Leaf margin shapes help plants adapt to their environment. Lobed or serrated edges increase the movement of air around the leaf, leading to more water loss but also allowing the leaf to take in more carbon dioxide for photosynthesis. Smooth margins help rainforest plants to shed rainwater rapidly.

Smooth edge with no serrations (teeth) or indentations

ENTIRE
Eucryphia sp.

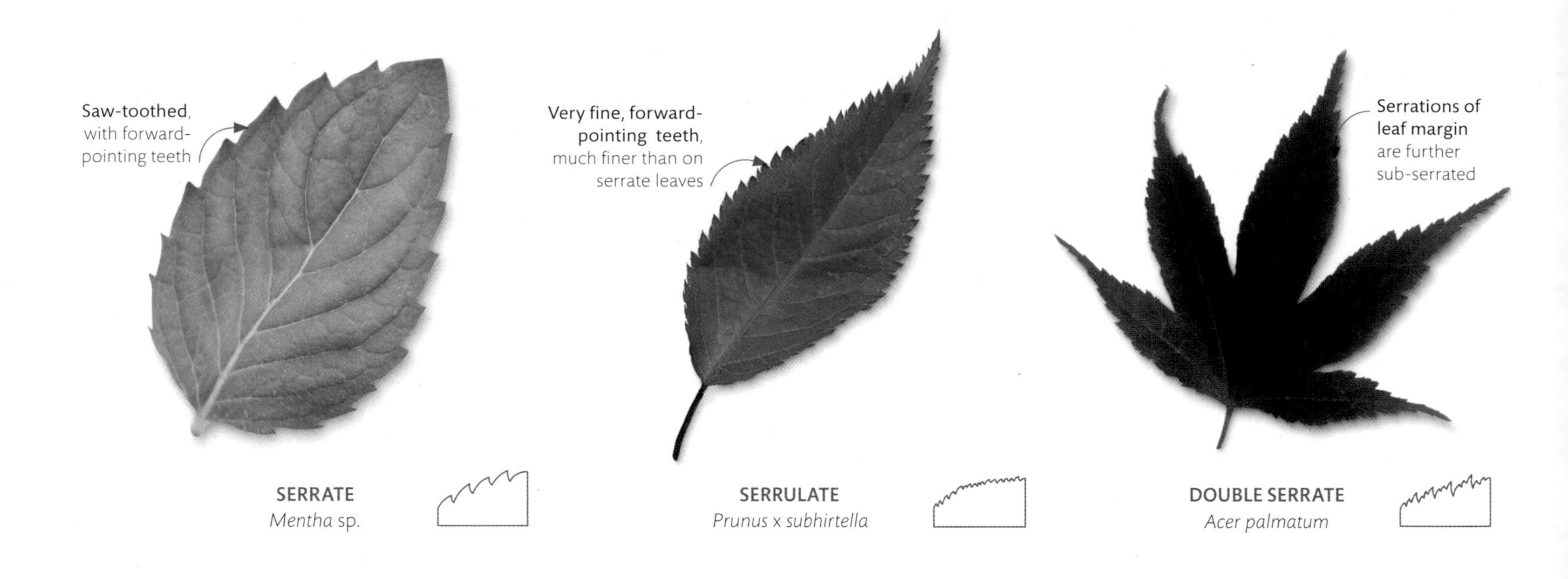

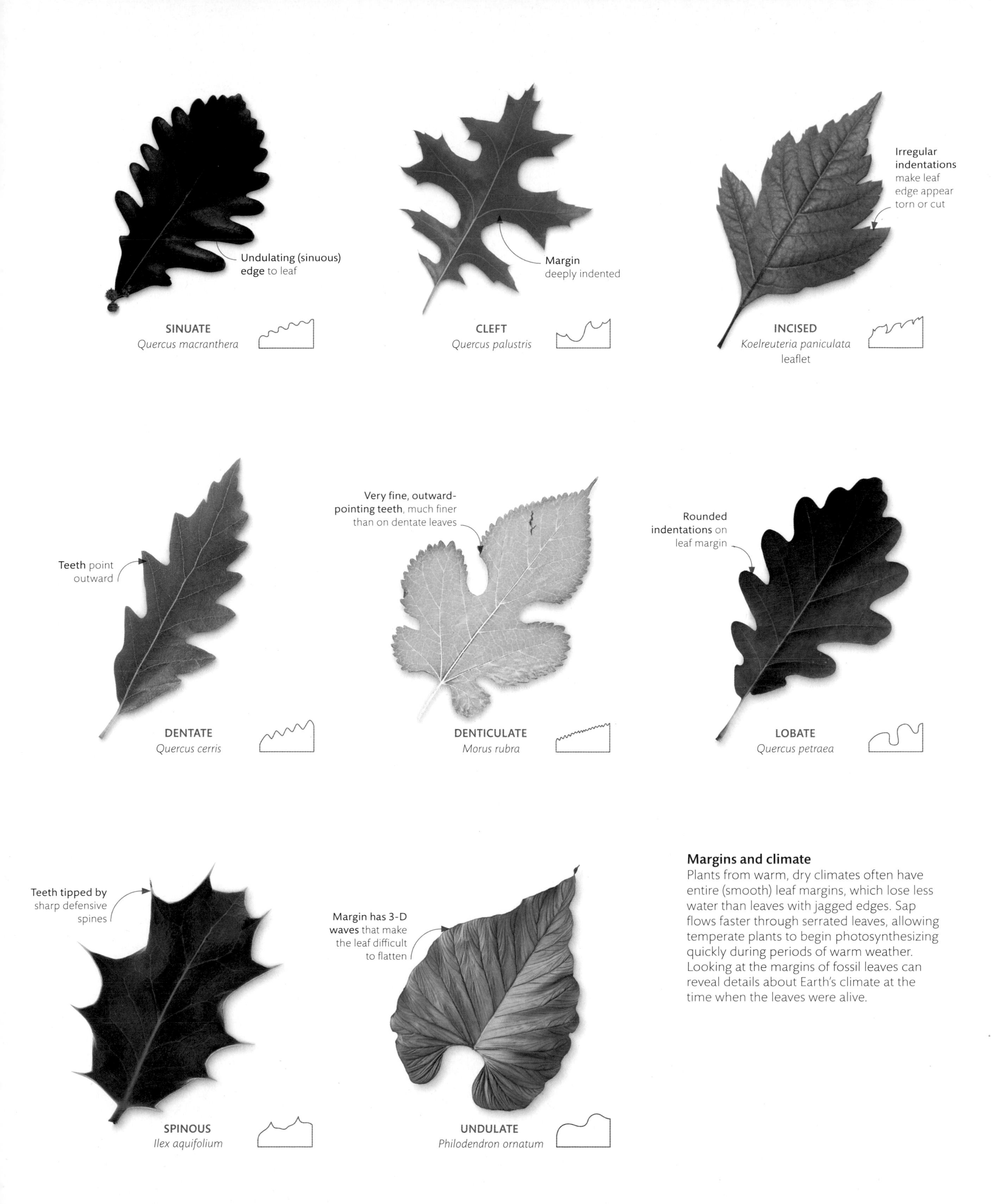

Margins and climate

Plants from warm, dry climates often have entire (smooth) leaf margins, which lose less water than leaves with jagged edges. Sap flows faster through serrated leaves, allowing temperate plants to begin photosynthesizing quickly during periods of warm weather. Looking at the margins of fossil leaves can reveal details about Earth's climate at the time when the leaves were alive.

hairy foliage

Deterring herbivores, protecting against weather extremes, and repelling competing plants with herbicides—plants achieve all this and more using hairlike structures called trichomes on their leaves, stems, and flower buds. Trichomes obstruct insects trying to feed or lay eggs, and may secrete toxins to defend themselves. Some plants with trichomes inject irritating chemicals into the skin of mammals, as a warning to stay away.

STACHYS BYZANTINA

MINT DEFENSES

The trichomes of many plants interfere with insect feeding, but some actively fight these pests. Mint (*Mentha* sp.) trichomes produce the essential oil menthol, which both repels insects and kills those that take a bite.

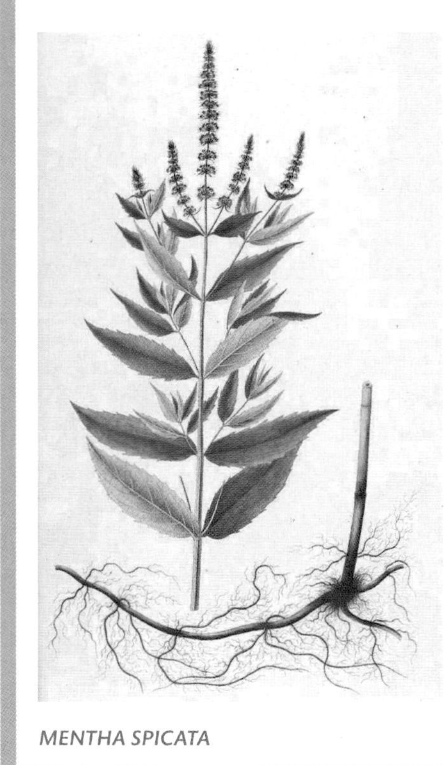

MENTHA SPICATA

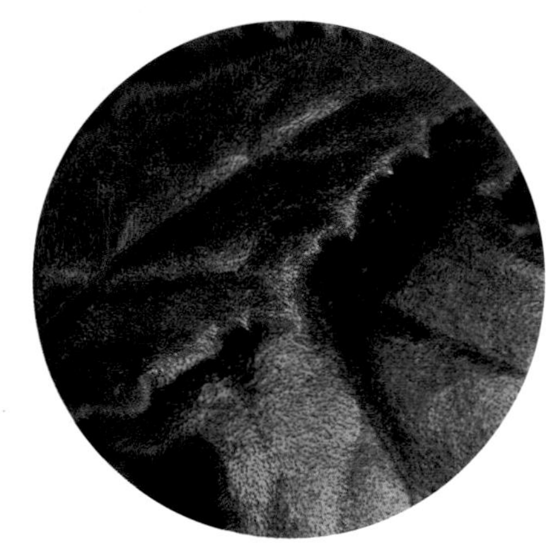

Velvet effect
The velvet plant (*Gynura aurantiaca*) has purple trichomes that contain anthocyanin pigments. These pigments prevent the usually shaded leaves from being damaged by flashes of strong sunlight that reach the forest floor.

A dense forest of trichomes makes it difficult for insects to move through the mass of hairs to feed on the leaf

Defense against the elements
Lamb's ears (*Stachys byzantina*) is covered by a silky layer of hairlike trichomes that help the plant cope with drought. The hairs trap moisture next to the leaves and deflect the wind, minimizing evaporation, while their silver color reflects excessive light and heat from the sun.

Fuzzy trichomes are harvested by some bee species, which use them to line their nests

Trichomes cover the leaves, stems, and flower buds, insulating them from frost and heat

Lamb's ears also have glandular trichomes. These excrete compounds with antimicrobial properties to help keep the plant disease-free

βάτος
Rubus sylvestris
83r

Easy identification
In Nicholas Culpeper's *The English Physitian* (1652), similar plant species were often placed side by side, in order to make identification easier. Here the corn marigold and the ox-eye daisy are depicted next to each other.

plants in art

ancient herbals

Herbals were books or manuscripts containing plant descriptions and information on their properties and medicinal uses. They were also used as reference guides for plant identification and botanical study. They were among the first books and literature ever produced, and some ancient examples contain the earliest known drawings and paintings of plants.

Herbals were probably based on the plant lore and traditional medicines of the ancient world. Some of the earliest examples, from the Middle East and Asia, date back to several thousand years BCE. Herbals became popular in Classical antiquity, the most influential being *De Materia Medica* (c. 50–70 CE), which was created by Pedanius Dioscorides, a Greek physician in the Roman army. Containing detailed information on more than 500 plants, it was copied extensively and used continuously for more than 1,500 years. It is not known whether the original version was illustrated, but the *Vienna Dioscorides*, considered to be the oldest manuscript version, features naturalistic and often minutely detailed paintings.

Woodblock and woodcut printing increased the scope for reproduction, but it was the invention of the printing press in the 15th century that led to a profusion of illustrated herbals and an improvement in the quality of the images. Although their popularity eventually diminished, herbals may be seen as forerunners of the scientific books with botanically accurate illustrations that replaced them.

Culpeper's herbal
This hand-colored botanical copperplate engraving is featured in *The English Physitian* by Nicholas Culpeper (1652). Affordable, accessible, and practical, it became one of the most popular and successful books of its type.

De Materia Medica
This page from Dioscorides' *De Materia Medica* illustrates a wild bramble (with the Greek word *batos* just above) identified as *Rubus sÿlvestris*. This copy of the manuscript was created in 1460, nearly 1,400 years after the original was first produced. It was part of the collection owned by renowned British botanist and naturalist Sir Joseph Banks.

> “... one of the rare types of manuscript with an almost continuous line of descent from the time of the ancient Greeks to the end of the Middle Ages.”

MINTA COLLINS, *MEDIEVAL HERBALS: THE ILLUSTRATIVE TRADITIONS*, 2000

Jade necklace
The tightly stacked leaves of the jade necklace plant (*Crassula rupestris* subsp. *marnieriana*) are small and rounded to reduce their surface area and minimize the evaporation of water, which they store within specialized cells. Their cuticles are often coated with a white waxy "bloom," which reflects damaging heat and light from the sun.

The tightly packed leaves resemble groups of lumpy stems

succulent foliage

Few plants can survive for long without water, but succulent plants store water in their thickened leaves or stems and are specifically adapted to conserve every drop. Not only do the leaves have a dense, waterproof, waxy cuticle, but their stomata (tiny pores) are often sunken, to decrease the airflow and increase the humidity around them. Unlike most plants, succulents open their stomata at night, to minimize loss of water by evaporation during the heat of the day.

CAM PHOTOSYNTHESIS

To conserve water, succulents use a type of photosynthesis called crassulacean acid metabolism (CAM). Instead of taking in carbon dioxide (CO_2) during the day, CAM plants open their stomata at night to reduce transpiration (see p.126). The carbon dioxide is stored as an organic acid compound, then moved to the chloroplast during the day and released for use in photosynthesis.

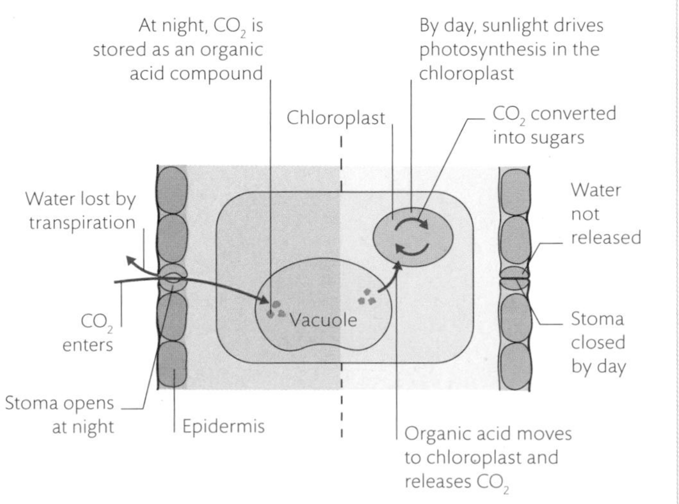

INSIDE A MESOPHYLL CELL

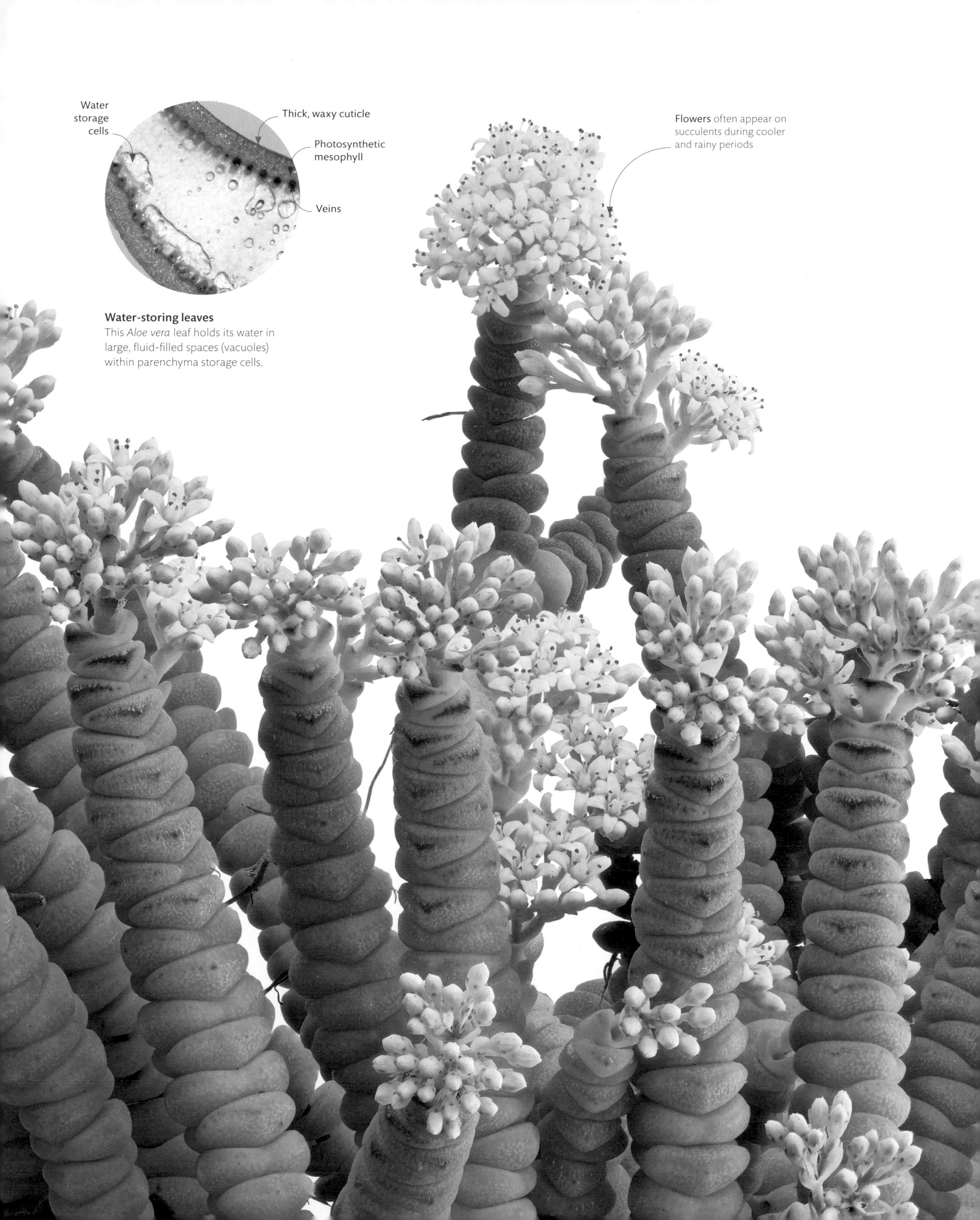

Water-storing leaves
This *Aloe vera* leaf holds its water in large, fluid-filled spaces (vacuoles) within parenchyma storage cells.

Self-cleaning leaves
The leaves of the sacred lotus (*Nelumbo nucifera*) are covered in microscopic bumps and a water-resistant waxy cuticle. Water droplets are repelled by this protective layer and quickly shed, but they pick up dirt particles as they run along the leaf, cleaning its surface and ensuring that light can reach the photosynthetic cells beneath. This useful property has been replicated in laboratories to create self-cleaning coatings.

waxy leaves

Plants first evolved in water, but about 450 million years ago they began to colonize the land. To prevent dehydration, they developed a waterproof, waxy coating called a cuticle over their leaves and stems. The cuticle also protects a plant from infection by microorganisms. Although it is translucent to let light in for photosynthesis, the cuticle also reflects excessive light and heat, to prevent damage to the plant.

IN CLOSE-UP
Waxy cuticles are made of water-resistant chemical compounds. These help to prevent water from evaporating from the leaf while also providing some protection against fungi and bacteria. The furlike covering on the euphorbia plant below are wax crystals, which often form on the outer surface of the cuticle.

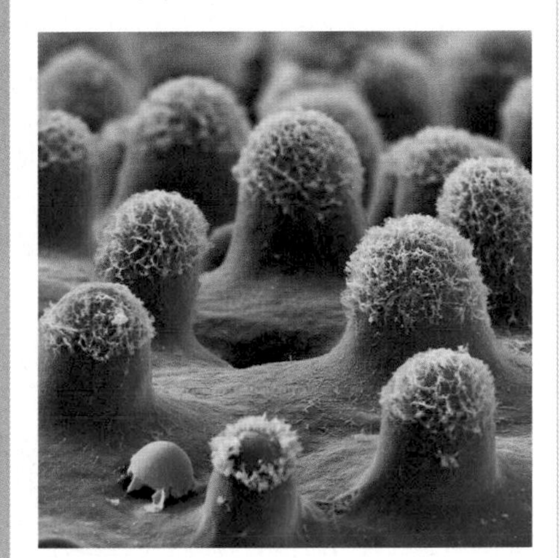
***EUPHORBIA* LEAF SURFACE**

The umbrellalike lotus leaves measure up to 2 ft (60 cm) across

The surface of a sacred lotus leaf is self-cleaning, keeping it free of dirt even in muddy habitats

A long petiole connects the leaves to roots at the bottom of the pond

Banishing dew
Unlike lily pads, lotus leaves often stand above the water on their slender petioles. As the large, balancing leaves vibrate, they cause the dew that forms between the water-repelling bumps on the leaf surface to run off.

Aloidendron dichotomum

the quiver tree

The name quiver tree was coined by the San people of southern Africa, who hollowed out this plant's branches and used them as quivers for their arrows. A large succulent, *Aloidendron dichotomum* is a very hardy species that can grow to 23 ft (7 m) in height and live for more than 80 years.

Native to southern Namibia and the Northern Cape region of South Africa, the quiver tree is essentially an aloe that grows as big as a tree. Like its smaller, fleshier relatives, it produces the rosettes characteristic of an aloe, but at the tips of its forking branches.

The quiver tree's branches are covered in a powdery white substance that acts like a form of protective sunscreen. As temperatures rise and the surrounding landscape sizzles beneath the fierce South African sun, the powder helps the quiver tree to maintain its internal temperature at a much more tolerable level.

In spring, long spikes with bright orange-yellow flowers emerge from each rosette of leaves. The spikes are like flags signaling to wildlife from far and wide. From bees to birds and even baboons, many animals arrive to feast on the quiver tree's nectar. Even when it is not in flower, the tree provides valuable nesting sites for birds among its sturdy branches. Mature quiver trees often boast huge colonies of sociable weaver birds, which weave large, intricate communal nests among the branches, taking advantage of what little shade they can provide.

Prolonged droughts have led to many quiver trees dying in the hotter parts of their native range. With climate change, such droughts are predicted to become more widespread and severe. The loss of the quiver tree would be an indication of serious changes in rainfall.

Forking habit
Although it grows into a tree, this plant does not produce wood. Instead, its stout branches are filled with pulpy fibers that store precious water. As a quiver tree grows, its branches fork repeatedly, branching in two each time.

Spikes of orange-yellow flowers produce copious amounts of nectar, which is vital to local wildlife

Succulent leaves
The quiver tree has typical aloelike leaves at the tips of its bright white branches. These succulent leaves are important for both photosynthesis and water storage.

silver leaves

In dry mountain climates with intense sunlight and high evaporation, many plants keep cool by having silvery leaves, which deflect light and heat. The color comes from a coating of translucent wax or hairs (trichomes) on top of the green leaf cells underneath. Both the wax and the hairs also help to reduce water loss: hairs raise the humidity around the leaf's surface, which minimizes evaporation, and the wax creates an additional layer of waterproofing.

EFFICIENT GAS EXCHANGE

The Tasmanian snow gum (*Eucalyptus coccifera*) has leaves that grow vertically, to minimize their exposure to the sun's heat. It also means that each leaf can have stomata (pores) on both sides and take in more carbon dioxide for photosynthesis, without losing too much water from evaporation.

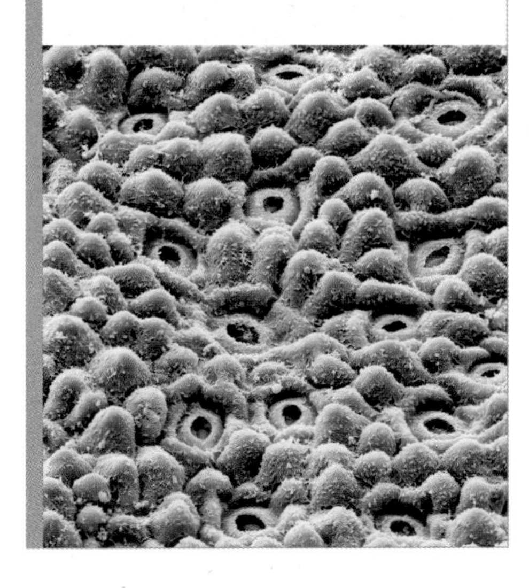

Wax is produced as soon as the leaf begins to form

The dense, waxy covering is difficult for insect pests to walk on

Both the leaf surfaces are coated in wax because the leaves are held vertically

Reflecting rays

The silver-leaved mountain gum (*Eucalyptus pulverulenta*) is coated with microscopic waxy tubes that give its leaves a silvery appearance. This waxy layer helps the plant reflect heat and keep cool. The stems of the florists' cultivar 'Baby Blue', shown here, grow horizontally from the main trunk. In this plant, from high mountains in southern Australia, the adult leaves are similar to the juvenile leaves, and are not replaced by narrow leaves as in most *Eucalyptus*.

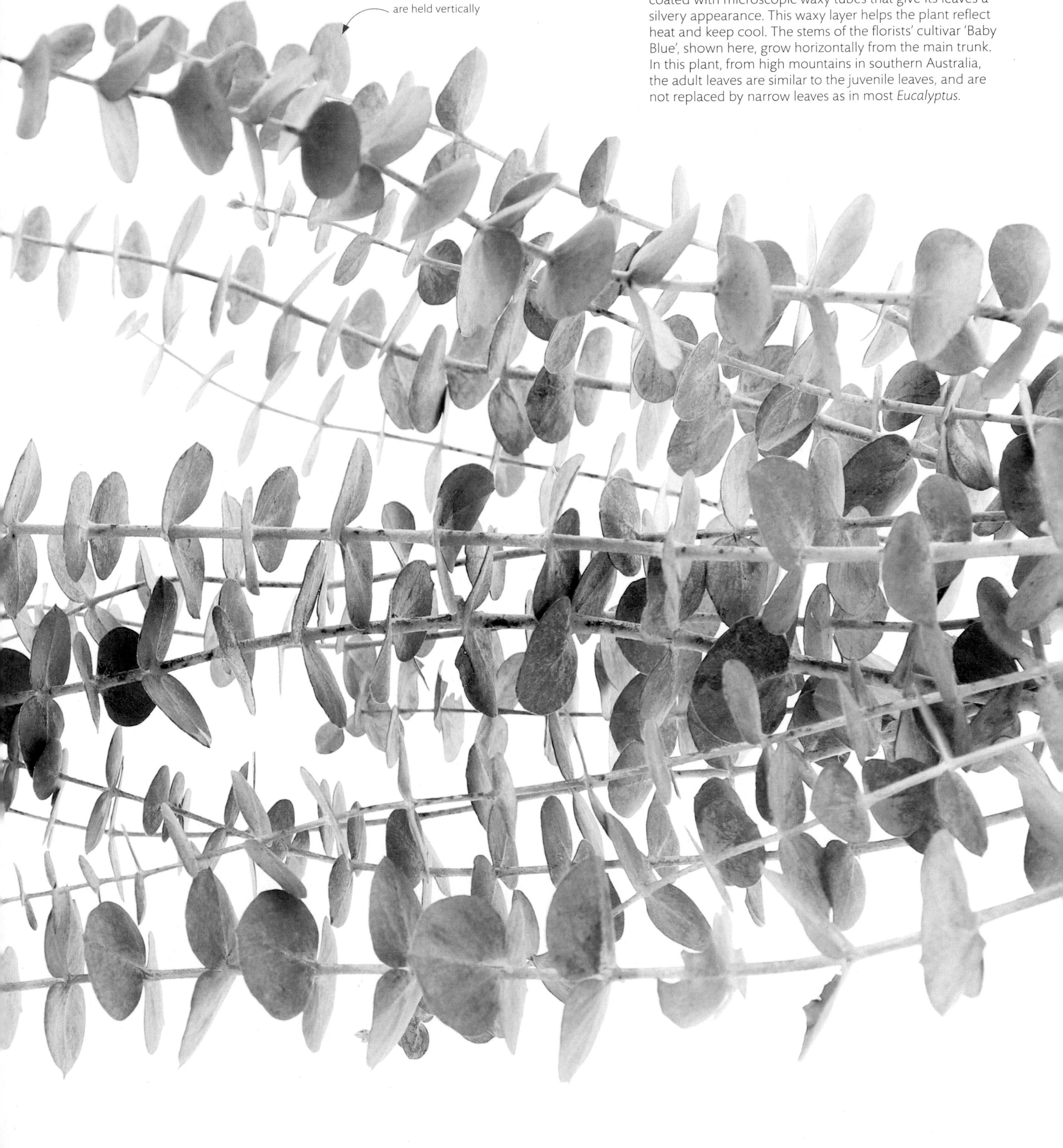

variegated leaves

Leaves that are two or more colors are known as variegated. Although common in gardens, these leaves are rare in nature because only the green parts of a leaf can photosynthesize. Some rainforest plants became variegated to avoid damage from sunlight flashing through the trees, or to mimic the effects of disease in order to deter plant-eaters. Most variegated garden plants are chimeras—the different colored parts of their leaves contain cells that are genetically different.

BEGONIA 'SILVER LACE'

BEGONIA 'ESCARGOT'

FITTONIA SP.

CALATHEA BELLA

ACER PLATANOIDES 'DRUMMONDII'

CODIAEUM VARIEGATUM

Mimicry

Angel wings (*Caladium bicolor*) is native to the forests of Central and South America. The white and red patches on its leaves cannot photosynthesize, but they make the leaf look as if it has been infested by leaf miner insects. This form of deception seems to be very effective: in a related species, variegated leaves like this were found to be up to 12 times less likely to be infested by leaf miners.

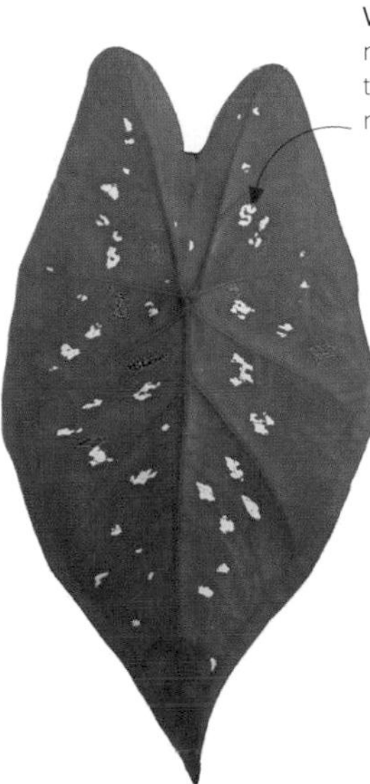

White marks mimic the tunnels of leaf miner larvae

CALADIUM BICOLOR

Variegated red leaves need more light than purely green leaves, as they photosynthesize less efficiently

AGLAONEMA SP.

Fake leaf-miner tunnels deter plant-eaters from attacking what appears to be an infested leaf, and an inferior food source

CALADIUM BICOLOR

Bred to be red
Native Japanese maples have green leaves, but some cultivars display red leaves all summer. The leaves turn an even brighter red in the fall.

Palmate (hand-shaped) leaves have 5, 7, or 9 serrated and pointed lobes

Acer palmatum

the japanese maple

Few leaves are as recognizable as those of the Japanese maple (*Acer palmatum*). Native to Japan, North and South Korea, and Russia, this attractive deciduous tree has become a mainstay in gardens worldwide thanks to its graceful shape, elegant leaves, and colorful foliage.

The Japanese maple is a relatively small, slow-growing tree that rarely exceeds 33 ft (10 m) in height, which means it is quite at home growing in the shade of its taller neighbors. In the wild, it grows in the understory of temperate woodlands and forests up to 3,600 ft (1,100 m).

One of the most intriguing features of this tree is how variable its appearance can be. Different native populations take on a range of forms, from diminutive shrubs to spindly trees. The leaves that have made it so famous are also quite variable in shape and color. All of this has to do with the species' DNA.

Japanese maples show a fairly high degree of genetic diversity, and seeds from the same plant can produce offspring that look wildly different from one another.

Plant breeders have taken advantage of this natural variety, creating more than 1,000 different cultivars since the 18th century. Much of that breeding has focused on producing bright red foliage. Such vivid coloration comes from a group of plant pigments called anthocyanins. The pigments help to protect leaf tissues from exposure to too much UV radiation or harmful swings in temperature. There is some evidence that anthocyanins may also act as a deterrent to insect herbivores.

Japanese maples produce small red or purple flowers that are pollinated by wind or insects. These are followed by winged seeds called samaras (see p.314). Like little helicopters, the seeds catch the breeze and twirl off and away from their parents.

Small wonder
Some Japanese maples have upright forms; others are weeping and often develop a dome-shaped habit. In the fall, the leaves turn startling shades of yellow, orange, red, purple, and bronze before they are shed.

fall colors

As summer draws to a close, the shorter days and cooler temperatures alert trees and shrubs that they need to prepare for winter. Spectacular displays of color may follow when chemical changes within the leaves gradually transform them from green to brilliant shades of yellow, orange, and red.

Color combinations
Leaves are colored by chemical compounds. The most visible of these is chlorophyll, which makes leaves appear green. Carotenoids are responsible for yellow and orange hues, while anthocyanins create shades of red and purple.

Yellow and orange shades are caused by carotenoids, which degrade more slowly than green chlorophyll

Chlorophyll absorbs the red and blue-violet parts of sunlight but reflects green light, giving leaves their normal color

Green chlorophyll degrades in bright light, but it is constantly replenished by new chlorophyll during the growing season

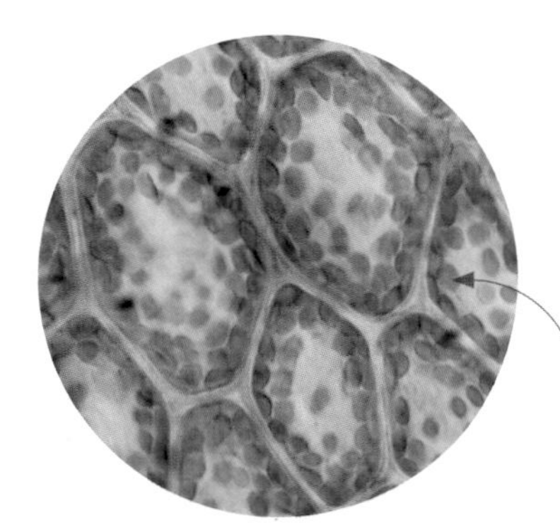

Chloroplasts are where photosynthesis takes place, using light energy captured by chlorophyll

Inside a leaf cell
Leaf cells contain several pigments, but green chlorophyll dominates during spring and summer, when photosynthesis is at its height. The chlorophyll is found in special cell structures called chloroplasts, which produce the plant's food.

Yellow carotenoids are always present, but they become more visible when chlorophyll is broken down in fall leaves

HOW COLORS CHANGE

In the fall, the plant produces hormones that tell its leaf tissues to die. The chlorophyll in the leaves breaks down more quickly than the carotenoids, so the leaves turn yellow. Meanwhile, red-purple anthocyanins are actively produced in large cell compartments called vacuoles. They give the leaf an orange hue, which turns to red as the last of the carotenoids are lost.

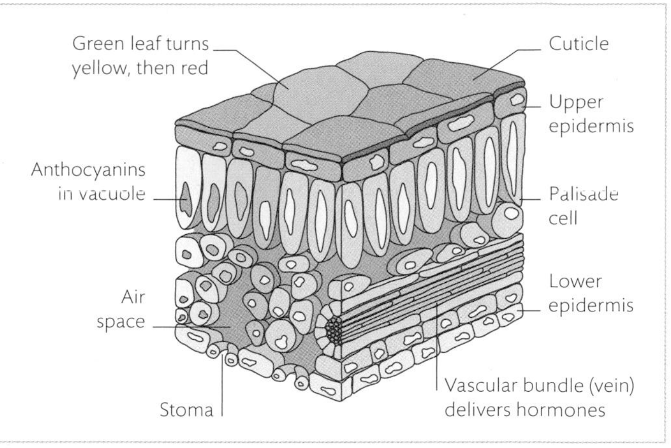

Common carotenoids, such as beta-carotene, look orange because they absorb blue and green light and reflect red and yellow

Red colors occur when the tree sap is acidic; purples occur if the sap is alkaline

Anthocyanins may act as a natural sunscreen that protects leaves until their nutrients have been recycled for use in other parts of the plant

Red-purple colors occur when leaf starches break down into sugars, which react with other chemicals to form anthocyanins

CHEMICAL COCKTAIL

Stinging nettle trichomes contain an arsenal of irritating chemicals, including formic acid, histamine, serotonin, and acetylcholine. These compounds work together to make the pain and discomfort that they cause last much longer than each chemical could achieve individually.

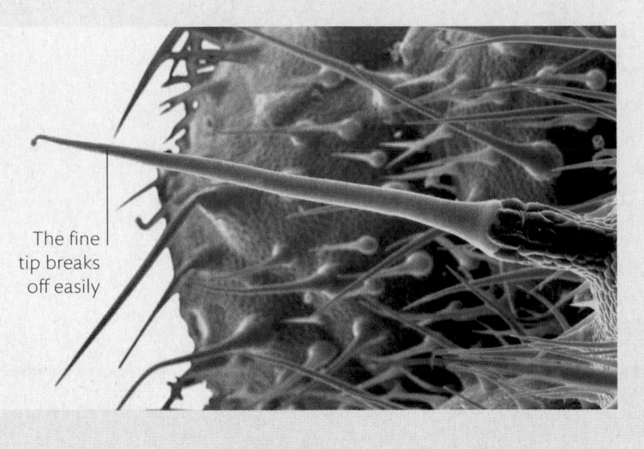

The fine tip breaks off easily

stinging leaves

Plants that sting, such as nettles, are covered in single-celled hairs (trichomes) full of chemicals that cause pain, inflammation, and irritation in animals unlucky enough to brush against them. The brittle trichome tip breaks off easily when touched, revealing a sharp hypodermic needle that immediately injects its toxic mixture under the skin of browsing herbivores. If nettle leaves are damaged by herbivores, they respond by producing more trichomes.

Nettle stems and stalks also carry stinging trichomes

Many more stinging trichomes are located along the veins on the underside of the leaf

Short, non-stinging trichomes provide some defense against insects

Stinging nettle
The sting of the common nettle (*Urtica dioica*) is irritating to mammals and some birds, but it does not affect insects. As a result, nettles are an important habitat for caterpillars and other insect larvae that might otherwise be eaten.

Nettle trichomes are brittle because they are formed from glasslike silica

Vachellia karroo

the sweet thorn

Formerly called *Acacia karroo*, this fragrant acacia is one of the hardiest trees in southern Africa. A highly adaptable species, it is equally at home in humid forest, on savanna, or in semidesert. Once established, the sweet thorn can handle almost any adversity—even wildfires.

The tree's daunting "thorns"—technically spines developed from leafy outgrowths (stipules) at the base of leafstalks—are up to 2 in (5 cm) long. Being dressed in spines is enough to put off many, but not all, grazing animals. Giraffes, in particular, have no problem wrapping their leathery tongues around its branches to feed on the fine, mimosalike leaves. The bark, flowers, and nutritious seedpods also provide sustenance to animals, as does the gum that oozes from wounds in the bark, a special favorite of green monkeys and lesser bush babies.

By no means a large tree—bigger specimens reach around 40 ft (12 m) in height—the sweet thorn is also relatively short-lived, with a maximum lifespan of 30 to 40 years. It is, however, able to cope with extreme conditions. As well as being frost-resistant, the sweet thorn can survive drought due to its long taproot, which enables it to draw on water reserves deep underground. When nutrients are scarce, it can generate its own supply of nutrients by using structures on its roots that house nitrogen-fixing bacteria.

Fast-growing and tolerant of many soil types, the sweet thorn is able to establish itself without shade or shelter, and is even impervious to fire. Seedlings that have survived their first year can be burned to a crisp, only to sprout new stems, thanks to the energy stored in their roots.

The sweet thorn's adaptability makes it an aggressive, invasive species when it is introduced outside its native range. The fact that few animals are brave enough to graze its well-protected foliage also helps it outcompete other vegetation.

Wall of thorns
Sweet thorns are leafless during the winter, when the density of their long, white thorns is clearly visible. These spines make it a favorite nesting spot for birds, as they deter all but the most persistent nest predators.

Each pom-pom inflorescence is made up of many individual flowers

Pom-pom display
In early summer, the sweet thorn's canopy erupts with hundreds of yellow pom-pom-shaped blooms. The tree's long flowering season offers a reliable source of pollen and nectar to bees, making this an important species for honey production.

leaf defenses

Running away is not an option, so plants have evolved many ways to deter would-be predators. Some have converted their leaves into sharp spines, which hurt animals that attempt to take a bite. Spines can be modified parts of the leaf, such as the vascular tissue, petiole, or stipules, while other plants have converted the entire leaf into defensive spines.

STAYING SAFE WITH SPINES

Acacia trees often grow large spines that provide a home to beneficial ants, which viciously protect their host from attack. Most cacti have converted all of their leaves into spines; they photosynthesize using their succulent stems.

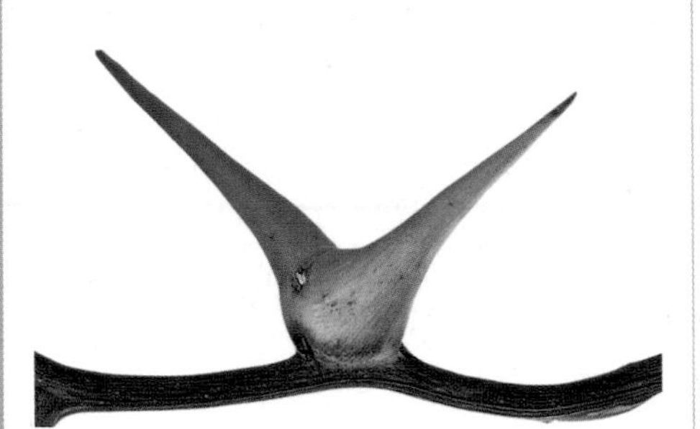

ACACIA (*ACACIA SPHAEROCEPHALA*)

CACTUS (*MAMMILLARIA INFERNILLENSIS*)

The flowers are unusual for a thistle, in that they resemble daisies when open

The leaf undersides are cobwebby, covered with soft, white hairs

Main stem leaves form broad, spiny "wings"

Upper leaf surfaces are almost shiny

Small, prickly leaves near the flowers are mainly for defense against herbivores

Bracts covered in spines protect the developing flower buds

Flowers are produced on short side branches off the main stem

Unprotected flowers are sometimes eaten by weevils; the leaves are usually free from pests

Spines are most common in plants from arid regions, such as this South African thistle, and the Saguaro cactus (*Carnegiea gigantea*).

South African thistle
The spines of this *Berkheya purpurea* are hardened extensions of the vascular tissue protruding from the leaf. As well as protecting the plant from herbivorous mammals, these spiny edges prevent caterpillars and other small pests from finding a safe place to start munching.

Surviving the storm
The coconut palm (*Cocos nucifera*) is extremely wind-tolerant, its pinnate leaves allowing strong winds to pass right through them. In heavy storms, fronds facing the wind may snap off at the base, but this does not usually damage the rest of the plant.

Fronds can reach 18 ft (5.5 m) in length

Leaves surround and protect the delicate growing tip at the top of the stem

A pinnate leaf has leaflets arranged in pairs along the midrib

leaves in extremes

During a hurricane, the leaves of many trees catch the wind like sails, and branches or trunks may snap. Palm trees, however, can be bent horizontally by extreme winds and still emerge largely unscathed. They survive storms thanks to their flexible stems and aerodynamic leaves. Most palms have featherlike pinnate leaves, with a strong, flexible midrib and leaflets that fold up in the wind to avoid major damage.

Leaflets of large fronds fold up like a fan to reduce the area exposed to gales
The trunk is flexible due to its vascular tissues that allow it to bend without breaking

Bismarckia nobilis

the bismarck palm

Offering welcome shade under its impressive canopy, the Bismarck palm is perhaps the most elegant and spectacular of all the fan palms. *Bismarckia nobilis* hails from the arid grasslands of northwestern Madagascar. It is one of the few endemic Madagascan species not currently in decline.

Named after the first chancellor of the German Empire, Otto von Bismarck (1815–1898), this tree is the sole member of its genus. *Bismarckia nobilis* is not the tallest of palms, but larger individuals can up grow to 60 ft (18 m), though it can take a century to reach such heights.

In its native land, the Bismarck palm must endure extremes of climate, so it has to be hardy. The dry season heralds punishing heat, relentless sun, little to no rain, and wildfires; the wet season brings lots of rain and high humidity. When the weather is driest and the sun fiercest, the tree uses its deep root system to access water far below ground, while the waxy coating that gives its leaves their silvery hues acts like sunscreen and protects the sensitive photosynthetic structures inside from an overdose of solar radiation. A stout trunk combined with leaves and growth tips positioned far above the reach of most flames enables the Bismarck palm to survive all but the very worst wildfires. To make the most of the rain when it does come, the curved frond petioles channel water down to the base of the trunk.

This palm is dioecious, meaning that individual trees produce either male or female flowers, never both. Pollination is achieved by insects and, if male and female trees are growing close enough together, via wind as well. Only the females trees will grow fruit, with each small flower developing a single fleshy, but inedible, stone fruit.

Massive fronds
Fan palms are so named for the shape of their fan-shaped leaf fronds. The Bismarck palm's rounded, silvery blue to green fronds can reach up to 10 ft (3 m) across. Each frond is made up of a spray of multiple stiff, sharp-edged leaflets.

Myriad small, creamy white flowers form on long stalks that eventually droop under the weight of the resulting fruit

***Bismarckia* inflorescence**
The Bismarck palm develops long, ropelike inflorescences made exclusively of either male or female flowers. Large clusters of fruits form on the female inflorescences.

The leaf is broad so the upper surface can collect the maximum amount of sunlight

Floating leaves provide shelter for young fish, insects, and amphibians, but—since they have little competition—the leaves can soon cover an area of water, harming other aquatic life by blocking light and oxygen

Firm anchors
Although the leaves of giant water lilies (*Victoria amazonica*) float on the water surface, they are firmly anchored to underground stems at the bottom of the pond. The formidable prickles along the stalk and leaf surface protect the leaves from being eaten by fish.

floating leaves

While most plants fight for space on land, some have turned to a life on water. Aquatic plants, such as the giant water lily, benefit from unlimited access to water and less competition for light and nutrients. Their floating leaves use pockets of air as buoyancy aids, trapped either in air chambers within the leaves or between dense hairs on their surface. The leaves of the giant water lily are so buoyant that they can support the weight of a human baby.

Purple tissues on the lower surface of the leaf keep it warm by absorbing excess sunlight passing through the photosynthetic cells above

Prominent veins

The long leaf stalk connects to a stem buried in the mud

Leaf underside
Water lilies lose a lot of water through evaporation. Taking up more water means absorbing dissolved toxic compounds such as heavy metals, which are safely stored in epidermal glands. Prominent veins are surrounded by thick-walled supporting cells that stiffen the structure of these broad leaves.

HOW WATER LILIES FLOAT

Large air chambers inside water lily leaves provide them with the buoyancy they need for a floating life. Rigid, star-shaped structures called sclereids within the spongy leaf tissues help to maintain the shape of the leaves. This, in turn, enables the leaves to use the water's surface tension to stay afloat.

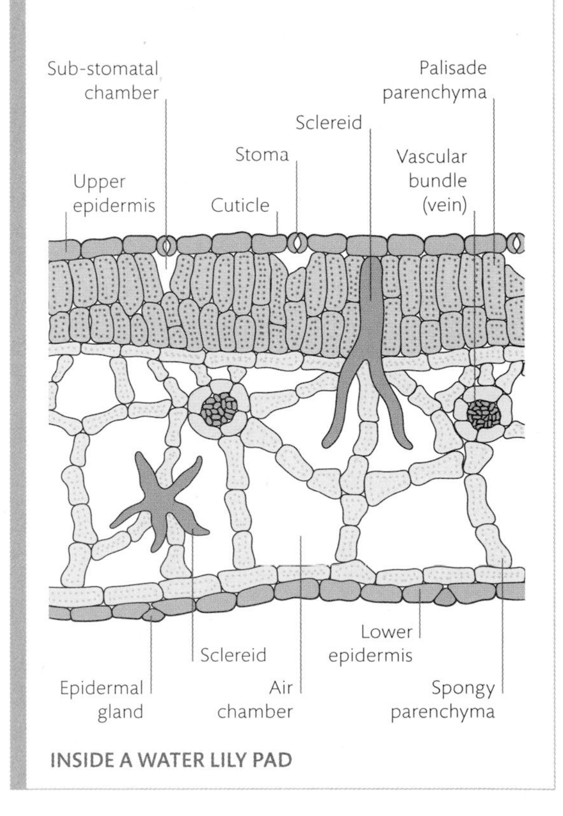

INSIDE A WATER LILY PAD

The waxy surface helps water drain easily into the central cavity

plants with pools

Tadpoles living in trees might sound bizarre, but some tree frogs lay their eggs in pools of water held between the leaves of plants such as this bromeliad, *Neoregelia cruenta*, which grows on trees in the rainforest. These tiny pools, or phytotelmata, can also support insects, nematodes, and even small crabs. Phytotelmata can be found in crevices and cavities between leaves, in *Nepenthes* and *Sarracenia* pitcher plant traps, tree hollows, and within bamboo stems.

Wide, concave leaves funnel rainwater toward the central cup to ensure the plant has a steady water supply

leaves that **eat**

Carnivorous plants capture and consume prey to obtain nutrients lacking in the soil. They have developed ingenious ways of ensnaring their victims. Venus fly traps snap shut to imprison insects, aquatic bladderworts create a partial vacuum to suck in swimming creatures, and the sticky hairs of sundews fold around prey, holding it tight as it is digested. Insects drown in the liquid at the bottom of pitcher plant cups, while corkscrew plants guide their unsuspecting quarry into a digestion chamber.

Pitcher plants
There are two main pitcher plant families, the Sarraceniaceae and the Nepenthaceae (shown here). Insects that fall into a pitcher plant cup are slowly dissolved. Phosphorus and nitrogen are the main nutrients that pitcher plants absorb from their digested prey. Instead of being eaten, the larvae of some insects are adapted to live in the water inside pitchers.

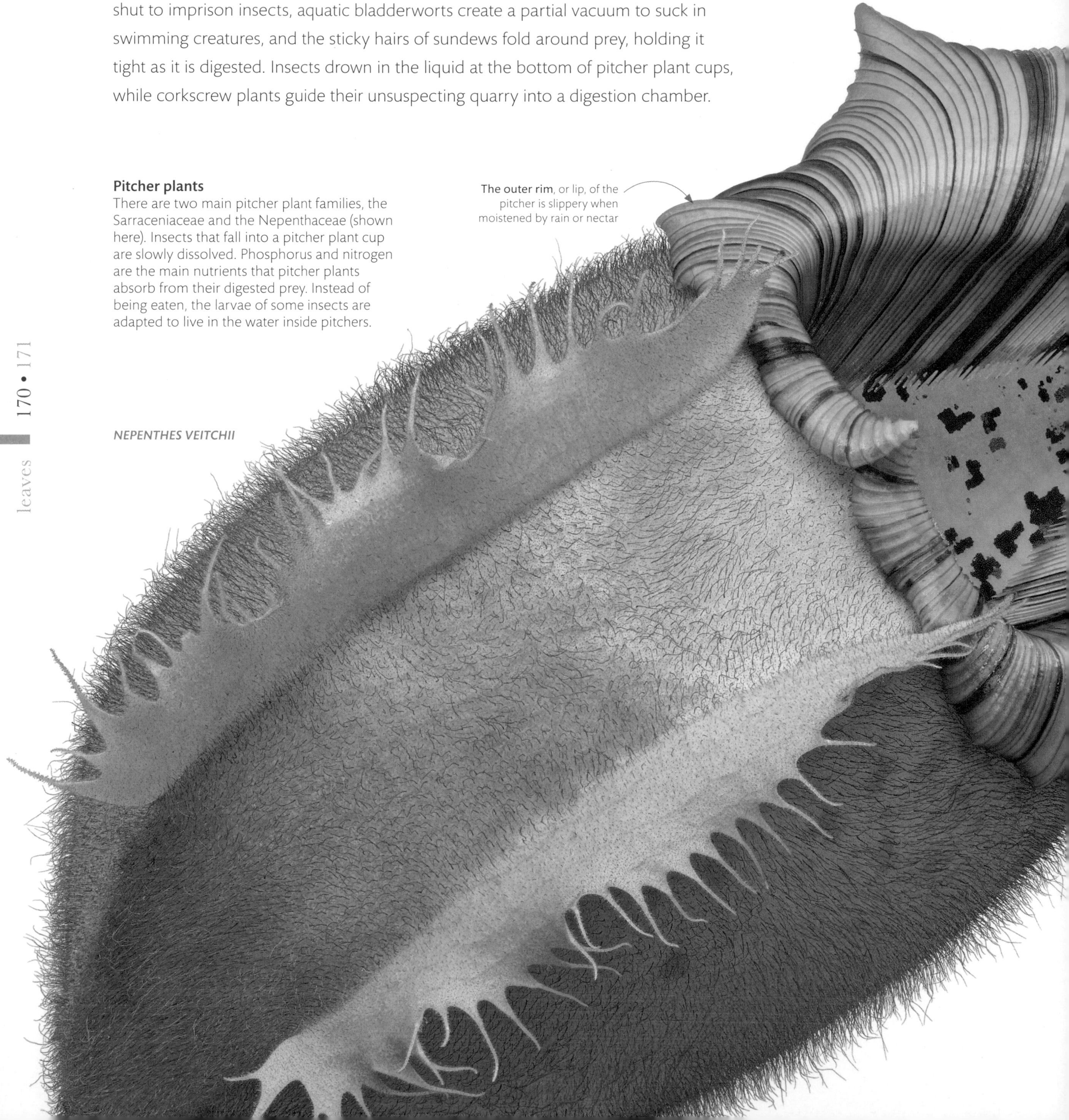

The outer rim, or lip, of the pitcher is slippery when moistened by rain or nectar

NEPENTHES VEITCHII

Bright colors, strong scents, and sweet nectar are all used by pitcher plants to attract insect prey

The lid (operculum) prevents rain from filling the pitcher and diluting its digestive enzymes

Nepenthaceae produce pitchers from tendrils at the tips of otherwise normal leaves; pitchers of Sarraceniaceae are formed by narrow, cone-shaped leaves

When the lip of the pitcher is wet, insects find it almost impossible to keep their footing

The tendril is an extension of the leaf's midrib; it may loop around a support to keep the pitcher steady

Trap trickery

The lip (peristome) of a pitcher is not always slippery. During "safe" periods, scouting ants discover the trap's nectar and leave to inform other worker ants. By the time a group of ants returns, the peristome has regained its slipperiness, so the pitcher captures many ants at once.

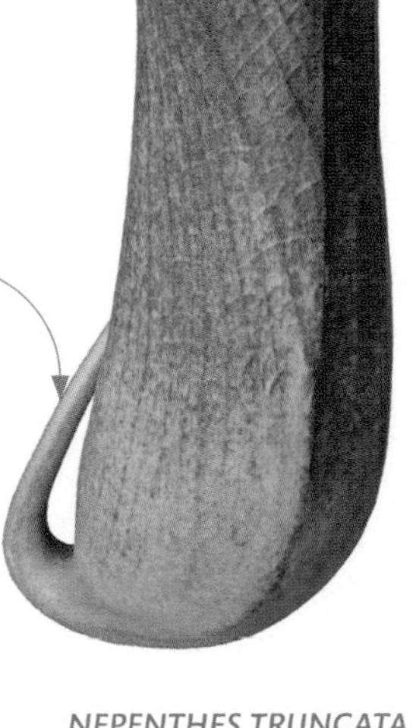

NEPENTHES TRUNCATA

INSIDE A PITCHER PLANT

Struggling insects that fall into the pitcher are unlikely to find a foothold on the waxy sides of the trap. The exhausted prey eventually drops into the liquid at the bottom and is digested by a variety of enzymes secreted by the pitcher's digestive glands.

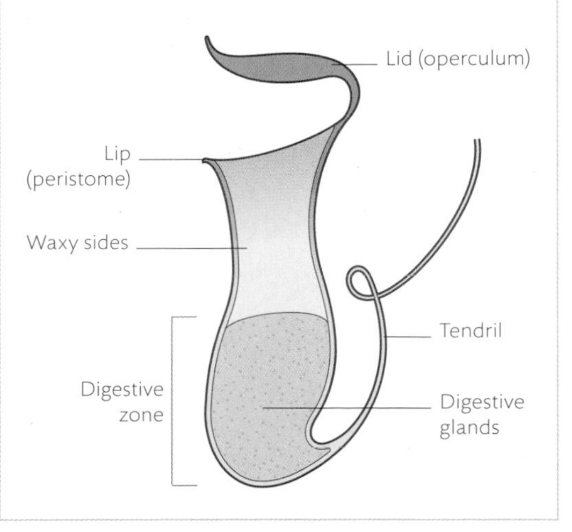

Drosera sp.

sundews

Armed with hundreds of sticky glands on each leaf, sundews are an insect's worst nightmare. Lured by the promise of sweet nectar or refreshing dew, visiting insects rapidly become ensnared by gluey drops of mucilage as the leaves curl around their flailing bodies and proceed to digest them.

The idea that there were plants with a carnivorous diet shocked many early naturalists. To Carl Linnaeus, it was an affront to the natural order of the divine plan. Not all views were so negative, though. Charles Darwin took great joy in studying sundews. Writing to a colleague, he went so far as to say: "at this present moment I care more about *Drosera* than the origin of all the species in the world."

The carnivorous appetite of sundews evolved in response to the nutrient-poor conditions of their typical habitats, which include bogs, swamps, marshes, and fens. With nitrogen in short supply, sundews have foregone symbiotic relationships with soil fungi and instead trap their own nitrogen-rich insect prey. Sensing the movement of insects arriving in search of nectar, their sticky glands swiftly close in on their prey. The more the victim struggles, the more securely it is held.

In most sundews, the process is completed by the leaf wrapping itself around the insect. The leaf then releases digestive fluids that break down the prey's body; other glands absorb the resulting juices.

All sundews produce attractive flowers far above their leaves, so as not to devour their pollinators. While sundews may seem exotic, they can be found growing on every continent except for Antarctica. Australia is the center of diversity for this genus, boasting 50 percent of all known species. Many species have become beloved, albeit curious, houseplants.

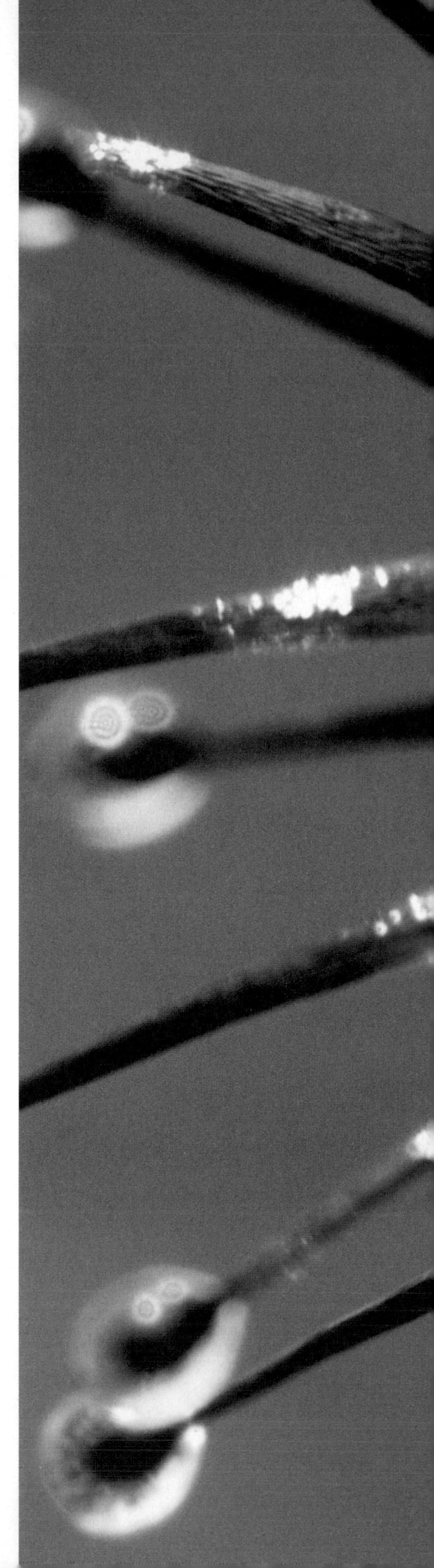

Deadly rosette
There are over 190 *Drosera* species, and they take many shapes and forms, from pygmy rosettes to tuberous climbing vines. The glistening, sticky droplets at the end of their "tentacles" give these plants the common name of sundews.

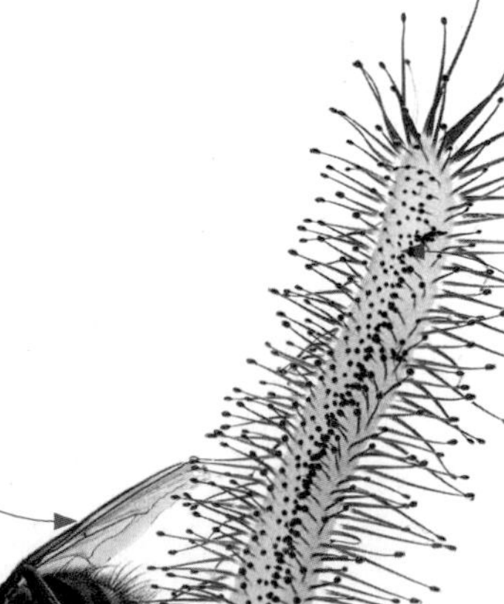

A trapped insect is ensnared by yet more sticky droplets as it tries to escape

The long, thin leaf will curl around the insect and digest it

Sticky glands sit atop tentaclelike trichomes (hairs) that are mobile and sensitive

Cape sundew with prey
Native to South Africa, the Cape sundew (*Drosera capensis*) is one of the most commonly cultivated sundews. It produces multiple pink flowers from late spring through early summer.

Plantlets are produced by genes involved in seed development

Mother of thousands

Kalanchoe daigremontiana is known as "mother of thousands," since it produces many tiny plantlets on its leaves. Once the plantlets form roots, they drop from the parent plant onto the soil below. Producing such prolific "babies" means it can quickly take over a suitable habitat.

Kalanchoe **plantlets** can colonize habitats more quickly than seedlings that grow from seed

The plantlets line up along the margins of the leaves

KALANCHOE DAIGREMONTIANA

VEGETATIVE REPRODUCTION

While seeds develop from flowers fertilized by the pollen of another plant, vegetative reproduction involves just one parent—so plants produced vegetatively are genetic clones of their parent. Many plant species produce new clones from their roots, tubers, or shoots, and most plants are able to grow from cuttings taken by gardeners; these, too, are forms of vegetative reproduction. In fact, under the right conditions, almost all plant tissues are capable of regenerating entire new plants.

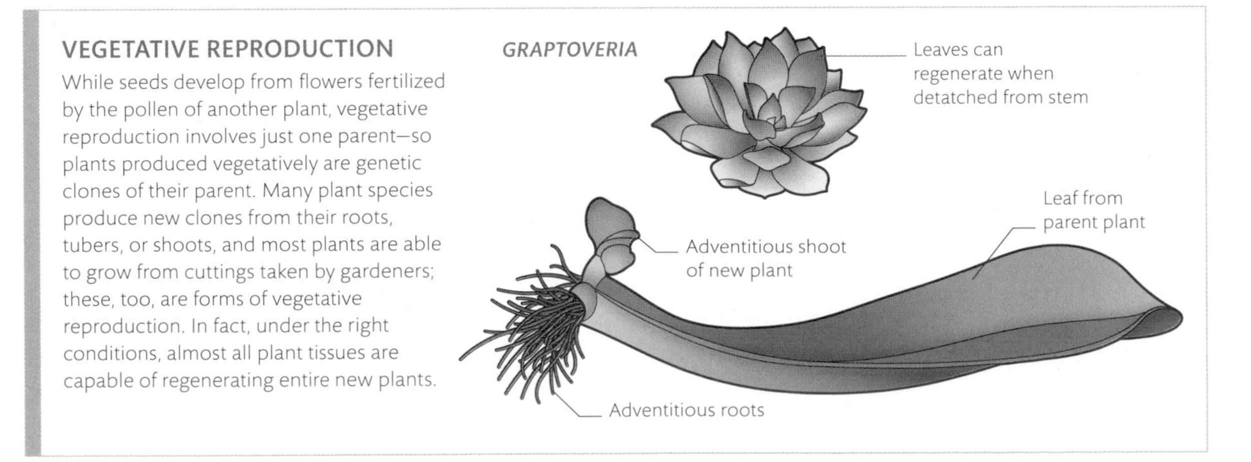

leaves that **reproduce**

For plants, finding the perfect home is a matter of pure chance. Their seeds are blown, washed, or carried away from the parent plant, and could end up anywhere. When they find a suitable environment, some plants make copies of themselves (clones) to spread across their new home. Some of the most fascinating examples of this are plants that create tiny plantlets on their leaves. The plantlets are nurtured by the parent plant until they are big enough to survive on their own.

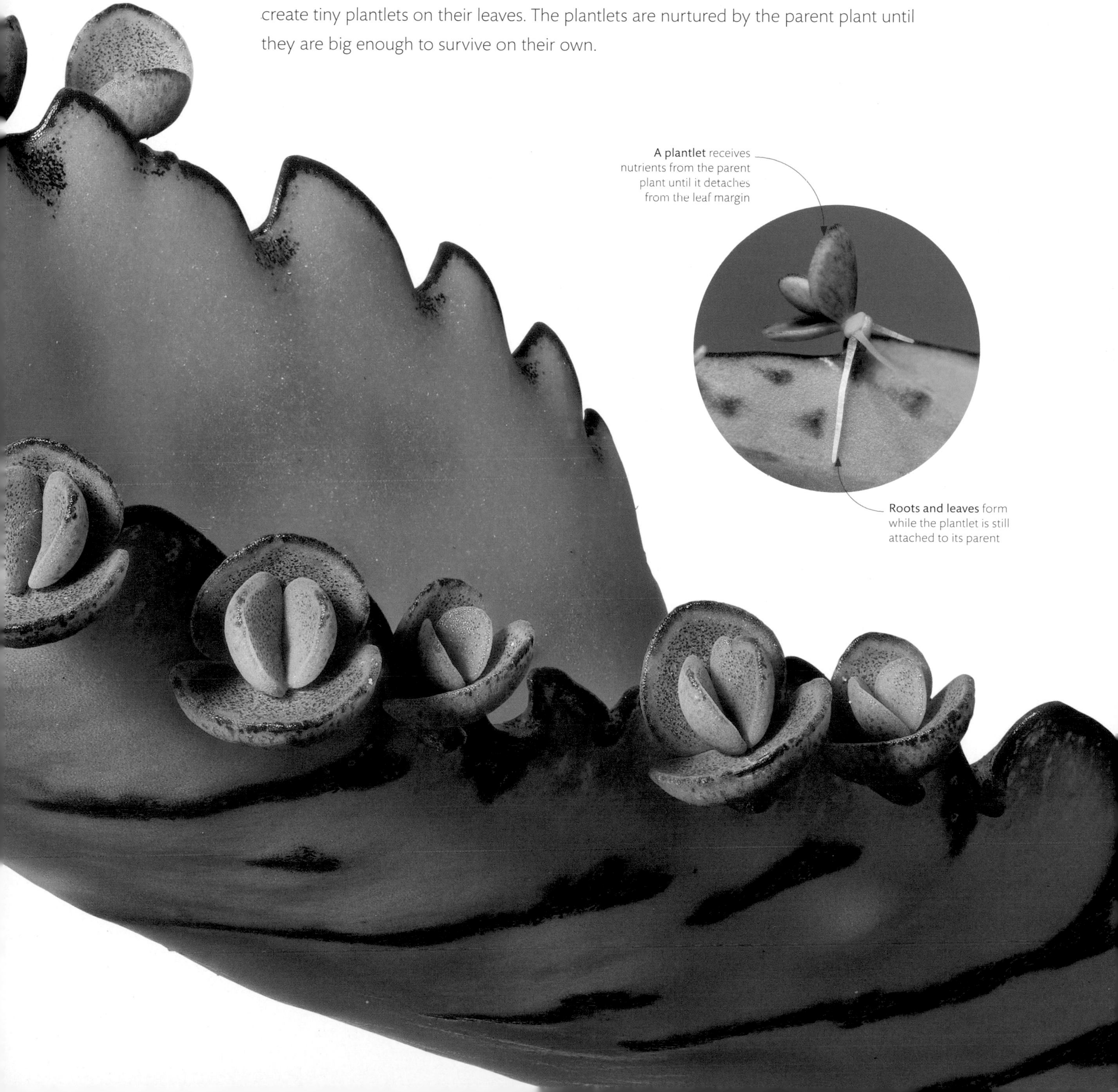

A plantlet receives nutrients from the parent plant until it detaches from the leaf margin

Roots and leaves form while the plantlet is still attached to its parent

NOURISHING BRACTS

The feathery bracts of the stinking passionflower (*Passiflora foetida*) exude a sticky substance that traps insects that might otherwise eat the flowers or fruit. The plant is protocarnivorous: it partially digests the trapped insects for nutrients.

PASSIFLORA FOETIDA

Sharp spines protect the developing flowers of the wild cardoon, but domesticated artichokes lack such protection

Spiky bracts
The cardoon (*Cynara cardunculus*) is a close relative of the globe artichoke. Its dramatic inflorescences are protected by an arrangement of thick, spiky bracts known as the involucre, which defends the soft developing floral tissues against both insect and mammalian herbivores. Each flower produces a single seed, or achene, topped by the hairy pappus (modified calyx) that aids dispersal by the wind.

Bracts are modified leaves, but they do not look like the large, cleft foliage typical of cardoons

protective bracts

Beneath many flowers and inflorescences are modified leaves called bracts. The function of bracts is twofold: some showy bracts mimic colorful petals to attract pollinators (see p.179), while others provide a protective barrier around the developing flowers or fruit, shielding them from attack by herbivores or the elements. Hairy bracts can deflect wind and heat, while spines deter animals from taking a bite.

Hundreds of these individual purple flowers make up each inflorescence

The bracts contain antimicrobial compounds to defend against fungi and bacteria

The fleshy bases of the bracts form the edible part of artichoke hearts

Silvery bracts, stalks, and leaves help cardoon plants to reflect heat and excess light

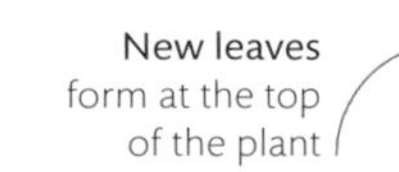

leaves and spines

Many plants produce spines from leaves or parts of leaves, such as the petiole (stalk) or the leafy outgrowths known as stipules. The main function of spines is to defend the plant against herbivores. Some species, such as cacti, have converted all their leaves into spines; with a reduced leaf surface area, they lose less water from evaporation.

Crown of spines
Despite its common name, the crown of thorns plant (*Euphorbia milii*) actually produces spines rather than thorns (which grow from shoots). Spines, modified from the stipules at the base of the plant's leaves, help to defend its succulent stem against herbivores. The older leaves fall off as the plant grows, leaving a spine-covered stem with only a few new leaves at the top.

MODIFIED STIPULES

Many plants form stipules—outgrowths at the base of the petiole. They are most common in eudicots, which produce a pair of stipules at the base of each leaf, but single stipules occur in some monocots. Stipules have various adaptations allowing them to perform specific functions: some plants use them for photosynthesis; others as tendrils for climbing; and some plants enjoy extra protection with scaly or spiny stipules.

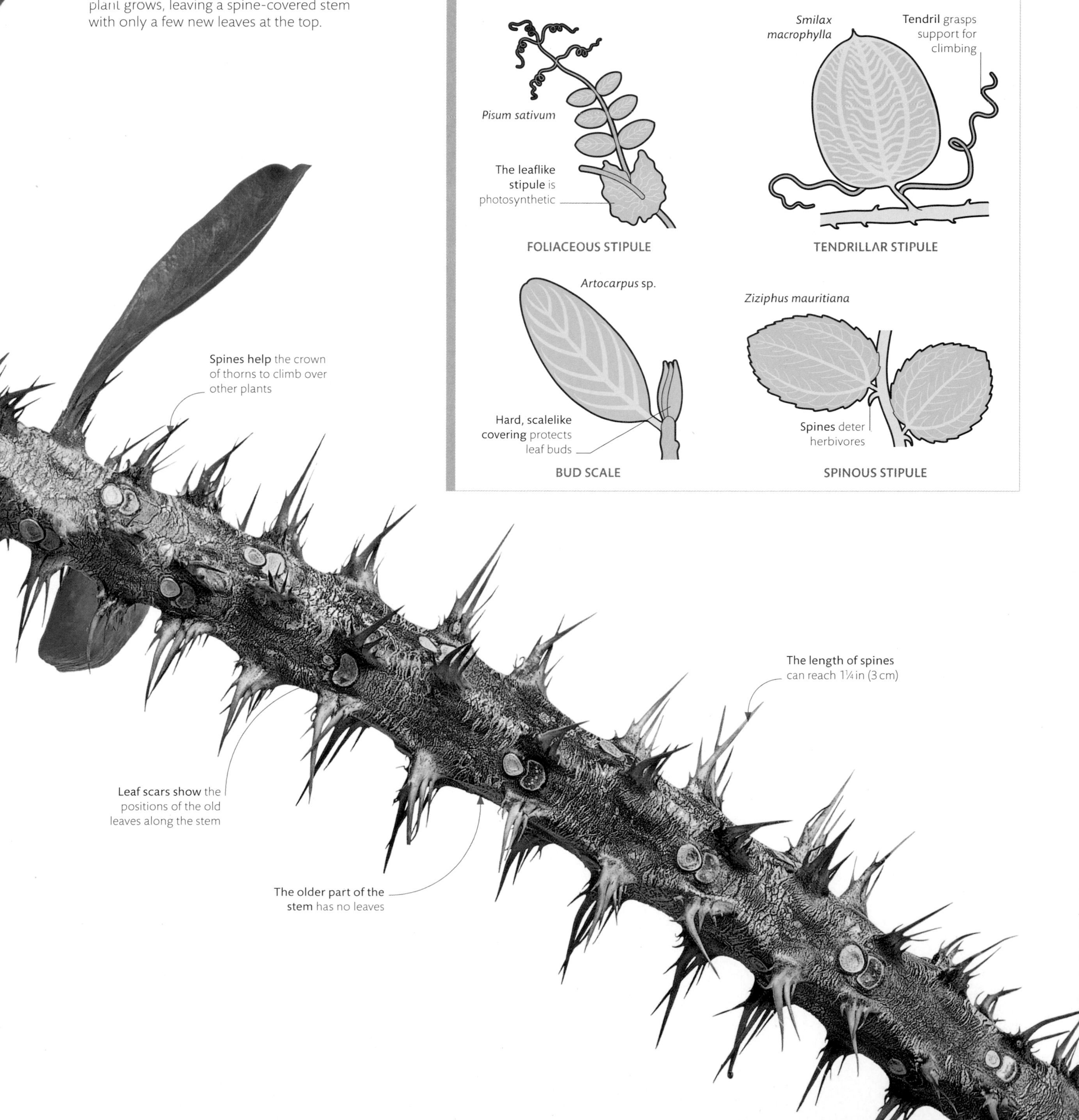

flowers

flower. the part of a plant from which its fruit or seed develops, consisting of reproductive organs (stamens and pistils) that are often surrounded by colorful petals and green sepals.

parts of a flower

About 90 percent of plants produce flowers, ranging from tiny, almost microscopic grass blooms to alien-looking giants that measure 3 ft (1 m) or more across. The most familiar are individual "complete" or "perfect" flowers, so-called because they contain both male and female reproductive organs within a single bloom.

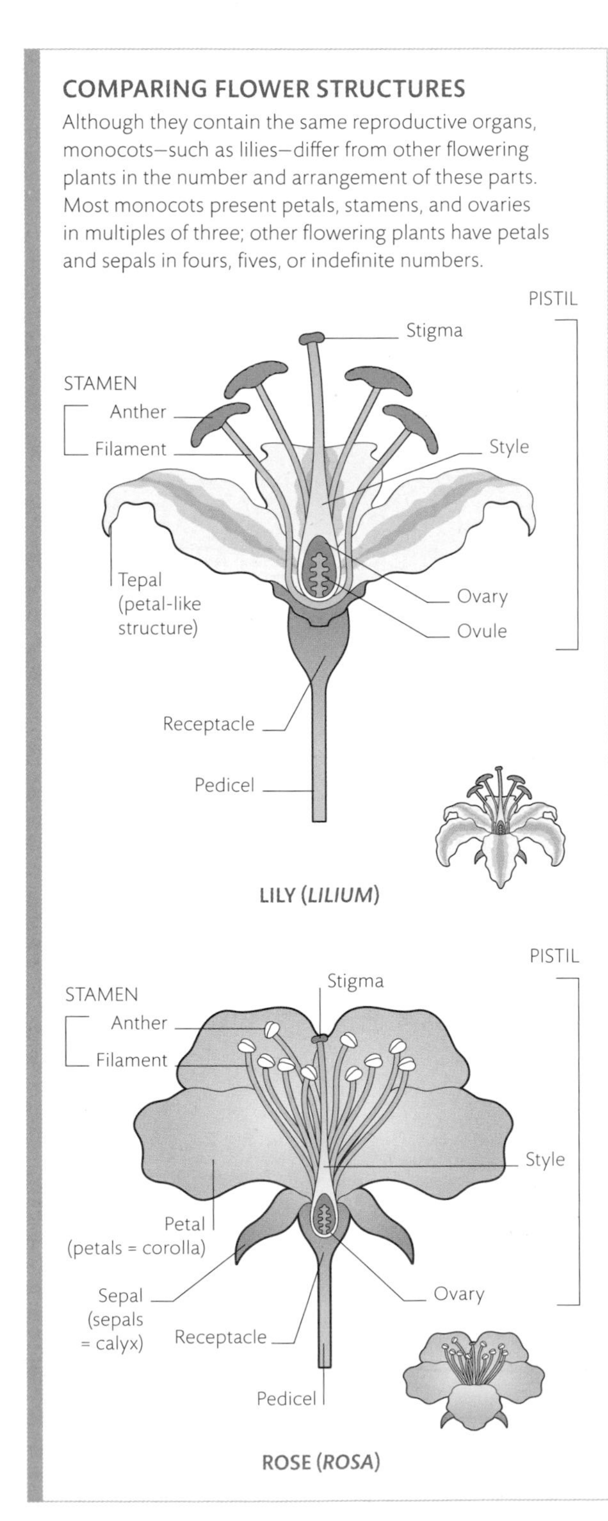

Simple flowers
A simple flower, such as this fuchsia, is comprised of male stamens and a female pistil surrounded by sepals and petals. Compare this with a compound ray and disk flower (see p.218).

A male anther, supported by the filament, produces pollen

Sepals enclose the flower head, peeling back when it blooms

***FUCHSIA* SP.**

Inside flowers

Dissecting a flower reveals its structure and helps with identification. In fuchsias, such as the one here, the petals, sepals, and stamens occur in groups and multiples of four. Slicing through the ovary and the receptacle (the thickened part of the stem where a flower's organs attach) reveals the true extent of the stigma's length, as well as the flower's ovules.

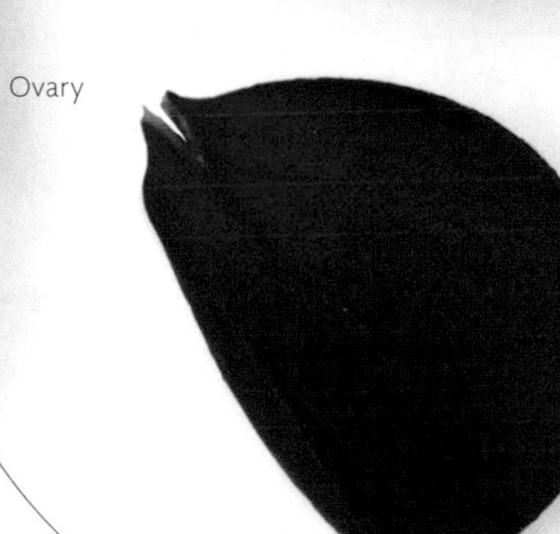

ancient flowers

Most flowering plants are defined as either monocots or eudicots, but some do not fit into either group. These are the so-called "primitive" species (basal angiosperms), which account for less than five percent of flowering plants. The closest living relatives of the first flowering species of plants on Earth, they include the magnolia family, or Magnoliids.

The outer flower consists of tepals—undifferentiated sepals and petals—arranged in whorls

Early flower buds
Unlike many species of flowering plants, magnolia flower buds emerge after being encased in bracts rather than protective sepals.

ANCESTRAL BLOOM

Flowering plants first emerged around 247 million years ago, when water lilylike plants with simple flowers that produced both pollen and ovules appeared. Gradually, these diverged into trees and woody plants, herbaceous land plants, and aquatics. The ancient woody plants have survived more successfully than the others, and have since evolved into many different tree and shrub families. Japanese star anise (*Illicium anisatum*) is one example. With its reduced number of ovules merged into a starlike fruit, it is now thought to belong to one of the first groups of woody plants to evolve.

ILLICIUM

Family resemblance
Botanists believe that the earliest flower looked much like a southern magnolia (*Magnolia grandiflora*). Its flowers contain numerous spirally arranged male and female organs, and produce conelike seed receptacles when fertilized.

flower shapes

Botanists categorize flowers in various ways: noting how their reproductive organs are arranged, for example, or whether certain structures are present. Shape, however, is one of the most useful classifications. Noting whether or not a flower head is symmetrical, then examining the arrangement of petals (the corolla), provides the best starting point.

shape of corolla

ROTATE
Lycianthes rantonnetii

Crown-shaped outgrowth of corolla

CORONATE
Narcissus 'Jetfire'

symmetry

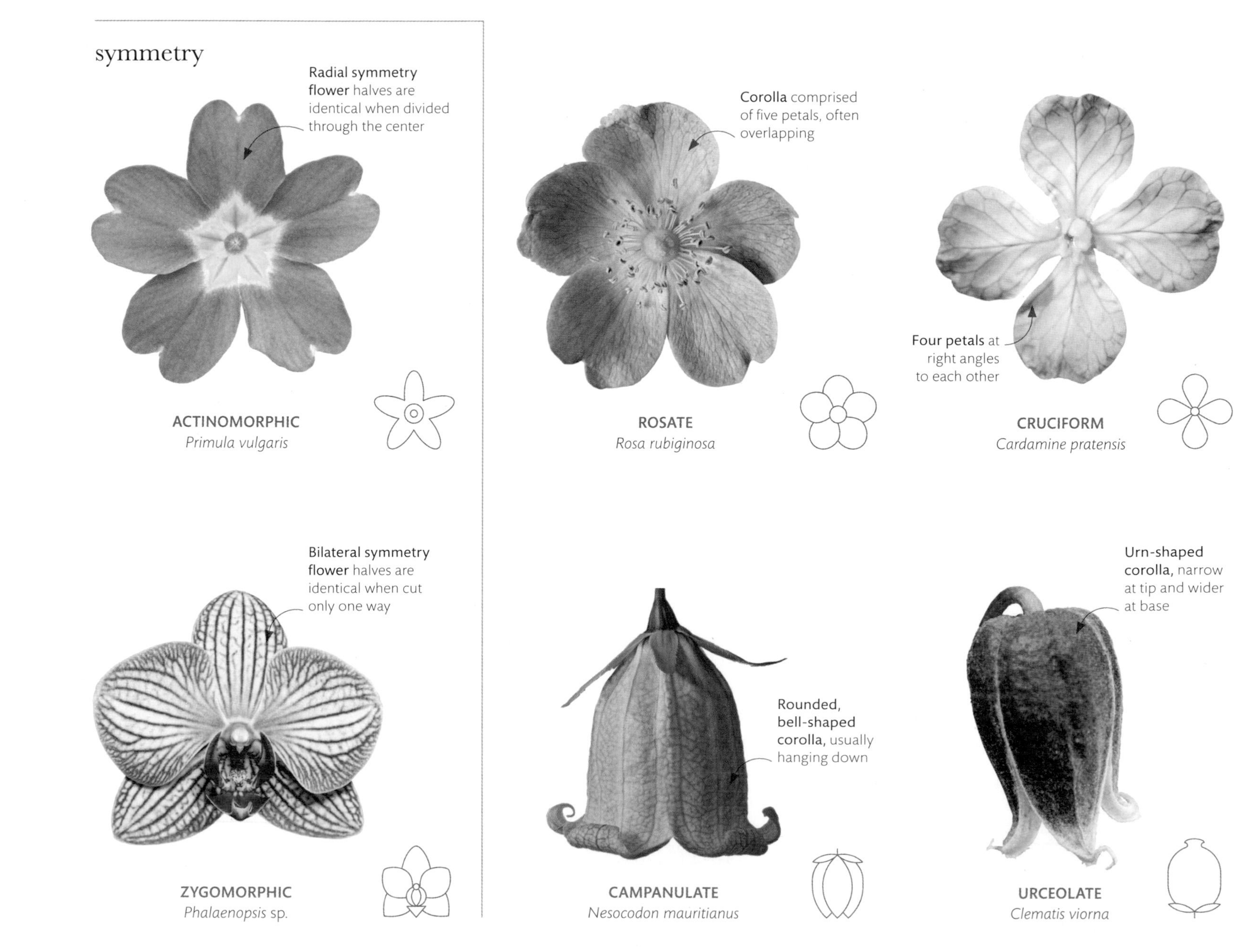

ACTINOMORPHIC
Primula vulgaris

ROSATE
Rosa rubiginosa

CRUCIFORM
Cardamine pratensis

ZYGOMORPHIC
Phalaenopsis sp.

CAMPANULATE
Nesocodon mauritianus

URCEOLATE
Clematis viorna

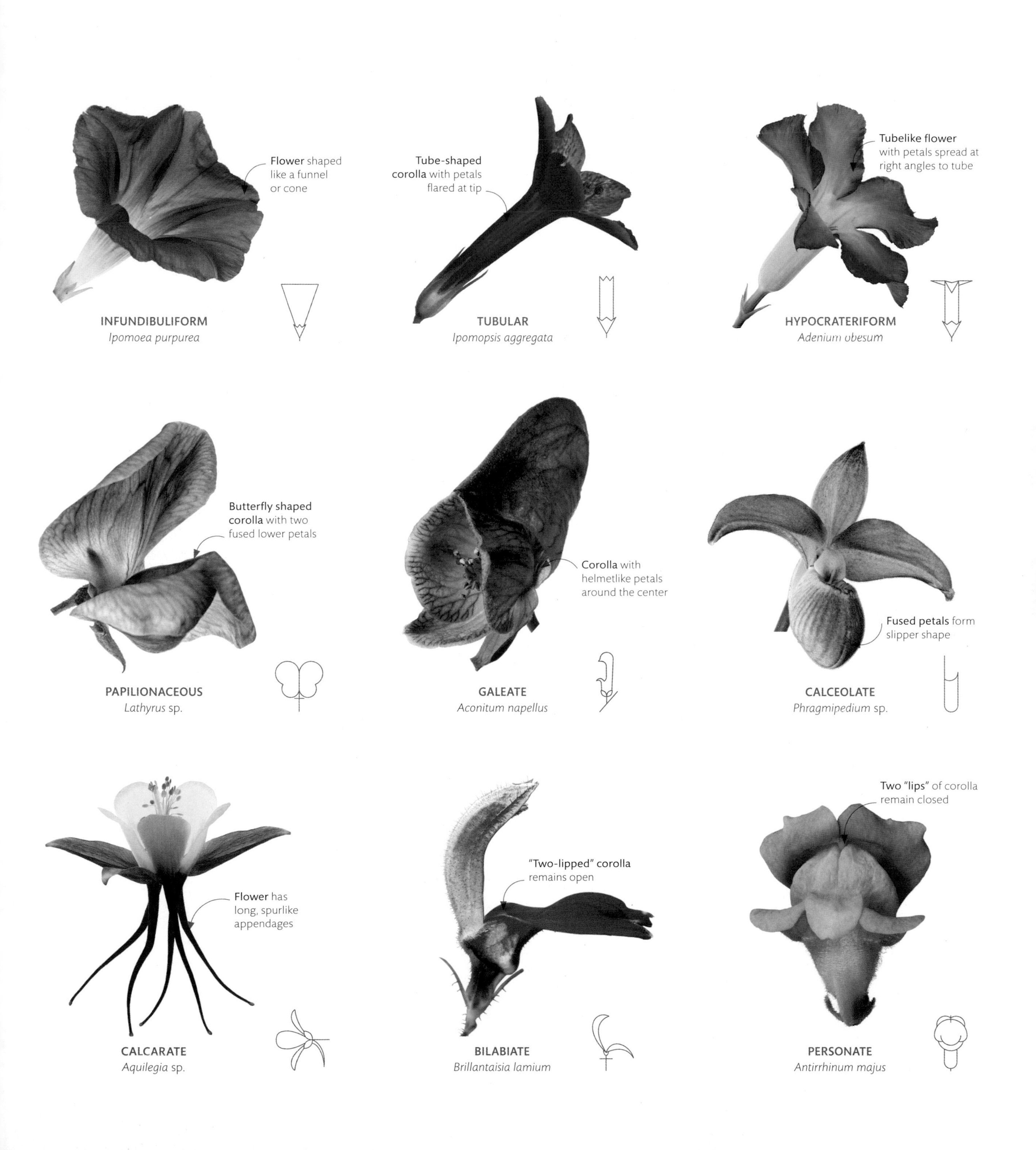

Flower shaped like a funnel or cone
INFUNDIBULIFORM
Ipomoea purpurea
Tube-shaped corolla with petals flared at tip
TUBULAR
Ipomopsis aggregata
Tubelike flower with petals spread at right angles to tube
HYPOCRATERIFORM
Adenium obesum
Butterfly shaped corolla with two fused lower petals
PAPILIONACEOUS
Lathyrus sp.
Corolla with helmetlike petals around the center
GALEATE
Aconitum napellus
Fused petals form slipper shape
CALCEOLATE
Phragmipedium sp.
Flower has long, spurlike appendages
CALCARATE
Aquilegia sp.
"Two-lipped" corolla remains open
BILABIATE
Brillantaisia lamium
Two "lips" of corolla remain closed
PERSONATE
Antirrhinum majus

HERBACEOUS PERENNIALS

Botanically, any plant that flowers and sets seed year after year, from the mighty oak to the humble geranium, is called a perennial. In horticulture, however, the term refers mainly to non-woody plants. The phrase "herbaceous perennial" denotes any soft-stemmed plant, such as a peony, whose stems die back completely in the fall or winter, in contrast to evergreen perennials, such as hellebores.

PAEONIA PEREGRINA

Purple stigmas protrude from the center of the fully open bloom

The sepals unfurl, revealing green filaments topped by purple anthers

Nigella flowers bloom 10–12 weeks after seeds are sown

flower development

Flowers develop whenever a flowering plant is ready to reproduce. Life cycles range from weeks to years, depending on the species. Annuals germinate, flower, reproduce, and die within a year or less. Biennials spend a season developing from seed, rest over winter, and bloom the following spring, living for roughly two years. Perennials bloom every year and live for three years or more. Some trees—woody perennials—can live for centuries.

The sepals begin to fold back, away from the maturing stamens and stigmas

The stigma withers and curls to form part of the multichambered seedhead

The anthers shrivel and die, eventually falling off

Life of a nigella flower
Annuals such as this *Nigella papillosa* 'African Bride' germinate, set seed, and die within a single growing season—no live root, stem, or leaf is left behind, and the plant only survives by its seed.

DAHLIA 'DAVID HOWARD'

GLOBE THISTLE (*ECHINOPS BANNATICUS* 'TAPLOW BLUE')

DAYLIGHT HOURS

Genetic changes triggered by exposure to sunlight make leaves produce a floral hormone called florigen, which "tells" a plant when it is time to flower. Plants make more or less florigen in response to daylight, and some species need more daylight than others.

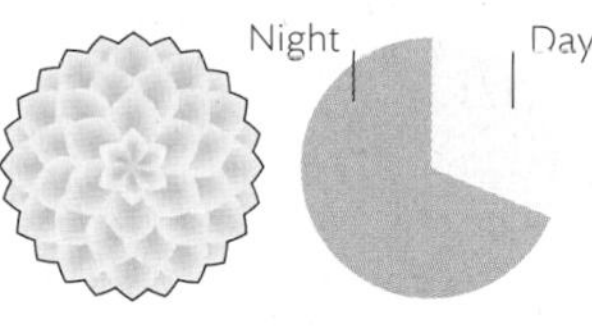

8 DAYLIGHT HOURS

Short-day plants need 8–10 hours of daylight, followed by 14–16 hours of darkness; if the darkness is interrupted at all, they do not bloom. Examples: poinsettias, dahlias, some soybeans.

Night | Day

14 DAYLIGHT HOURS

Long-day plants need 14–16 hours of continuous daylight, followed by 8–10 hours of darkness; the daylight is more critical than the dark. Examples: globe thistle, lettuce, radishes.

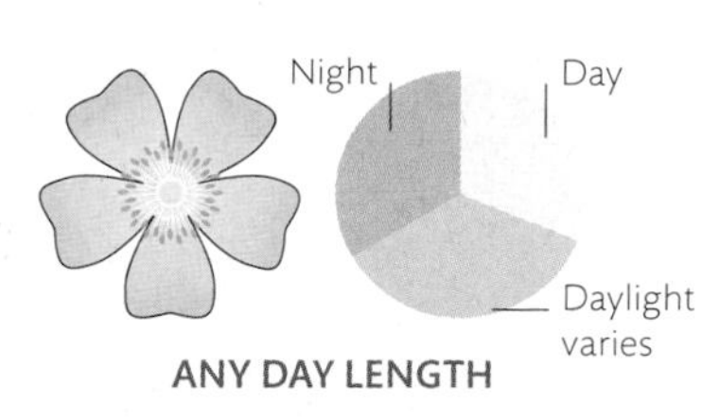

ANY DAY LENGTH

Day-neutral plants are the most flexible flowering plants and bloom if exposed to anything from 5–24 hours of continuous daylight. Examples: sunflowers, tomatoes, and some peas.

flowers and seasons

Throughout the world, plants respond to the changing seasons. Regardless of which hemisphere they live in, most plants germinate, grow, and reproduce in spring and summer, when warm weather means that pollinators are plentiful. They release seed in late summer and fall, when their growth slows down in preparation for winter. While these responses are partly a response to seasonal changes in temperature, the most crucial trigger is the changing amount of daylight, known as photoperiodism. Only plants at the equator experience days and nights of equal length all year.

DOG ROSE
(*ROSA CANINA*)

Blooming times
Plants need differing lengths of daylight to bloom, and their flowers open at different times of year. Globe thistle flowers open in midsummer, whereas dahlias do not open until much later in the year. Some roses flower only in early summer, whereas others, such as most cultivated roses, may flower throughout the summer.

> “ Beautiful scenes throughout the year are what you should remember. ”
>
> CHINESE PROVERB

Flowers and poetry
A dainty sweep of magnolia blossoms is one of 12 ink and color leaves inscribed with poems from an *Album of Fruit and Flowers*, the work of Qing Dynasty artist Chen Hongshou (1598–1652). Early flowering magnolias are celebrated in China as "the flower that welcomes spring." Legend has it that, at one time, only the Emperor was allowed to grow the tree.

Chinese hwamei
Bird and flower painting was one of the many admired skills of Yuan Jin (active 1857) in the Qing Dynasty. This ink and color work features the Chinese hwamei, or melodious laughing thrush, amid viburnum blossoms.

plants in art

chinese painting

The strokes of ink and color on silk and paper in Chinese flower painting have much in common with Chinese calligraphy. Scholars, trained from an early age in the art of handwriting, used calligraphic brushstrokes in their paintings. Flower painting was regarded as "silent poetry" and poetry as "painting with sound;" over time, the two became combined in artworks that represent the song of nature.

Floral thoughts
Ming Dynasty artist Chen Chun (1483–1544) excelled in the study of ancient poetry, prose, and calligraphy and brought this wealth of knowledge to his art. He regarded flower painting, such as this spontaneous collection of spring peach blossoms and *Ziziphus jujuba*, as "idea writing."

Chinese flower painting originated from Buddhist banners decorated with flowers that were brought into China from India in the first century CE. The art form reached its peak in the Tang Dynasty (618-907) and has endured across the centuries.

The "four treasures" of traditional Chinese painting and calligraphy are ink stones, ink sticks, brushes, and paper. Artists use combinations of four basic techniques: outlines filled with colors, "boneless" washes, ink lines, and sketching ideas. The fine tip of a Chinese brush produces an infinite variety of strokes depending on which part of the brush is used and the pressure on paper or silk.

Plant subjects have a distinct character in the eye of the artist and a symbolism recognized throughout Chinese culture. In bird and flower painting, a specific genre associated with the Taoist philosophy of harmony with nature, certain birds and flowers are paired for their symbolism: the crane and pine tree, for example, both represent longevity.

Lateral buds mature after terminal buds

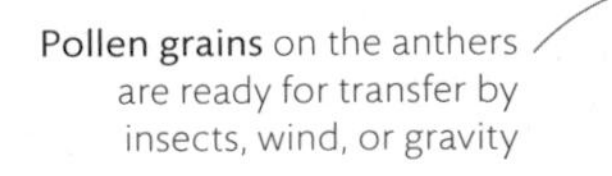

Open petals invite pollinators to land on flowers, transferring pollen between blooms

Self-pollination
Plants such as St. John's wort (*Hypericum pseudohenryi*) reproduce through the transfer of pollen from one flower to another on the same plant. This can also occur within the same flower. Both of these processes are described as self-pollination.

CROSS-POLLINATION

Cross-pollination occurs when pollen from an anther of one flower is transferred to the stigma of another flower on another plant of the same species. Once transferred, the pollen grains release a tube that pushes down the style into the ovary. This delivers male sperm cells to fertilize female egg cells within the ovary.

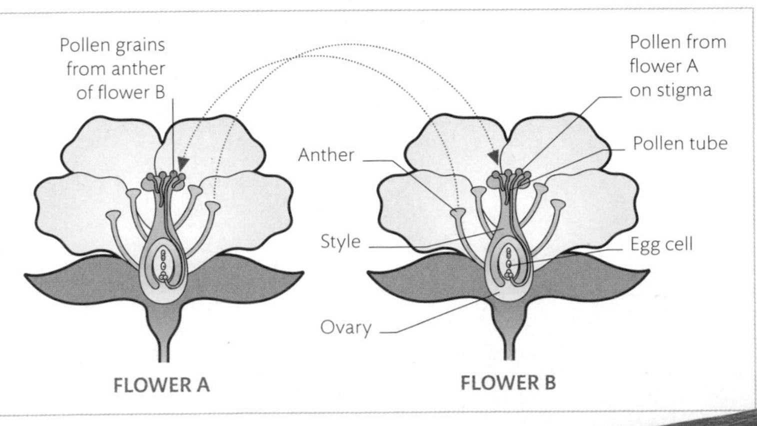

flower fertilization

The formation of flowers indicates that plants are ready to produce—or "set"—seed and pass on their genes. Fertilization occurs when sperm-bearing pollen is transferred to the flower's female reproductive organs, or pistil, where it fuses with the female reproductive cells (the ovules) to create seeds.

The anthers shrivel and the stamens droop after a flower is fertilized

A fertilized ovary changes color, transforming into a red, ripening seedhead

Petals fold back and wither around the fertilized ovary, a sign the flower is starting to set seed

The stigma loses stickiness and turns brown on the ripening seedhead

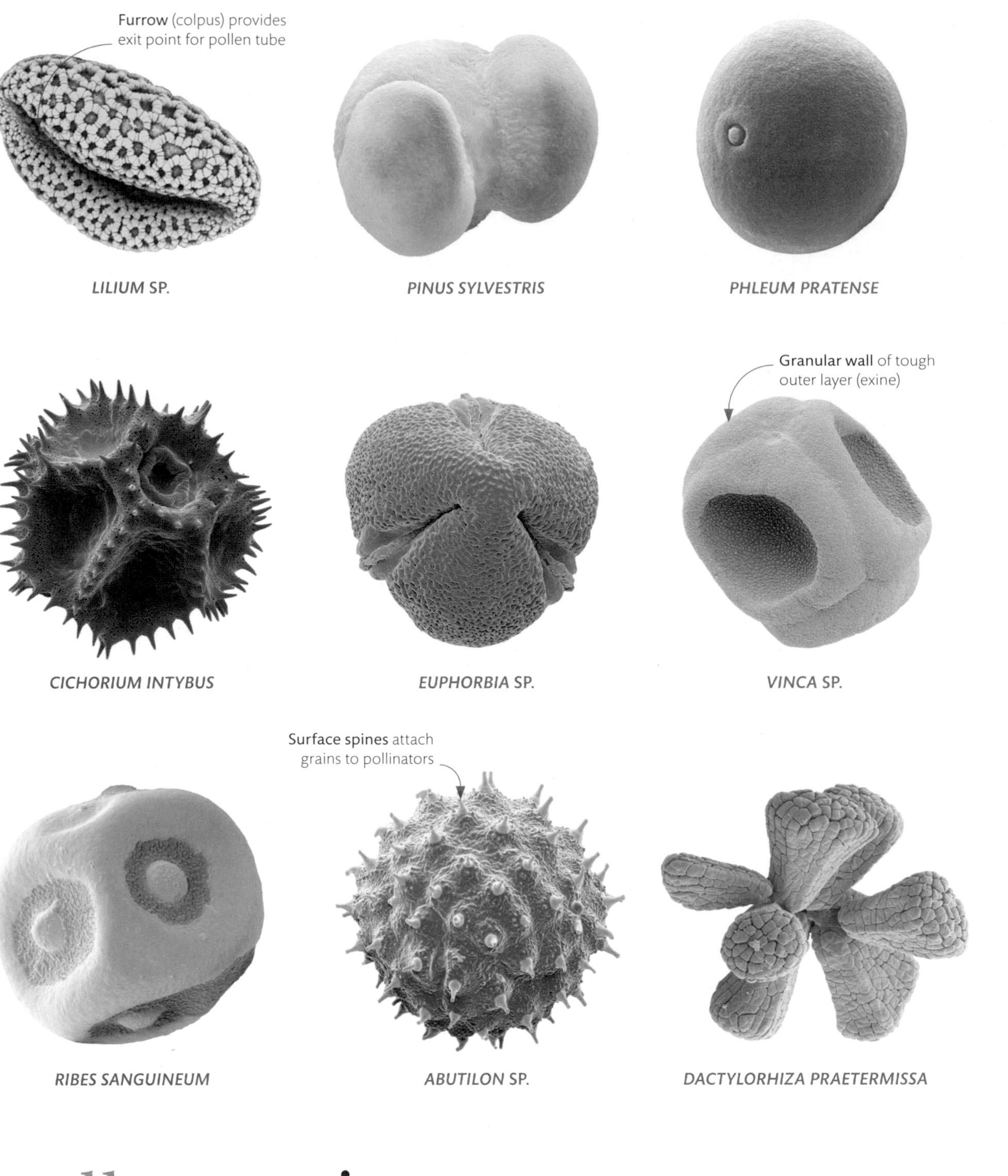

pollen grains

Although they look like dust to the human eye, pollen grains vary greatly in shape, size, and texture. Scanning electron microscopy reveals spheres, triangles, ovals, threads, and disks, among other shapes. Grain surfaces can be smooth, sticky, spiny, striated, meshed, or grooved, and marked by pores or furrows.

Plentiful pollen
A honeybee can fill its leg baskets with about 0.0005 oz (15 mg) of pollen in a single foraging trip to flowers like this cactus.

Spatial constraints
The blooms of species such as Japanese lantern (*Hibiscus schizopetalus*) are located far apart. Hanging pendantlike from the ends of branches, they are easily seen by bird and insect pollinators.

A slender pedicel up to 6in (15cm) long supports each flower

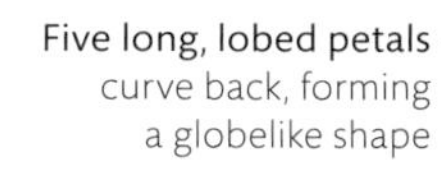

Five long, lobed petals curve back, forming a globelike shape

PIN AND THRUM

Primroses feature two incompatible flower types to reduce the possibility of self-pollination. The stigmas in a "pin" flower are at the top of the floral tube, but are located inside the tube in a "thrum" flower. Pollination between a pin and a thrum is more likely to result in fertilization than pin-to-pin or thrum-to-thrum pollination.

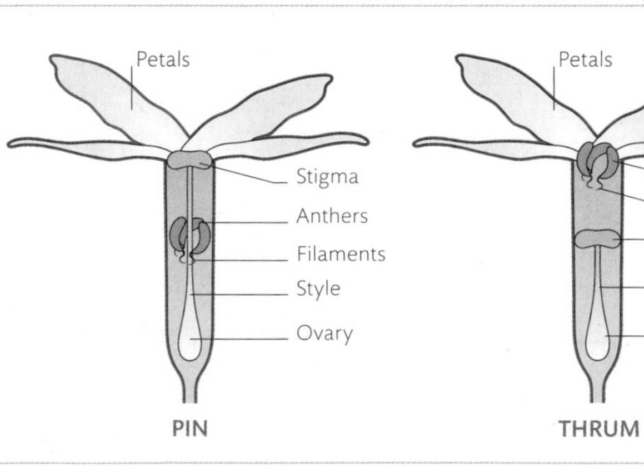

The underside of petals shows the veining and blotching typical of many *Hibiscus* species

Green flower buds dangle from pedicels sprouting from leaf axils on upper stem

Joint, where the pedicel connects to the peduncle

Self-avoidance

Most *Hibiscus* shed their pollen before the styles are receptive. They coat the legs and undersides of pollinators with pollen, which is carried to another bloom. When the styles become receptive, they curve up to receive pollen from another flower.

The staminal column extends from the center of the flower and is twice as long as its petals

Pollen-loaded anther

Curved style

encouraging diversity

Many plant species have evolved to favor pollination from other plants rather than from their own flowers, as cross-pollination usually results in stronger seeds and healthier plants that are more resistant to disease. One way in which a plant can reduce the chance of self-pollination and increase the possibility of cross-pollination is by keeping the reproductive organs of its flowers as far apart as possible, as is the case with the Japanese lantern.

Large, vibrantly colored petals attract pollinators such as bumblebees
The winged structure pushes the bee to the right, then to the left nectary deep within its center
Two large, oval leaflets attach to the winged stems

managing pollen

Flower shape can influence how plants release and receive pollen, and some plants have evolved blooms that make it difficult—or impossible—to pollinate themselves. Asymmetrical flowers such as pea blossoms allow only the strongest insects to gain access; once inside, the internal structure ensures that pollen is received and dispersed in two separate stages—avoiding self-pollination.

Tendrils allow the plant to scramble up and over varied terrain

Two-stage pollination
Everlasting pea blooms (*Lathyrus latifolius*) have two nectaries. When a bee forces its way inside, the petals direct it first to the right, then to the left, releasing the style, which has a hairy, brushlike area on which the pollen settles, below the stigma. The stigma touches the bee, collecting pollen from other flowers. When the bee moves to the other side, the brush pats it again, this time to transfer its own pollen for transport to other plants.

一種
千葉鋸歯あり
紫色淡紫辺の物
一種
單辨鋸歯
ありて紫色
淡紫辺の物

Mount Fuji with Cherry Trees in Bloom
This polychrome woodcut by Hokusai, made around 1805, celebrates spring with a view of the snowy peak of Mount Fuji seen through cherry blossoms and mist. The luxurious print, known as a *surimono*, is on thick paper and includes metal pigments such as copper and silver powder.

plants in art

japanese woodblock prints

Chrysanthemum
At the peak of his career as a landscape artist, Hokusai turned his attention to flowers, and produced a set of ten stylized woodcut prints known as the "Large Flowers." This detail from a larger work captures the detail of a chrysanthemum's many petals.

A mainstay of Japanese art for centuries, woodblock printing was at a peak of popularity in the 19th century. The prints were made with water-based inks that enhanced color, glaze, and transparency, and their bold, simplified forms and subtle colors were the perfect medium for capturing the beauty of the Japanese landscape and its native flora.

As the power of the last Shogunate declined and travel restrictions were lifted, Japan's botanists were drawn to Western scientific methods. Iwasaki Tsunemasa (1786–1842), a young shogun with a passion for the natural world, spent time with German scientist Philipp Franz von Siebold, who was based in the Dutch East Indies (now Indonesia). Iwasaki, also sometimes known as Kan-en, roamed the countryside, bringing back specimens to draw, paint, and name for his epic work, *Honzo Zufu*, a botanical atlas featuring 2,000 plants.

Perhaps the best-known artist of the Edo period is Katsushika Hokusai (1760–1849). He learned the art of woodblock printing, known as *ukiyo-e* (the art of the floating world), as a young apprentice and went on to excel in paint and print in every genre. Late in life he wrote: "At seventy-three, I began to grasp the structures of birds and beasts, insects and fish, and of the way plants grow. If I go on trying, I will surely understand them still better by the time I am eighty-six, so that by ninety I will have penetrated to their essential nature."

" Illustrations must be made with all possible skill and accuracy. If they are not, how then will it be possible to differentiate plants which closely resemble each other? "

IWASAKI TSUNEMASA, PREFACE TO *HONZO ZUFU*

Honzo Zufu
Botanist Iwasaki Tsunemasa collected plants and seeds from the countryside and grew them in his garden so that he could record their fine detail in his artworks. This vibrant woodblock print of an opium poppy is taken from his *Honzo Zufu* (Botanical Atlas). The first four volumes were printed in 1828, and the entire work of 92 volumes was finally printed in 1921.

Male and female plants
Tiny, unisexual flowers are often smaller than their bisexual counterparts, but tend to mature faster. By producing a whole host of anther-bearing flowers, a male holly (*Ilex aquifolium*, shown left) increases its chances of attracting insects to carry pollen to female trees—which are the only ones to produce berries.

FEMALE FLOWERS OF HOLLY (*ILEX AQUIFOLIUM*)

HOLLY BERRIES ON FEMALE TREE

unisexual plants

In the animal kingdom, separate, individual males and females are the norm when it comes to reproduction, but in the botanical world, when one plant bears flowers of only one sex, it faces several challenges. While such dioecious plants, which include many trees, avoid self-fertilization, they rely completely on pollen being transported successfully from male to female—often over quite a distance.

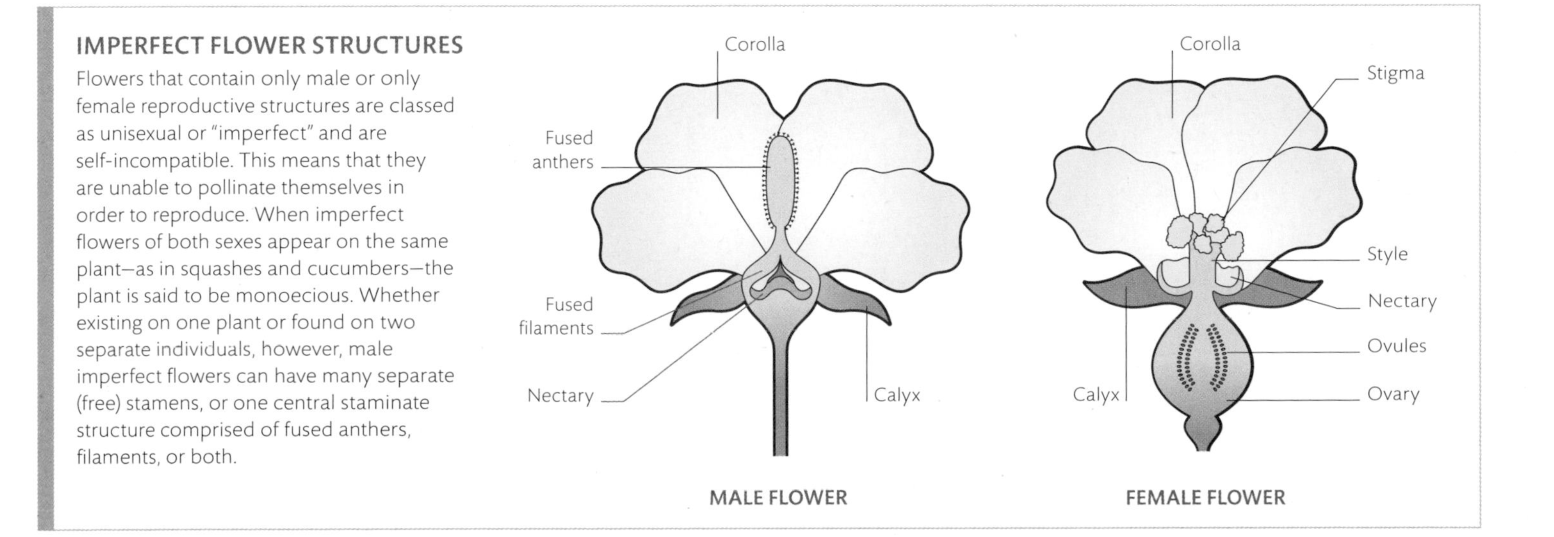

IMPERFECT FLOWER STRUCTURES

Flowers that contain only male or only female reproductive structures are classed as unisexual or "imperfect" and are self-incompatible. This means that they are unable to pollinate themselves in order to reproduce. When imperfect flowers of both sexes appear on the same plant—as in squashes and cucumbers—the plant is said to be monoecious. Whether existing on one plant or found on two separate individuals, however, male imperfect flowers can have many separate (free) stamens, or one central staminate structure comprised of fused anthers, filaments, or both.

The sterile zone, or appendix, is comprised of sterile male flowers, which may play a part in attracting pollinators

incompatible flowers

Many species avoid pollinating themselves by means of a careful arrangement of structures in the case of bisexual flowers, or by having the unisexual male and female flowers on separate plants. Some inflorescences accomplish this even when they contain flowers of both sexes within one unit. Staggered maturity rates, leafy shields, and buffer zones ensure that these plants are largely cross-pollinated, creating a healthier genetic mix.

SEPARATION STRATEGY

Many arums, such as dumb canes (*Dieffenbachia sp.*), have sterile zones in the middle of the spadix, separating male and female flowers. These flowers never develop fully but perform the important role of helping to prevent pollen from fertile males reaching fertile females.

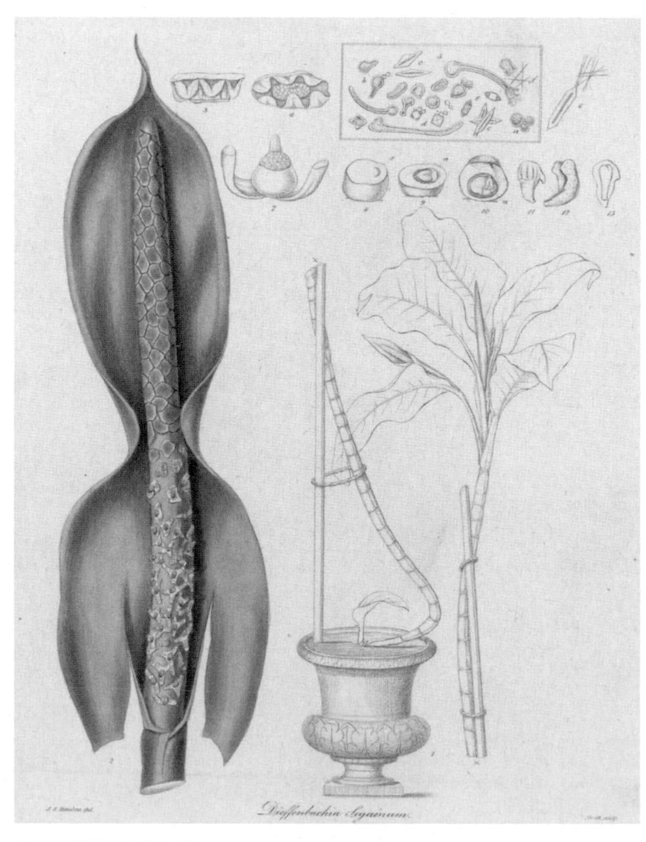

***DIEFFENBACHIA* SP.**

MALE FLOWERS

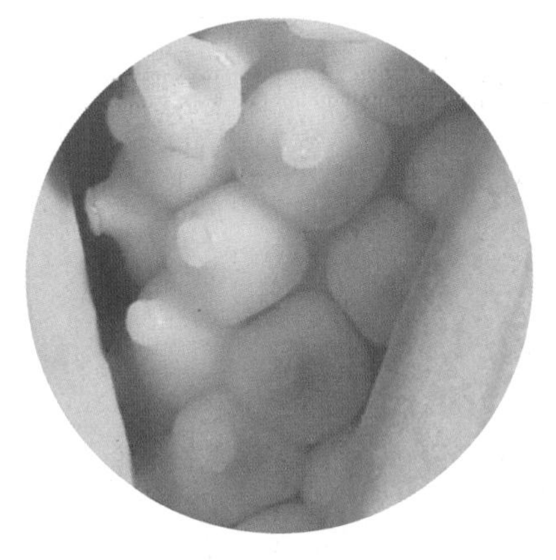

FEMALE FLOWERS

Female flowers are enclosed by the spathe and mature 1–2 days before the male flowers, lowering the risk of self-pollination

All in the timing

Plants in the arum family (Araceae), such as *Pseudodracontium lacourii*, produce a spadix (a columnar inflorescence), that contains both fertile male and female flowers and sterile ones (usually male) in a separate zone at the top or center. The female flowers reside at the base, wrapped in a sheathlike spathe, and the male flowers only release pollen when the females have already been pollinated by beetles.

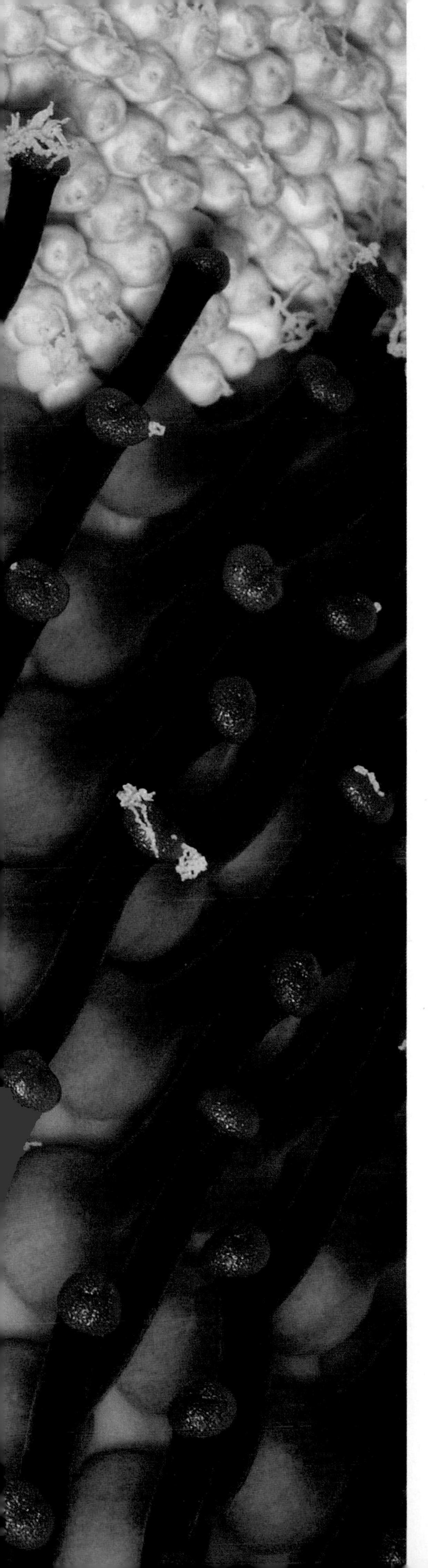

Amorphophallus titanum

the titan arum

This colossus of the Sumatran rainforest holds the record for the largest unbranched inflorescence in the world. The huge size of the inflorescence is matched by its powerful odor, which has been likened to that of rotting flesh—hence its nicknames of "corpse plant" and "carrion flower."

Appearances are deceptive with the titan arum: what looks like a 10 ft (3 m) tall flower is really an inflorescence called a spadix surrounded by a frilly tissue known as a spathe. The flowers themselves are deep inside, at the base of the spadix.

This plant is remarkable not just for its inflorescence: it also generates its own heat. As the inflorescence matures, energy stored in a massive underground organ called a corm is used to warm the flowers to about 90°F (32°C). The corm, which weighs up to 110 lb (50 kg), is the largest known in the plant world. It is thought that producing all this heat helps the plant to disperse its foul scent throughout the dense rainforest to attract pollinators.

In essence, pollination for the titan arum relies on a ruse. The smell of rotting meat draws insects such as flesh flies and carrion beetles in search of a meal and a place to mate and lay their eggs. They find none of these rewards, and instead are doused with pollen. With luck, they fall for the trick again the next evening and pollinate another titan arum bloom.

The inflorescence itself only lasts 24 to 36 hours before it collapses. It takes so much effort to produce that individual plants can only bloom every three to ten years. When not in flower, the titan arum exists as a single, enormous leaf, roughly 15 ft (4.5 m) tall. As impressive as the titan arum is, this plant is at risk of becoming endangered if the deforestation of its habitat continues at the current rate.

Titanic bloom
The flowers of the titan arum are arranged in dense clusters, with male flowers on top (white) and female flowers at the bottom (red). The overall structure of the titan arum acts like a chimney, channeling the fetid smell up into the air column to spread it far and wide.

The fleshy, hollow spadix retains the heat produced underground

A frilly spathe surrounds and protects the flowers at the base of the spadix

Meaty allure
Hundreds of tiny flowers are hidden behind the titan arum's huge, petallike spathe. The burgundy coloration of the spathe is thought to mimic the appearance of decomposing meat.

Ladies first
Amazon water lilies have relatively short lives. They open at dusk, heating up and exuding a pineapple scent that is irresistible to scarab beetles of the *Cyclocephala* genus. The petals then close on the beetles, imprisoning them until the next evening.

flowers that change sex

Most flowering plants have hermaphroditic flowers containing both male and female parts. These separate sex organs usually mature alongside each other during the blooming cycle. In some hermaphrodites, however, the reproductive parts mature in such separate, distinct stages that the flowers effectively change sex—either from female to male (protogyny) or from male to female (protandry). The Amazon water lily (*Victoria amazonica*), which is pollinated by beetles trapped within its flowers, has flowers with female sex organs that mature before the male ones.

CAPTIVE POLLINATORS

The beetles that pollinate Amazon water lilies are rewarded with high-calorie starchy pads attached to the flowers' carpels, which the beetles feed on during their incarceration. While bumbling from one carpel pad to another, they transfer pollen collected from other male-phase flowers to stigmas inside the floral prison, then get dusted with more as the female flower changes into a male one.

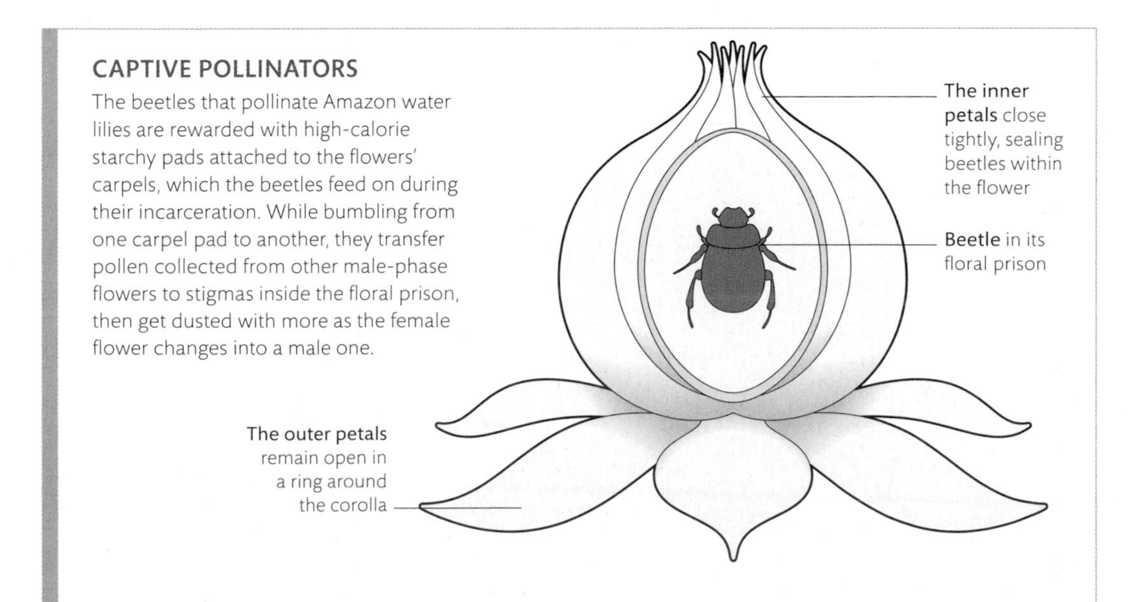

Undergoing change
Changing from white to pinkish purple signals that a giant water lily flower is entering its male phase. All of the beetles trapped inside the flower spend the day being dusted with pollen shed by the water lily's maturing anthers. When the flower opens on the second evening, the beetles fly away, seeking more female flowers. The flower then sinks below the water's surface, its task accomplished.

Bell-shaped **flowers** turn pink as they mature

Drumstick allium
Allium sphaerocephalon features round or egg-shaped umbels. In an umbel, the inflorescence stalk, or peduncle, has a wider, rounded end. This stalk serves as a "platform" for numerous flowers, whose stems, or pedicels, are all the same length.

UMBEL

Anthers shed most of their pollen within 1–2 days of opening, before the stigma is fully developed. Once mature, a stigma remains receptive for several days, and may receive pollen from lower, younger flowers

Pollen is transferred to the uppermost flowers as bees move from younger to older blooms

inflorescences

Many of the most striking floral displays occur in plants with an inflorescence, where numerous flowers are produced on a single stem. This is especially true of ornamental alliums. From a distance, these members of the onion family give the impression of bearing one large bloom, but a closer look reveals that each "bloom" consists of numerous tiny flowers, or florets. If a single inflorescence's florets open at different times over a period of days, or even weeks, and each floret produces pollen capable of fertilizing its neighbors on the same plant, the opportunity for self-pollination is immense.

Petals change color to guide pollinators to the flowers that offer the best rewards

Flourishing individuals
Many-flowered inflorescences cram multiple sources of nectar and pollen into a small space, allowing pollinators to feast while utilizing less energy in moving from flower to flower. The blooms often mature at different rates, which actively encourages pollinators to return; this strategy promotes both self-pollination on the same inflorescence and cross-pollination between different plants.

Flowers at the base may open as much as two weeks later than the uppermost blooms

The uppermost flowers often open first; as they age, flowers may increase the amount of nectar they produce to encourage more visits from insects

inflorescence types

Inflorescences are defined by how their flowers are arranged around main and lateral stalks, called peduncles and pedicels. Inflorescences can be either determinate or indeterminate. Determinate inflorescence stalks end in a single flower, while indeterminate ones terminate in a vegetative bud. Once a terminal flower bud forms, growth in that direction stops for determinate flowers. Indeterminate inflorescences, by contrast, continue to grow and can produce flowers at varying stages of maturity on the same inflorescence. Here are some examples of the many types.

determinate inflorescences

SOLITARY
Tulipa sp.

indeterminate inflorescences

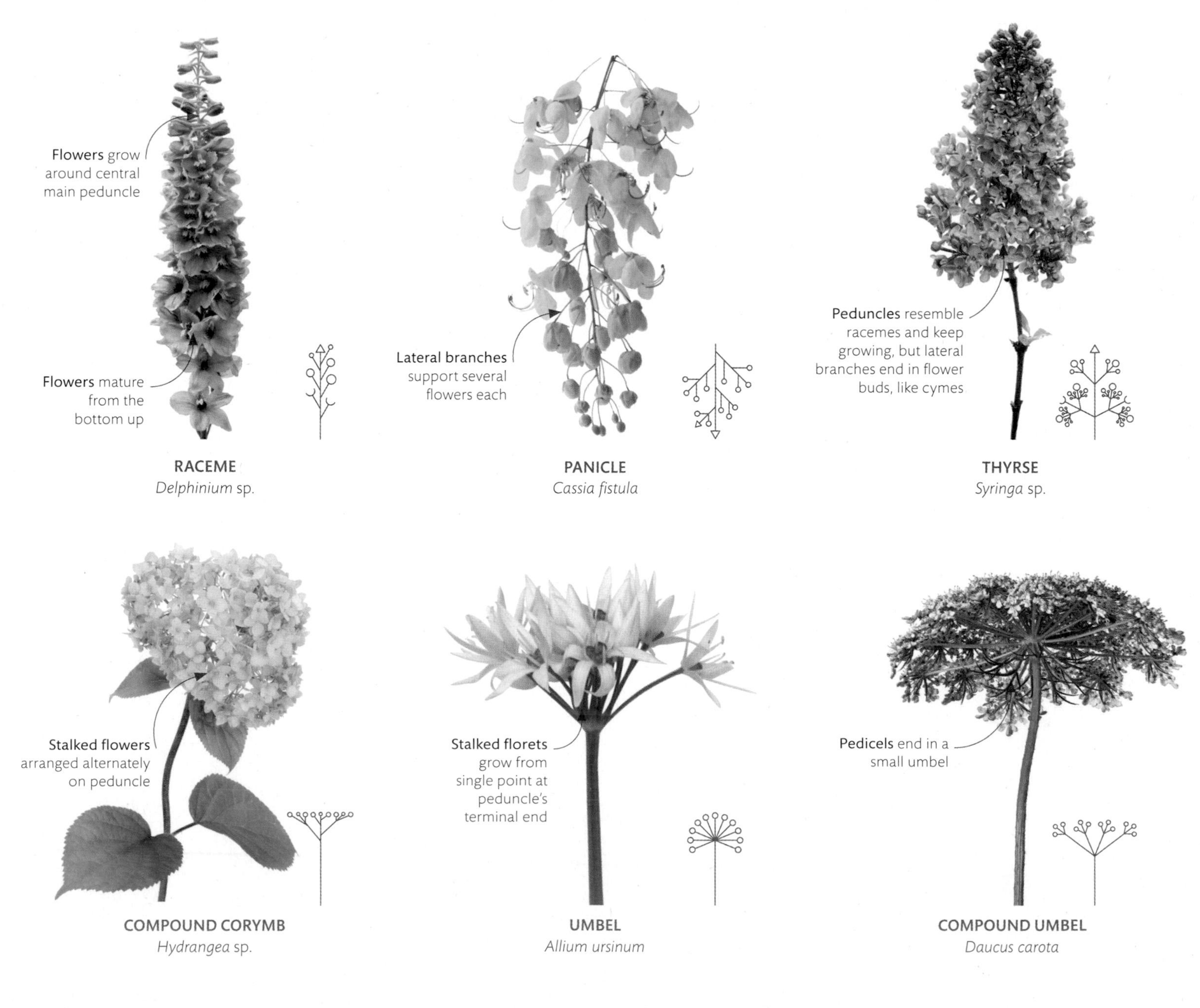

RACEME
Delphinium sp.

PANICLE
Cassia fistula

THYRSE
Syringa sp.

COMPOUND CORYMB
Hydrangea sp.

UMBEL
Allium ursinum

COMPOUND UMBEL
Daucus carota

Terminal flower typically opens before its lateral blooms

Flowers grow from central point, forming simple cyme

CYME
Dianthus chinensis

Secondary branching made up of simple cymes

COMPOUND CYME
Ranunculus acris

Alternate pedicels on each side create a zigzag flower pattern

SCORPIOID CYME
Iris sp.

Stalkless (sessile) flowers attached directly to peduncle

SPIKE
Callistemon sp.

CATKIN
Alnus glutinosa

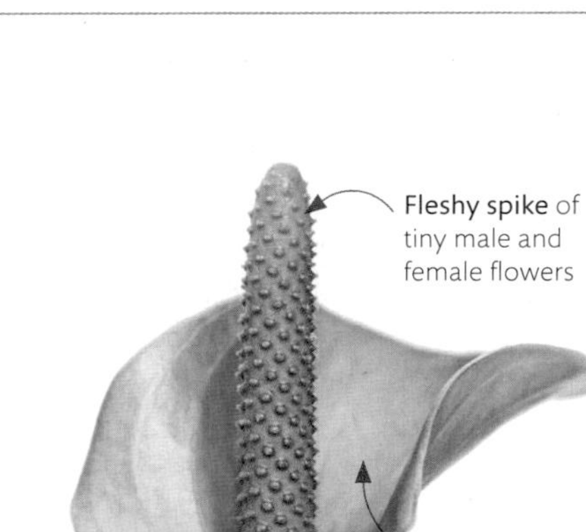

SPADIX
Anthurium sp.

CAPITULUM
Taraxacum sp.

CAPITULUM
Echinacea purpurea

VERTICILLASTER
Stachys palustris

The whorl of bracts folds back, allowing the head to expand

Central disk flowers begin to color and swell as ray flowers elongate

As ray flowers unfold, the straplike shape of each individual petal is revealed

ECHINACEA PURPUREA 'MAXIMA'

How composite flowers open
Composite blooms open from the outside in. Disk flowers expand and change color as they mature, but only open once the surrounding rays have completely unfurled.

rays and disks

The daisy family (Asteraceae) is one of the largest groups of flowering plants. They have a distinct floral structure—each apparently "single" flower head is comprised of tiny blooms, known as ray and disk florets. Some, like dandelions, are composed of petallike rays; some, like thistles, contain only tubelike disks; and some, such as echinacea, have both ray and disk florets.

FLOWER STRUCTURE

While their petals (ligules) differ greatly, ray and disk blooms both have fused anthers that form a cylindrical shape, and a pappus—hairlike bristles that have replaced the sepals, or calyx, found in typical flowers—that aids in seed dispersal.

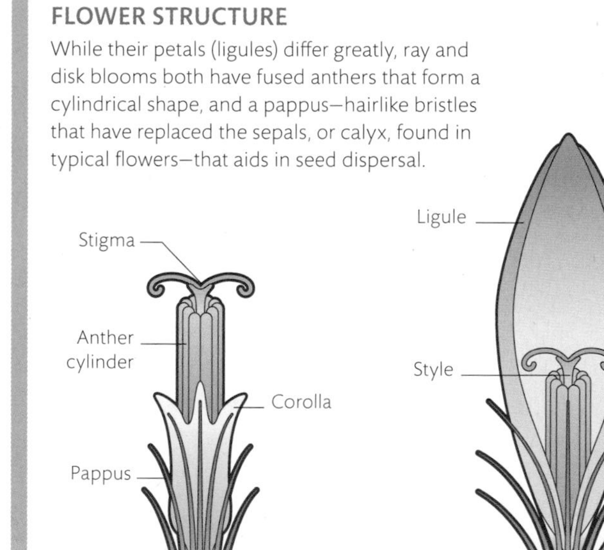

Many-flowered bloom
In echinacea heads, maturing disk flowers enlarge to form a rounded center. Oval, pinkish ray flowers—which may be sterile, but help attract pollinators—flatten and flex backward away from the disks. The tubular corollas blush orange red, turning darker as they open to reveal forked, pollen-coated styles, each surrounded by five small, pointed edges.

Outer disk flowers turn from green at the base to red-orange toward the tip

Helianthus sp.

the sunflower

Sunflowers have been cultivated since at least 2600 BCE, not only for their bright yellow blooms that echo the sun, but also for their highly nutritious seeds. Originally native to the Americas, the sunflower has since spread across the globe.

Sunflowers are known for their habit of tracking the sun across the sky. Contrary to popular belief, this heliotropism only occurs as plants are developing. Both the leaves and the flower buds track the sun's path to maximize their exposure to its life-giving rays. Once the flowers have bloomed, this daily motion stops and the flowers generally orient themselves toward the east. In this way, they are able to take advantage of the sun's heat as soon as it rises over the horizon, which increases pollinator visits as well as the speed at which their seeds develop.

Big yellow inflorescence
Each flower in the inflorescence's center, called a disk flower, produces a single seed. Most of the disk flowers shown here have yet to open. The more seeds a plant produces, the better the chances that it will have offspring the next year.

What looks like a single flower is really an inflorescence composed of many tiny flowers. The flowers of the inflorescence mature from the outside in, providing plenty of opportunities for pollinators throughout the blooming period.

The genus *Helianthus* has some 70 different sunflower species, mostly annuals or biennials. The most commonly encountered sunflower is *Helianthus annuus*, which has been selectively cultivated for centuries to grow a single, outrageously large inflorescence on top of a long, bristly stalk. Wild sunflowers look quite different, producing many branching stalks, each ending in a much smaller inflorescence.

Some sunflowers are allelopathic: they produce a chemical cocktail that inhibits other plants' growth. By poisoning plants around them, the sunflowers limit the competition that they face and increase the amount of seeds they can produce.

Each "petal" is made up of the fused petals of an individual ray flower

False petals
The sunflower's bright yellow "petals" are actually sterile flowers, called ray flowers, that exist only to attract pollinators to the fertile disk flowers at the center of the inflorescence.

Spring catkins
Among more noticeable wind-pollinated flowers are catkins, produced by trees such as white oak (*Quercus alba*) and hazel (*Corylus* spp.). Most catkins are formed of male flowers. The merest breeze releases clouds of pollen, which fertilizes the female flowers.

Two glumes, sheathlike lower or basal bracts, enclose the base of each spikelet

Each floret is held within two bracts: the lemma and palea

The palea is the short, inner bract

The longer bracts are called lemma

Spikelets have one or multiple florets

Shimmering seed heads
The northern sea oat (*Chasmanthium latifolium*) is a tall, clump-forming grass native to the woodlands and inland waterways of central and eastern North America. It produces dangling inflorescences that are typical of many wind-pollinated plants. The plant is sometimes called "spangle grass" because of its oatlike seedheads that shimmer in the sunlight.

The culm, or main stem, allows the large, drooping spikelets maximum exposure to breezes

Fertilized florets become more rigid before they release their seed

wind-pollinated flowers

Easily overlooked, most wind-pollinated flowers use breezes to carry their pollen, so they do not need to draw attention to themselves with showy petals. Many plants that use wind pollination, such as grasses and hazel catkins, hide their flowers in special bracts that protect the reproductive organs until they emerge for just long enough to be fertilized.

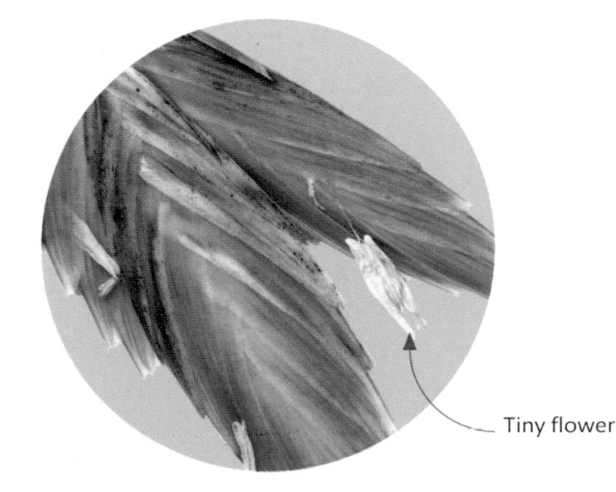

Seizing the day
From spring to fall, male and female flowers emerge from green spikelets of beach-dwelling sea oats (*Uniola paniculata*). Florets open only once in the early morning, then quickly close.

Anther

INSIDE GRASS FLOWERS

Grass florets have specialized bracts known as lemmas and paleas instead of petals. These are pushed apart when the lodicule at the base swells, opening the flower. Anthers and stigmas extend beyond the flower in order to release and receive pollen grains on the wind.

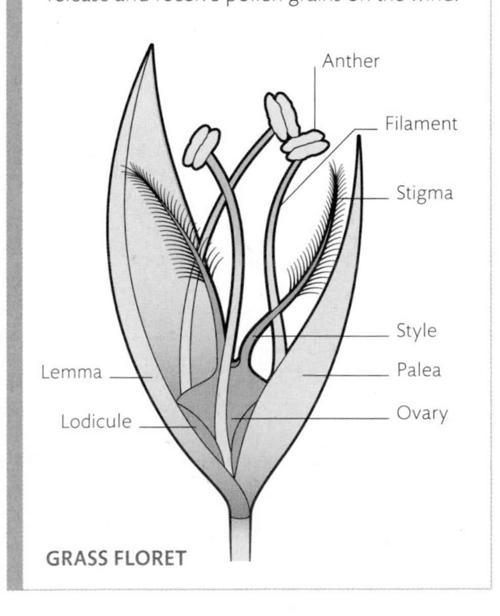

GRASS FLORET

Catching the breeze
Fluffy, wheatlike tufts give *Cenchrus longisetus* its common name of feathertop, but the cloudlike blooms are for more than show. As well as long anthers that release pollen, the feathery structure of its stigmas allows them to capture pollen grains from passing breezes, maximizing the chances of fertilization from other grass blooms.

grass flowers

Despite comprising the third-largest plant group on Earth, grasses are rarely associated with flowering. This is partly due to size—individual flowers of low-growing species are often so tiny that they are imperceptible to the naked eye—and partly due to structure. Grass flowers are wind-pollinated, so instead of brightly colored blooms, they feature fuzzy, tuftlike heads, usually on long stems, and rely on air currents for pollination and seed dispersal.

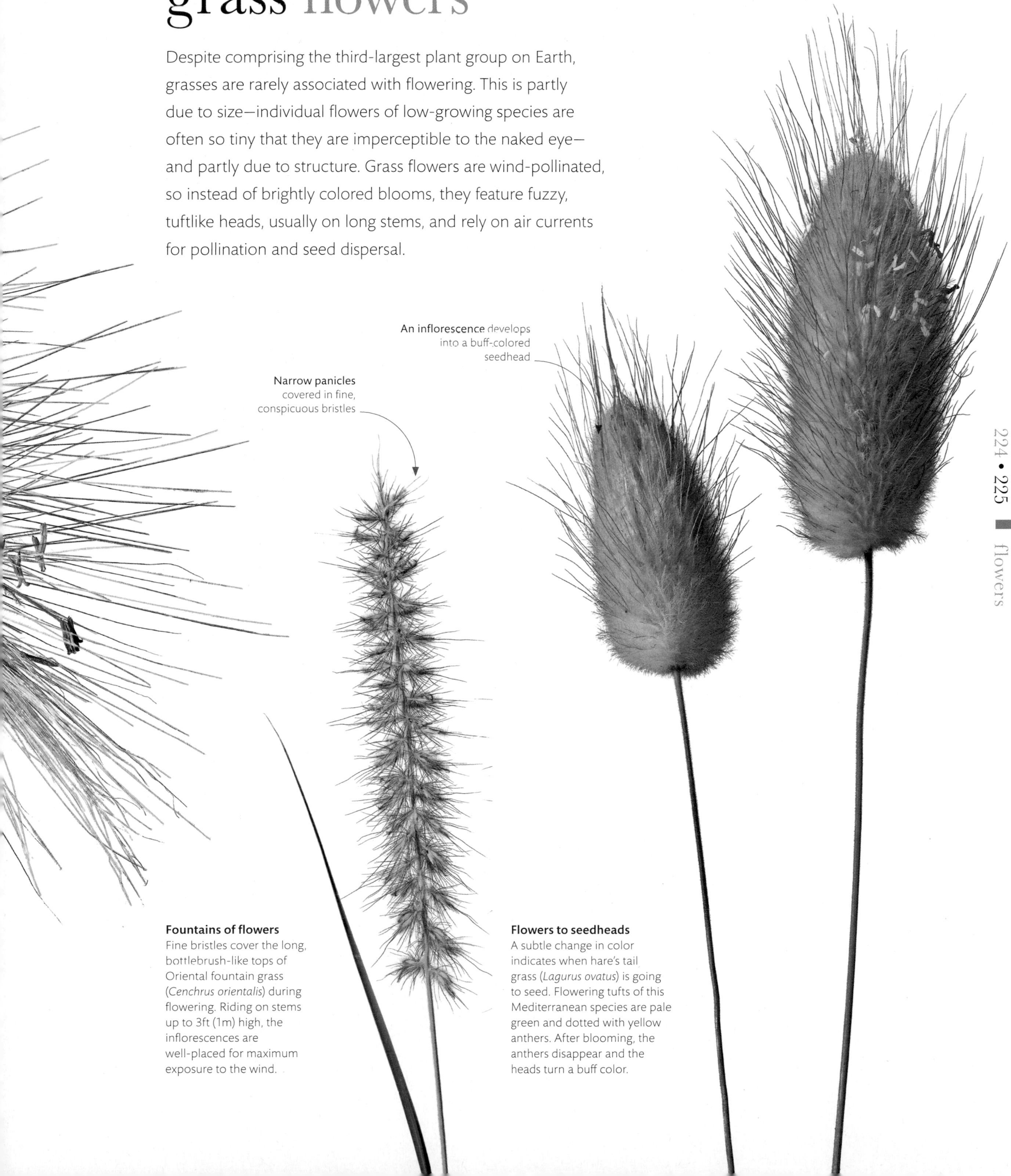

An inflorescence develops into a buff-colored seedhead

Narrow panicles covered in fine, conspicuous bristles

Fountains of flowers
Fine bristles cover the long, bottlebrush-like tops of Oriental fountain grass (*Cenchrus orientalis*) during flowering. Riding on stems up to 3ft (1m) high, the inflorescences are well-placed for maximum exposure to the wind.

Flowers to seedheads
A subtle change in color indicates when hare's tail grass (*Lagurus ovatus*) is going to seed. Flowering tufts of this Mediterranean species are pale green and dotted with yellow anthers. After blooming, the anthers disappear and the heads turn a buff color.

Nectar-seeking geckos crawl over flowers, collecting sticky pollen on their bodies as they go

Vital relationship
In certain habitats, reptiles serve as key pollinators. In Mauritius, the ornate day gecko laps nectar from the blossoms of ox trees (*Polyscias maraisiana*), and pollinates this endangered species in the process.

The bell-shaped bloom hangs down, so only adept climbers can reach the sweet rewards

NESOCODON MAURITIANUS

Pale blue petals ensure the blood-red nectaries stand out to attract the plant's gecko pollinators

Unusual color
Most nectar is colorless, relying on scent to attract pollinators, but the blue Mauritius bellflower (*Nesocodon mauritianus*) is an exception. To increase its chances of reproduction, it releases scarlet fluid from blood-red nectaries. The color red is very attractive to the ornate day geckos that pollinate these rare bellflowers in their rocky habitat.

flowers and nectar

Nectar is a plant's ultimate bribe. The sweet, sticky liquid produced by glands called nectaries attracts a wide range of pollinators. While nectaries also appear on stems, leaves, and buds, they are most closely associated with flowers, where nectar serves as a reward for pollination. Sugars such as sucrose, glucose, and fructose are the main constituents of nectar along with trace amounts of amino acids and other substances. The type, volume, and even color of the nectar a flower makes differ from species to species—tailored to the tastes of the animals that pollinate them.

FLORAL NECTARIES
Within a flower, the three most common positions for nectaries are at the base of the ovary, the base of the stamens (specifically the filament), and at the petal bases. All of these locations require pollinators to push past the reproductive parts of the flower to access the nectar—a design that fosters pollination. However, nectar-exuding glands may also occur on other parts of the ovary, the anthers, stamens, pistils, stigmas, and on petal tissue

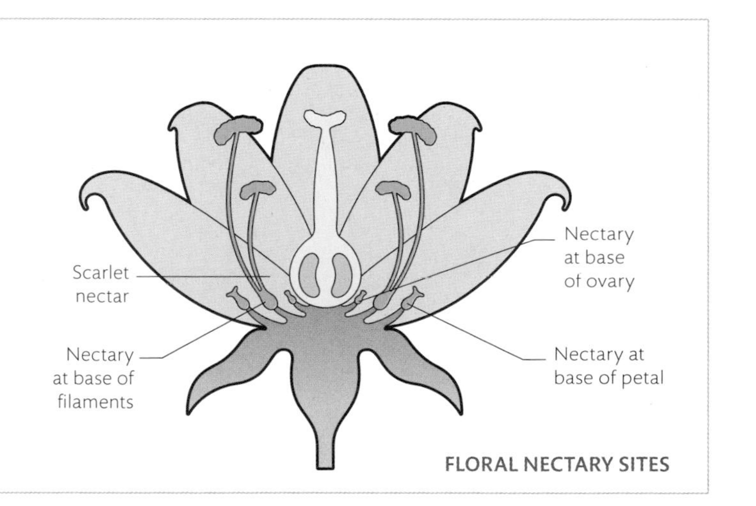

FLORAL NECTARY SITES

BOWL-SHAPED BLOOMS
Wide, open flowers such as poppies are ideal for flying insects, particularly bees, which can land easily on individual, bowl-shaped heads. The ability to alight easily and the exposed floral parts mean that pollinators expend less energy. This benefits the flowers because it means that more poppies are visited by pollinators—and are therefore fertilized—in a shorter amount of time.

ICELANDIC POPPY (*PAPAVER NUDICAULE*)

designed for visitors

Floral color and scent undoubtedly play crucial roles in attracting the attention of pollinators, but the shape of a flower can also determine just who those pollinators might be. Birds—except hummingbirds—need perches, whereas insects such as bees need landing platforms. Providing these features not only encourages the right visitors—it also enables some flowers to guide the right pollinators into their reproductive structures at the right moment.

Landing pads
Landing stages take many forms. The spongy domes of plume thistles (*Cirsium rivulare*) enable butterflies and bees to grip onto them.

Accommodating insect needs
The sap of giant hogweed (*Heracleum mantegazzianum*) is caustic to humans, but its huge, umbrella-shaped flowers provide vast supplies of nectar and pollen for butterflies, flies, and beetles. These insects can probe each flower while comfortably "seated."

buzz pollination

Some 20,000 plant species attract pollinators using high-protein pollen as a lure. To increase the chance of passing on their genes, most of them, including potatoes and tomatoes, have distinctive structures that allow only specific insects access to their pollen. The insects, in turn, have evolved ingenious ways of collecting their reward. In buzz pollination, for example, certain bees transfer vibrations to flower heads to shake out pollen grains.

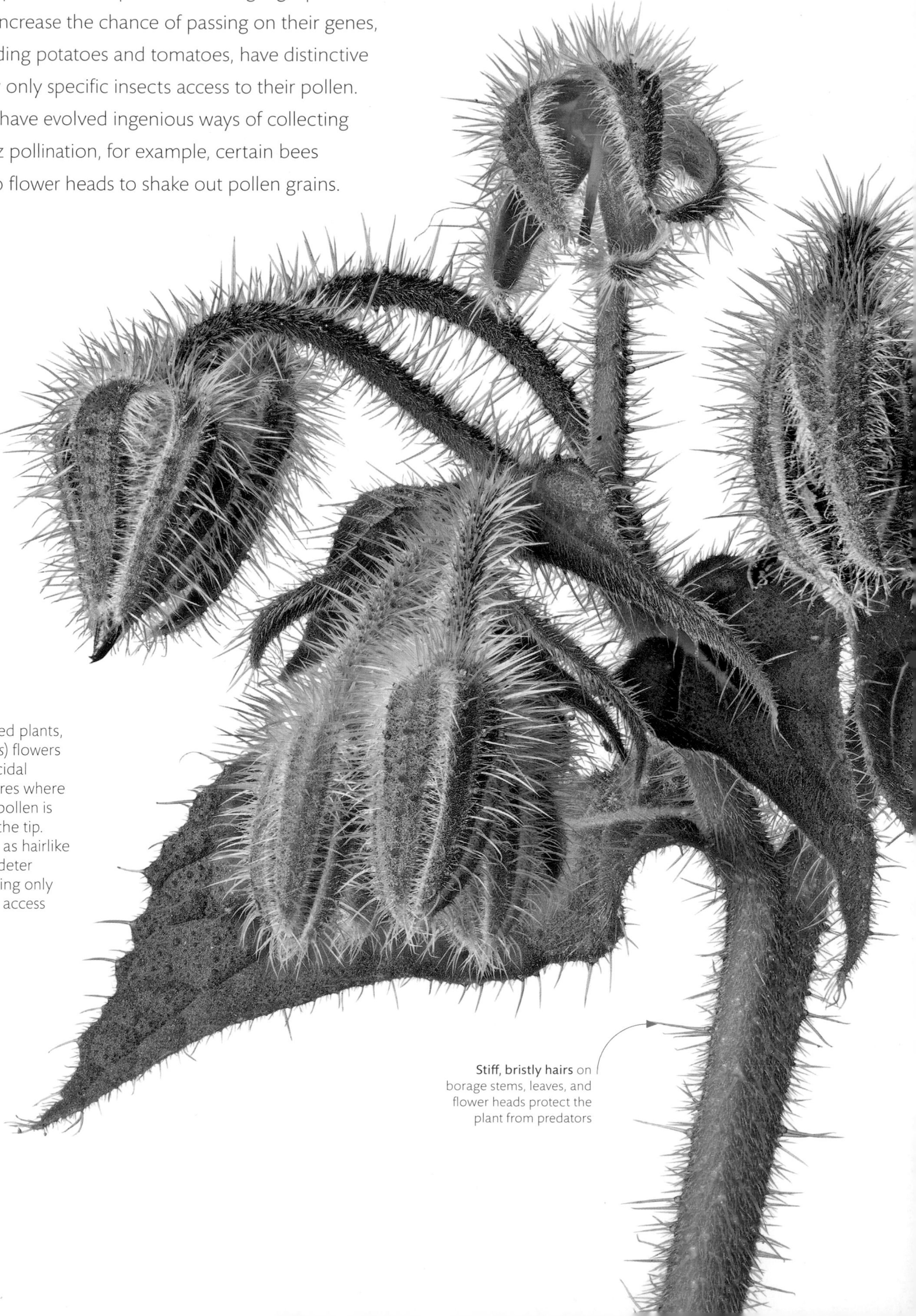

Pollination by design
Like other buzz-pollinated plants, borage (*Borago officinalis*) flowers store pollen within poricidal anthers, tubelike structures where only a small amount of pollen is released at a time from the tip. Other adaptations, such as hairlike trichomes, may help to deter unwanted visitors, allowing only desired pollinators clear access to the blooms.

Stiff, bristly hairs on borage stems, leaves, and flower heads protect the plant from predators

Bumblebees latch on to the anther with their jaws

Bumblebees gather pollen into basketlike corbiculae, but some sticks to their bodies and is transferred to another flower of the same species

Buzz-pollination in action
As a bee vibrates its flight muscles, the motion shakes the flower's anther, making it release a small cloud of pollen. This "sonication" exposes the bee to a force equivalent to 30 times that of gravity.

An anther cone protrudes from the flower's center, attracting specific pollinators while discouraging others

RESTRICTING ACCESS TO POLLEN

The flowers of buzz-pollinated plants share common features, even if they are only distantly related. Their protruding, central anther tubes have short filaments, mainly arranged in closed cones, and pollen is released via a slit or pore in the tube tip.

SOLANUM SP.

DICHORISANDRA SP.

RAMONDA SP.

DODECATHEON SP.

***Jimson Weed*, 1936**
The blossoms of the jimson weed (*Datura wrightii*), a common desert plant that thrives on roadsides and wasteland in the US, are magnified in Georgia O'Keeffe's largest floral canvas. O'Keeffe was immensely fond of the plant, in spite of the toxicity of its seeds, and captured its pinwheel growth habit in this exuberant composition, charged with energy and movement.

> “... they will be surprised into taking time to look at it—I will make even busy New Yorkers take time to see what I see of flowers.”
>
> GEORGIA O’KEEFFE

plants in art

radical visions

As modernist artists sought new techniques to reflect the urban mechanized landscape that surrounded them in the early 20th century, the faithful and realistic representation of the natural world fell by the wayside. Their radical approach rejected centuries of representational art to focus on abstraction, introspection, and a return to primitivism. Over time, these new freedoms inspired artists to create modern works with an intense response to nature.

American artist Georgia O’Keeffe was already an early embracer of the modernist movement when she produced the floral paintings in the 1920s and ’30s that were to become her most iconic works. These large-scale close-ups on giant canvases played with perspective, drawing viewers into the heart of a rose, or their gaze upward to a towering Canna lily.

Modernism was shot through with Freudian psychology, and O’Keeffe’s folds of petals and open blooms were given burdensome interpretations of female eroticism. It was not what she intended. Her rationale was to capture the fine detail of plants and write it large, so that people could witness the everyday miracle of plant beauty at a glance.

Flower power resurfaced in the 1960s as a hippie symbol of peace; blooms with simple shapes, bold patterns, and bright colors became fixtures of contemporary design. Although Andy Warhol’s playful postmodern series of silkscreen prints entitled *Flowers* was a startling departure from his works based on commercial brands and popular culture, it was a perfect fit for the times.

Postmodernist art
In his pop art series *Flowers*, Andy Warhol experimented with swatches of colors for silkscreen prints based on a photograph of four hibiscus flowers. The colors of the flowers change from yellow, red, and blue, to pink and orange, or all white, against a backdrop of a bed of grass.

Ribbonlike petals fully unfurl on warmer days, then curl up tightly during the cooler nights

Scaleless leaf nodes will open in spring, with one leaf per node along the stem

Winter pollinator
Witch hazel nectar sustains many insects during colder months. The satellite moth (*Eupsilia transversa*), active from fall to spring, is an important pollinator of witch hazel.

The stems are generally flower- or fruit-bearing; one year's flowering stem bears next year's fruit

The satellite moth shivers to raise its body temperature so that it can fly in sub-zero conditions

EARLY FLOWERS

Hardy perennials, such as the giant snowdrop (*Galanthus elwesii*), are vital to a healthy ecosystem, because they are one of the earliest sources of food for bees. The giant snowdrop is well adapted to winter conditions and reacts quickly to the arrival of warm, dry weather, going dormant for the summer. Although originating in Turkey, this species is now common in gardens around the northern hemisphere.

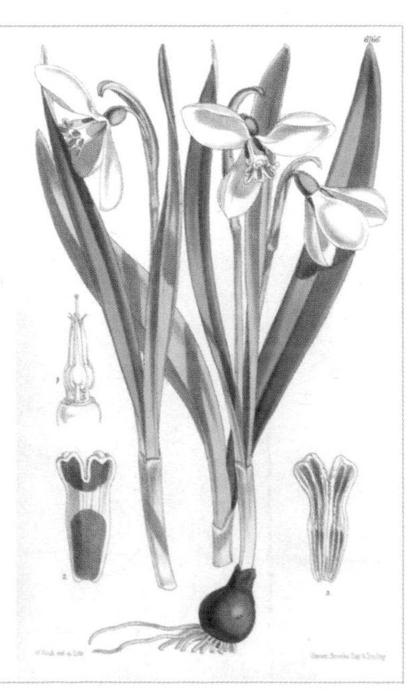

GALANTHUS ELWESII

winter blooms

During the fall and winter, competition among plants for pollinators drops, as most species have long since flowered and are entering a period of dormancy. Some plants, however, only bloom during the coldest months of the year. One of the hardiest is witch hazel (*Hamamelis* sp.), whose spidery, ribbonlike flowers survive temperatures as low as -4°F (-18°C), and daytime highs that are below freezing, yet still manage to bloom for weeks on end.

Protective bracts surround each flower, remaining in place after the petals fall

Next year's fruit
Native to North America, China, and Japan, witch hazels bloom from September through March or April, depending on the species. Any fertilized flowers will produce fruits, but these take a year to develop and usually appear on different stems from those that carry the current year's blooms. Mature fruits eject small black seeds, launching them up to 30 ft (9 m).

The flowers exude a delicate, spicy fragrance that attracts pollinating moths

Each flower, up to 2½ in (6 cm) long, extends from a 6½-ft (2-m) high stem
Anthers and stigma at the tip of a fused corolla
Nectaries on the ovary, at the base of the corolla, produce nectar full of sucrose

flowers for **birds**

Many flowers have evolved to attract bird pollinators. They share many common features, including a lack of fragrance, specific vivid colors, and the amount and type of nectar. Long-billed, long-tongued, hovering species, such as hummingbirds, prefer tubular flowers, while other birds, such as honeyeaters and sunbirds, visit blooms that provide convenient platforms on which to perch.

Hummingbirds insert their bills where the floral tube diverges from the petals

Invitation only
Hummingbirds hone in on red flowers, such as Chile's devil's tobacco plant (*Lobelia tupa*), shown at left. Multiple inflorescences at the ends of stalks, and nectaries hidden within floral tubes, ensure that only certain birds can reach them—and transfer pollen in the process.

Pollen presenters cover the flower head of *Grevillea*, giving it a "tangled" look

Maximizing transfer
Grevillea flower heads make the most of bird pollinators. While probing for nectar, the bills and heads of honeyeaters are brushed by pollen presenters, style tips bearing pollen transferred from a flower's own anthers before opening.

GREVILLEA 'COASTAL SUNSET'

PRIMATE POLLINATORS

The traveler's palm (*Ravenala madagascariensis*) seems to be adapted to large mammal pollinators. Several lemur species transfer pollen among plants as they prise apart the tough, protective leaves to take nectar by "paw-dipping," or by drinking directly from the flowers.

RED RUFFED AND RING-TAILED LEMURS

Each individual flower offers nectar to pollinators, transferring pollen grains to fur as the animals feed

The flower spike, 1–5 in (3–13 cm) long, supports hundreds to thousands of flowers that open in sequence

Sturdy flowers
Silver banksia (*Banksia marginata*) flowers develop on robust spike inflorescences that become repositories for the plant's woody seeds. With a mass of nectar-filled flowers opening in sequence, the spikes tempt birds by day and small mammals at night. Pollen transfers to fur just as it does to feathers, and the spikes support the weight of tiny nocturnal creatures as they search for food.

flowers for animals

Birds and insects pollinate many plant species, but mammals also play a crucial role in pollination. Many of these furry pollinators are small, nocturnal creatures such as mice, rats, and shrewlike sengis, lured by the sweet, energy-rich nectar and able to crawl over flowers without damaging them. Even larger carnivores, such as Cape gray mongooses, have been seen raiding flowers—and pollinating them in the process.

Underbelly fur
picks up pollen as the animal crawls over a flower spike looking for nectar

The tiny possums
weigh less than a hen's egg and are just 3½ in (9 cm) long

Pygmy possums

Prolific nectar and pollen feeders, Australia's pygmy possums help to maintain many habitats. These tiny marsupials pollinate banksias, eucalyptuses, and bottlebrush plants.

Flower bud

The buds emerge from notches on the edges of the leaflike stems. Buds usually start to unfold between 10 pm and midnight, in response to the falling night temperature.

Flower buds take about a month to reach maturity

Epiphyllum oxypetalum

the queen of the night

It seems counterintuitive to look for cacti in trees, let alone doing so in a wet, tropical forest, but that is exactly where one finds the queen of the night. Native to southern Mexico and much of Guatemala, this cactus has stunning blooms with an intoxicating scent that last for a single night.

The queen of the night lives an epiphytic lifestyle, making its home up in the forest canopy. All this species needs for its seeds to germinate is a little humus in a tree hollow or fork between branches. Despite its unusual appearance, the cactus shares many anatomical features with its more cylindrical relatives. What look like long, sprawling leaves are actually the main stems of the cactus. Because it spends its life clinging to the limbs of trees, its stems have evolved a flattened shape to help the cactus cling to these precarious surfaces. Its roots not only provide it with water and nutrients, but also anchor it in place to stop it from tumbling out of the canopy. The queen of the night is spineless. Spines serve to protect cacti from too much sun and the attentions of herbivores, but these are minor problems in the shaded, tropical forests where this species grows.

The name queen of the night refers to the plant's massive blossoms, which open only after dark. Extremely fragrant and brilliant white, the flowers are visited by nocturnal sphinx moths. The moths have only one chance to pollinate each flower: by the time the sun rises, the 10 in (17 cm) wide flowers have already wilted. If pollination has been achieved, the flowers are soon followed by bright pink fruits. Birds and other tree-dwelling animals relish the soft pulp inside the chubby little fruits. Seeds that pass through their digestive tracts are deposited on canopy branches, and a new generation begins.

Fragrant bloom

The flower's funnel-shaped center holds a mass of pollen-laden stamens and a long white stigma. The chemical that gives the flowers their wonderful aroma, benzyl salicylate, is commonly used as a fragrant additive to perfumes.

color attraction

The scent, size, and shape of flowers all play vital roles in appealing to pollinators, but color is undoubtedly one of the most important ways a plant has of attracting attention. The color preferences of insects and birds may sometimes seem strange to us—until we realize that their eyes, being structurally different, perceive a different color spectrum than ours. Bees and many other insects are able to perceive ultraviolet light, which helps them more easily identify the paths to nectar.

White flowers attract nocturnal moths and beetles, as well as butterflies and flies

Red and orange flowers are favored by birds

Pink flowers are preferred by butterflies and some moths

Yellow appeals to butterflies, bees, hoverflies, and wasps

Customized colors
Plants have evolved a rainbow of shades to match the visual preferences of their pollinators. While many colors are seen by both insects and birds, not all perceive them the same way—bees gravitate to purple, while some bird species are drawn to more vivid oranges and reds.

GREEN FLOWERS

Scientists believe that flowers evolved earlier than many insect pollinators, and that their blooms were colored green, like their surounding foliage. As relationships developed between plant and pollinator, and competition for pollinators intensified, plants began to adapt floral colors to attract certain species.

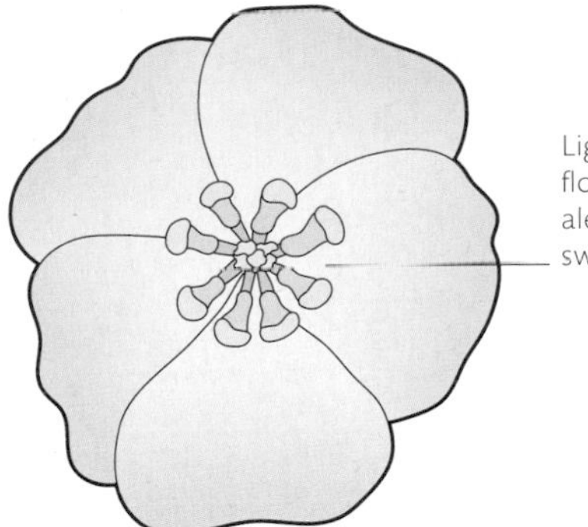

GREEN HELLEBORE (*HELLEBORUS VIRIDIS*)

nectar guides

The human eye views reflected light as various colors, but many pollinators see the world quite differently. Bees, in particular, perceive a specialized range of wavelengths, including the ultraviolet spectrum. This enables them to see features of a flower, such as lines, dots, and other patterns—invisible to the human eye—that guide them directly to where the nectar is. These "nectar guides" are all-important, not just for the bees but also for the plants, as the bees help the flower to disperse its pollen.

The smallest dots are farthest away from the nectary

Purple dots, along with violet and blue ones, are the three colors most likely to attract bees

The light background contrasts with the dark nectar guides

The large dots tell bumblebees that they are getting closer to the nectary

BEE'S-EYE VIEW

Bees lack the photoreceptors necessary to see colors like red, but their ability to discern ultraviolet light transforms seemingly plain surfaces into complex patterns. What we see as a plain yellow marsh marigold (*Caltha palustris*), a bee perceives as a light-colored bloom with a contrasting dark center—effectively a "bull's-eye" proclaiming "land here."

IN DAYLIGHT **IN ULTRAVIOLET LIGHT**

Follow the dots
The extraordinary mottling of the toad lily (*Tricyrtis hirta*) advertises its presence to nectar-seeking insects. Although the dots are also visible to humans in this instance, the increasing size of the dots is especially striking to bumblebees, the toad lily's main pollinators, because bumblebees prefer dots to linear nectar guides.

Immature
green bud

Nectary is deep inside
the flower, forcing bees
to pass its reproductive
parts to gain access

Glandular hairs
(trichomes) on the
stem deter unwanted
visitors from accessing
the nectaries

Alcea sp.

the hollyhock

With their tall racemes of big, showy flowers, hollyhocks are highly prized as garden plants. But like the blooms of many flowering plants, hollyhock blossoms bear markings that are hidden from human eyes and are only visible to pollinators able to see the ultraviolet (UV) part of the spectrum.

The 60 or so hollyhock species (*Alcea* sp.) belong to the mallow family (Malvaceae), making them distant cousins of plants such as hibiscus. In summer, hollyhocks produce large, funnel-shaped flowers. The visual appeal of these soaring blooms to the gardener is very different from what attracts pollinators able to detect UV light. Where human eyes may see a plain blossom, the UV-sensitive eyes of bees, the hollyhock's main pollinators, see bull's-eye markings around the center of the flower. The bull's-eye pattern is produced by special pigments that either reflect or absorb UV light. As well as bees, many insects, including butterflies, and even some birds and bats, can perceive UV light.

These markings, called nectar guides, are not unique to hollyhocks—many flowers have patterns that are only visible in UV. The markings vary widely, but they all serve the same function: like runway lights, they direct pollinators to a flower's stores of nectar and pollen. Both plant and pollinator benefit from this, as it means that insects spend less time looking for pollen and nectar, and flowers are pollinated more quickly. In fact, insects avoid mutant flowers lacking nectar guides. Some insects that hunt on flowers, such as crab spiders and orchid mantises, resemble nectar guides when seen in UV light. This clever disguise may help to lure insect prey toward the nectar—and their doom.

Fluorescing hollyhock
When a hollyhock is made to fluoresce under UV light, its bull's-eye pattern is revealed. As well as guiding pollinators to nectar, such markings may also help pollinators distinguish between flowers that look much the same to human eyes.

Hollyhock buds flower sequentially rather than all at once, helping to avoid self-pollination

The funnel-shaped flowers, around 4 in (10 cm) across, may be white, pink, red, purple, or yellow

Common hollyhock
Alcea rosea, the common hollyhock, can reach 8 ft (2.5 m) in height, and bears saucer-sized flowers along its upright stem. Native to China, it is now widely cultivated for its attractive blooms.

Emerging flowers are pale pink, signaling high acid and high nectar levels

Older flowers turn blue-purple, meaning less nectar and lower acidity

Floral acid test
The flowers of lungwort (*Pulmonaria officinalis*) bloom pink and turn blue-purple as they age. This color change is driven by acidity levels within the lungwort flower, which affect the colored pigments (anthocyanins). The flower's pH changes as it matures, so young, pink, nectar-rich flowers are more acidic than blue-purple ones.

Dark pink buds have the highest acidity levels

The reddish color fades on the petals of mature or pollinated blooms

A pinkish-red color indicates unopened buds or immature flowers with little or no available reward

color signals

While certain shades of flowers are known to be more or less attractive to different pollinator species, many plants take the use of color a step further. By varying the hues of individual flowers at a particular stage and age of development, plants such as honeysuckle (*Lonicera periclymenum*) not only attract the right pollinators, but can direct them toward maturing blooms that contain the most nectar or pollen rewards. In return, the plant receives more visits from passing insects, which results in a much higher proportion of fertilized flowers.

SUBTLE SIGNS
Some flowers deploy subtle color changes to indicate their status. On spring snowflakes (*Leucojum vernum*), tiny green spots change from green to yellow as the flowers mature. Thought to indicate whether or not the blooms have been pollinated, the spots on these early spring flowers attract bees to a much-needed food source.

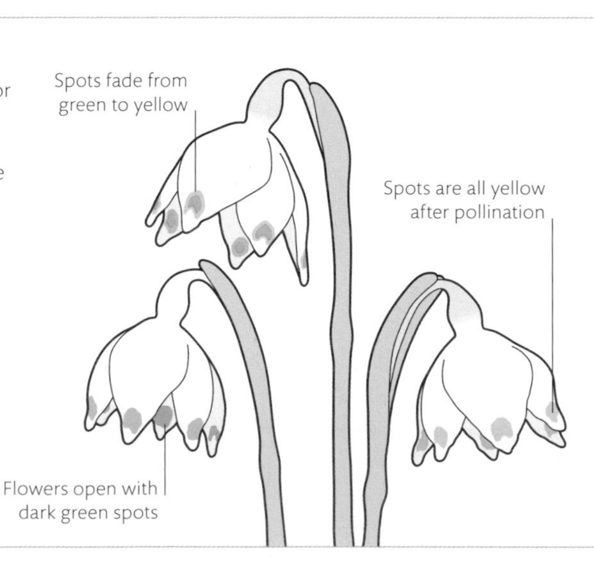

LEUCOJUM VERNUM

Communicating with color

A combination of scent and color indicates which honeysuckle flowers are worth visiting. Immature buds blush pinkish red, while white blooms offer the most pollen and turn yellow after pollination has occurred, when they still provide nectar for bees.

A dark pink to red color warns potential pollinators to avoid inner, unopened buds

White flowers are strongly scented to attract night-flying moth pollinators

Yellow, nectar-filled flowers attract long-tongued bees—which may also pollinate any neighboring white flowers

POLLINATION

A *Sarracenia* flower is structured to avoid self-pollination. Visiting insects push past the stigmas and transfer pollen as they enter the style chamber. As they drink nectar, they are coated in pollen from both the style and the anthers, then leave through gaps between the stigmas.

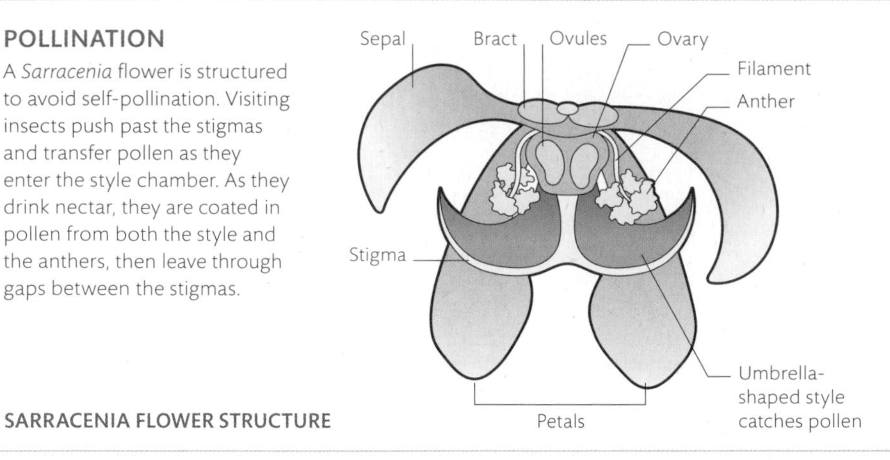

SARRACENIA FLOWER STRUCTURE

The curved sepals help shield the floral chamber, safeguarding the nectar and pollen

The leathery sepals remain on the flower long after its petals have fallen (as in the flower shown here); they are sometimes still present in winter

The floor of the style is covered with tiny hairs, which pass on both pollen and nectar to pollinators

The sepals gradually change color in many *Sarracenia* species as the flowers age

The flower hangs upside down, due to the weight of the umbrella-shaped style

The style curls around the developing ovary

Unique style
Single *Sarracenia* flowers develop on separate stems from the plant's carnivorous pitchers, which are often closer to the ground. To reduce the risk of ingesting a valuable pollinator, the flowers bloom in spring, well before the pitchers become active in summer. The style may be a strange shape to prevent self-pollination, but it does allow the plant to hybridize (cross pollinate) easily.

restricted entry

Carnivorous pitcher plants attract insects to ingest, but they also need pollinators to come and go in order to reproduce. To do this, the pitcher plant's flowers are physically separated from its deadly traps, not only by space but also by time, because they bloom before the traps become active. Their unique structure also controls the way in which pollinators enter and exit the flowers.

NEW WORLD PITCHER PLANTS

Pitcher plants belong to the Sarraceniaceae family. This group of three genera—*Darlingtonia*, *Heliamphora*, and *Sarracenia*—contains 34 species, and many of them are highly endangered. All grow in boggy areas with poor soil—which is why they need nutrients from the insects they entrap.

SARRACENIA DRUMMONDII

fragrant traps

Many plants use floral fragrance to attract pollinators. Some, however, go a step further and emit irresistible scents that lure insects to flowers in order to trap them inside for "forced pollination." Hundreds of orchid species—including nearly 300 greenhood orchids (*Pterostylis* spp.)—employ this method to ensure cross pollination, and access a wider gene pool.

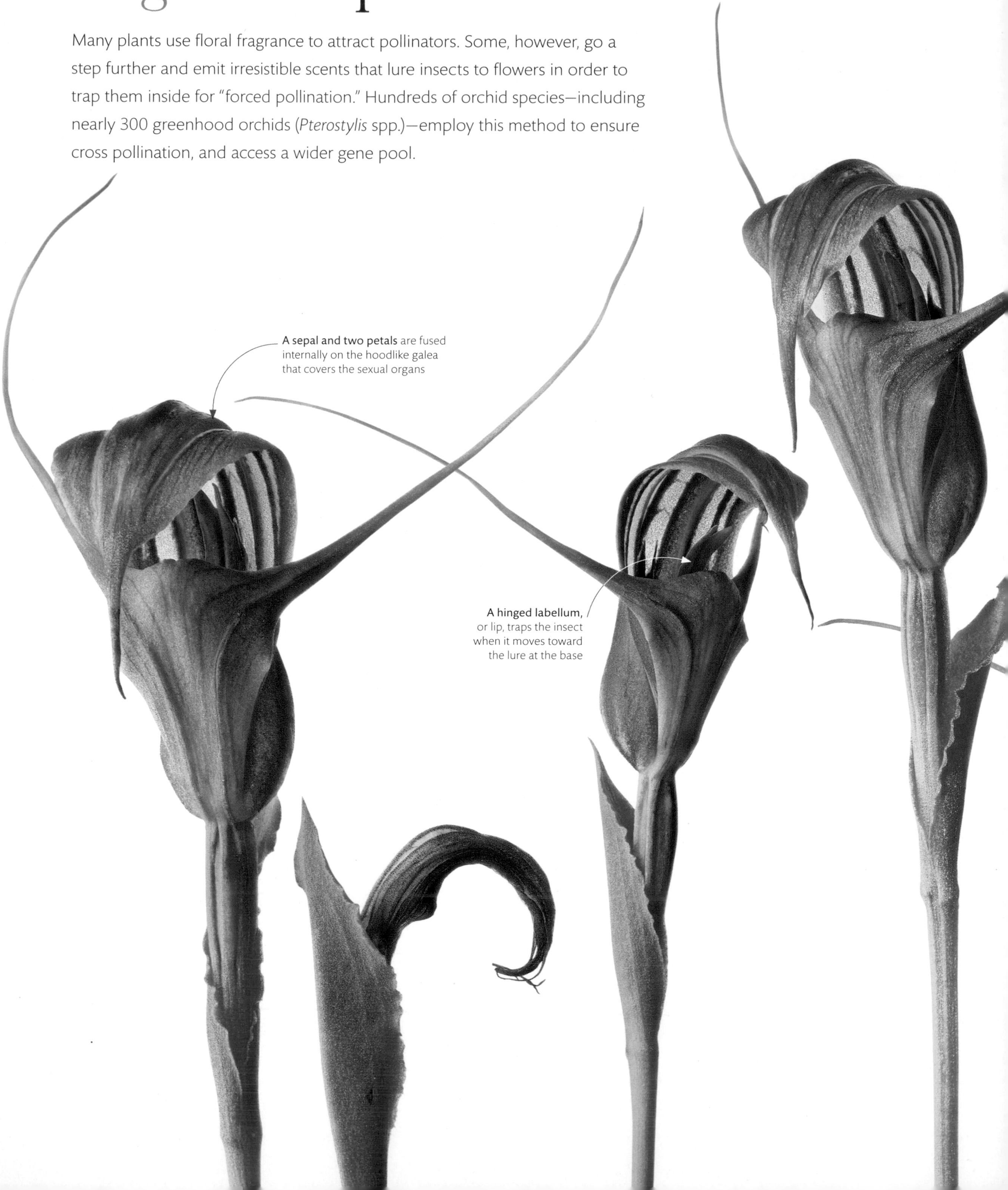

A sepal and two petals are fused internally on the hoodlike galea that covers the sexual organs

A hinged labellum, or lip, traps the insect when it moves toward the lure at the base

Two fused sepals form the front of the orchid, ending in elongated "points" on either side of the galea

Translucent stripes on the galea let in filtered light, guiding insects to the back of the flower

TRAPPING MECHANISM

When a gnat begins crawling along the labellum of a greenhood orchid, this lip flexes and tips the insect inside. This traps it within the gynostemium—a reproductive structure found in several plant families that consists of fused stamens and pistils. Once in this column, the insect can only exit by squeezing past the anthers, which press a mass of pollen, known as a pollinium, onto its back. The insect carries this to the next orchid, pollinating that flower in the process.

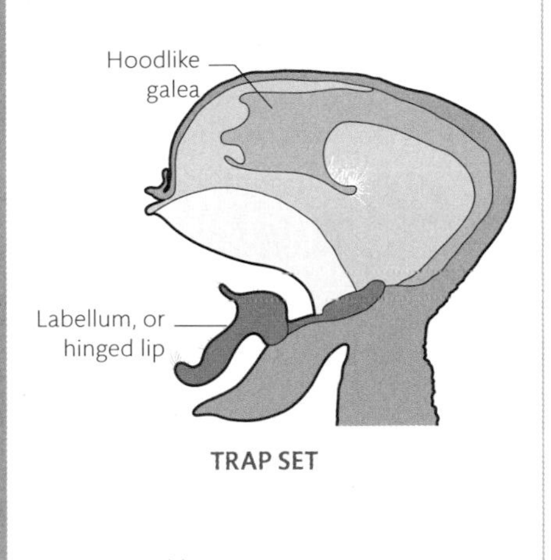

TRAP SET

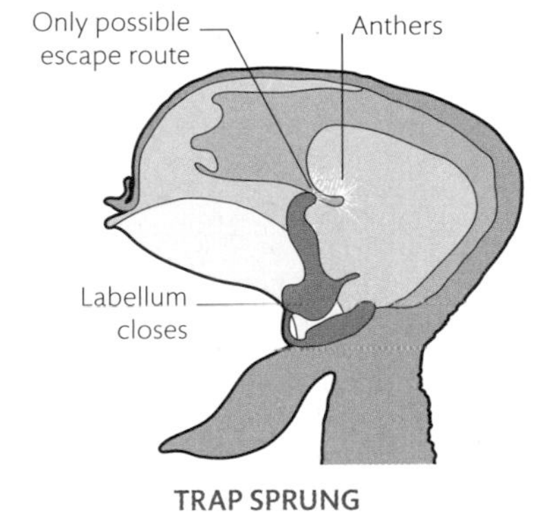

TRAP SPRUNG

Chemical attraction

Greenhood orchids (*Pterostylis* spp.) are native from Malaysia to Australia and New Zealand, but *P. tenuicauda* is endemic only in New Caledonia. The majority of insect visitors to these flowers are male fungus gnats (Mycetophilidae). It is thought that, to attract them, the scent given off by the blooms mimics the pheromones of tiny female flies.

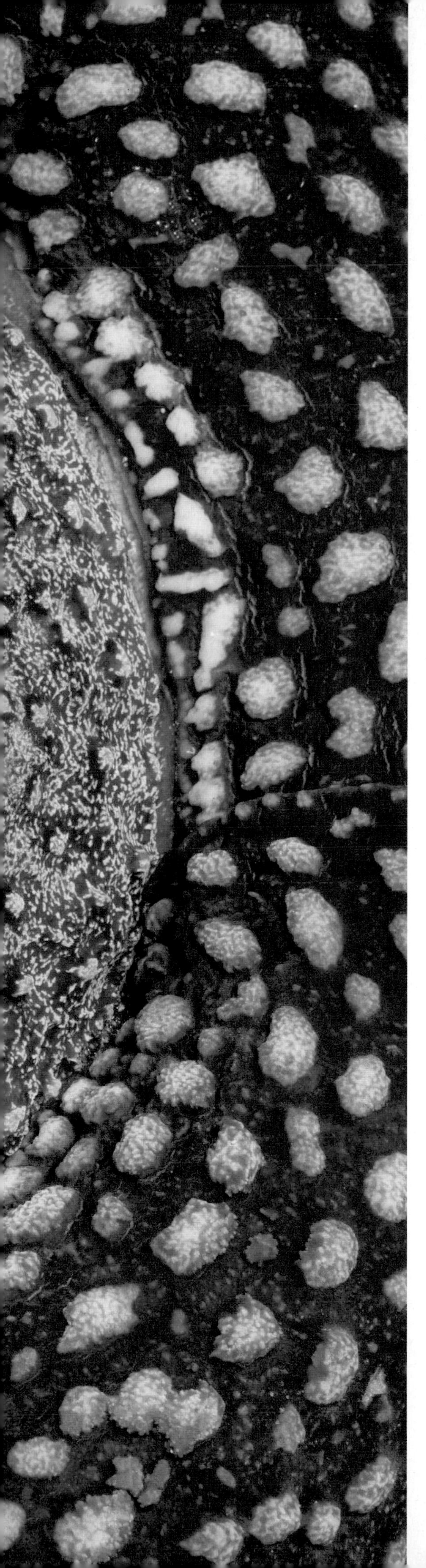

The red coloration and textured surface are thought to mimic the look and feel of decaying flesh

Short-lived grandeur
Corpse flower blooms may have evolved to such a huge size to make them easier for pollinators to find. For all its grandeur, the blooming of the corpse flower is an ephemeral event, lasting barely a week.

Rafflesia arnoldii

the corpse flower

At 3 ft (1 m) across and weighing 24 lb (11 kg), the corpse flower is the world's largest single flower. Despite its size, this rare flower is usually smelled before it is seen. Native to the diverse rainforests of Sumatra and Borneo, the plant mimics rotting meat in both fragrance and appearance.

The vast bloom is all that is visible of the *Rafflesia arnoldii*. This parasitic species has no stem, no leaves, and no roots. Invading the food-conducting vascular tissues of forest vines, it mainly consists of threadlike tissues living in and around its host's cells, from which it obtains the nutrients and water it needs. The corpse flower cannot live without its host, so it rarely has any serious effect on the vine's health.

At flowering time, tiny buds form on the vine stem and gradually swell to look like large purple or brown cabbages. The flower buds take up to a year to develop, during which time they are sensitive to disturbance. Flowers are either male or female, so they need to appear relatively near each other in order to reproduce. The principal pollinators of *Rafflesia arnoldii* are carrion flies. Lured into male flowers by the false promise of rotting meat, the flies become covered in gooey blobs of pollen. When they visit female flowers, they are corralled into a narrow crevice that forces them to brush against the stigma and transfer the sticky pollen. The rarity of these plants means that the odds of a male and female plant being in flower at the same time and within flying distance of one another are slim; sexual reproduction does not happen often.

This species relies on intact forest for survival. Deforestation may be pushing the species to the brink of extinction, but its rarity and cryptic lifestyle make it hard to properly assess numbers in the wild.

Mystery plant
Within the central cavity of the corpse flower is a disk covered with projections whose function is uncertain. The anthers and stigma lie below this disk. The plant's goopy pollen dries on the back of the flies and may remain viable for weeks.

special relationships

Mutualism—an association in which two different organisms each benefit from the actions of the other—is an intrinsic part of the botanical world, from fungi that nourish forest root systems to the pollination of flowers by animals. Over time, highly specialized relationships have evolved, resulting in structural changes on the part of flowering plants, and changes in the behaviour of the animals that depend on them.

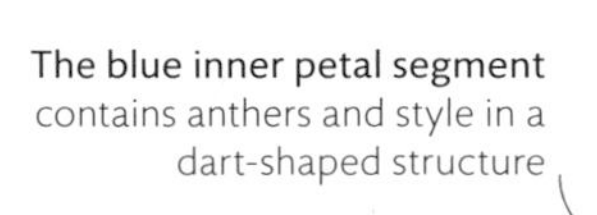

Bird of paradise
The South African *Strelitzia reginae* has evolved an inflorescence that resembles an exotic bird's head. Commonly known as the bird of paradise plant, its vivid pointed flower parts are specifically adapted to pollination by birds.

Whitish anthers protrude from the top of each blue petal segment

Threadlike pollen also gathers within the channel of the fused blue inner petal segment

A hard, beaklike spathe protects 4–6 flowers as they emerge one at a time

Pollen perch
The robust spathe and fused petal "dart" of *Strelitzia reginae* make an effective perch for feathered pollinators. The most common is the Cape weaver bird. As it presses down the dart to access the nectary, long strands of pollen collect on its feet to transfer to the next *Strelitzia* it visits.

Cape weaver birds keep their feet remarkably still as they press down petals, preventing flowers from self-pollinating

FIGS AND FIG WASPS

More than 700 fig species are pollinated only by wasps that enter the figs to lay their eggs. Figs are specialized inflorescences known as syconia—sacklike structures filled with simple flowers. The fig wasps themselves can reproduce only by laying eggs inside the syconia.

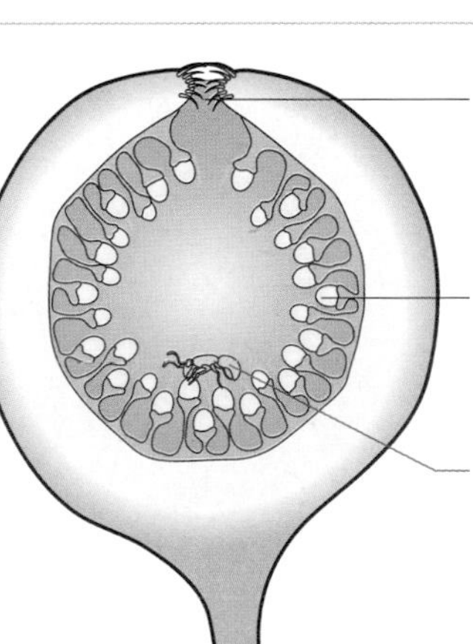

Opening near apex of fig (ostiole); wasp loses wings as she enters here

Flowers line the inside of the syconium

Wasp lays eggs down floret styles, transferring pollen to stigmas from her forelegs

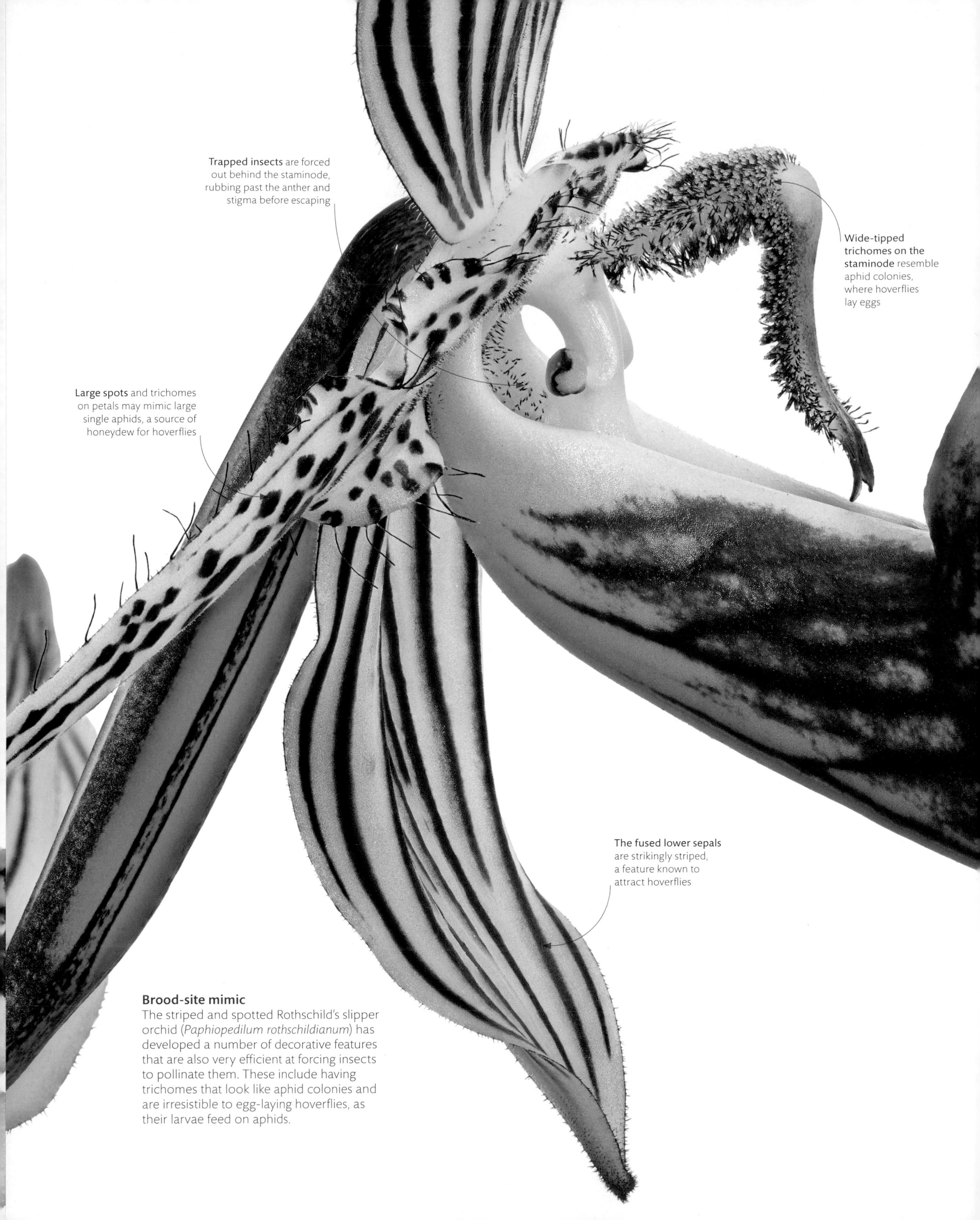

Trapped insects are forced out behind the staminode, rubbing past the anther and stigma before escaping

Wide-tipped trichomes on the staminode resemble aphid colonies, where hoverflies lay eggs

Large spots and trichomes on petals may mimic large single aphids, a source of honeydew for hoverflies

The fused lower sepals are strikingly striped, a feature known to attract hoverflies

Brood-site mimic
The striped and spotted Rothschild's slipper orchid (*Paphiopedilum rothschildianum*) has developed a number of decorative features that are also very efficient at forcing insects to pollinate them. These include having trichomes that look like aphid colonies and are irresistible to egg-laying hoverflies, as their larvae feed on aphids.

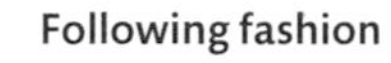

Following fashion
These two *Paphiopedilum* hybrids sport features similar to that of the Rothschild's slipper orchid, attracting pollinators with enticing stripes, and spots that can be mistaken for aphids.

PAPHIOPEDILUM HYBRIDS

After landing, hoverflies often fall into the labellum, or lip, and become trapped

designed to deceive

Some plants reward pollinators, but others operate through deception, fooling pollinators with the false promise of a sweet reward. In the case of slipper orchids, lures vary from spots or hairs that mimic aphids appealing to predators, to holes that resemble tunnels, which attract bees looking for nest sites. Many visiting insects become trapped with only one way out, which forces them past sex organs, pollinating flowers for no reward.

The pouch-shaped labellum—actually a third petal—resembles a slipper

Elongated, spotted petals, set almost horizontally, increase the plant's "advertising space" to attract pollinators

CLOSED FLOWER STRUCTURE

The corollas of *Lamprocapnos* flowers are much smaller than those of many of its relatives in the poppy family—they are also inconspicuous, and may be colorless. They completely enclose the anthers and stigmas, which are so tightly packed together that they may touch. This makes it easy for pollen to be transferred within the flower, allowing self-pollinated seeds to develop.

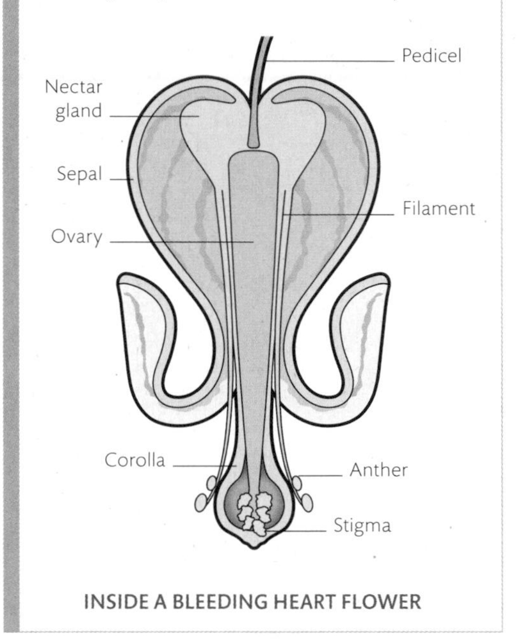

INSIDE A BLEEDING HEART FLOWER

The terminal bud of the raceme's main stem (peduncle) keeps growing

Heart-shaped blooms
The flowers of bleeding heart (*Lamprocapnos spectabilis*) may be pollinated with or without the help of insects. The flowers produce nectar, attracting bumblebees, but the close proximity of the male and female parts inside the tubular corolla makes self-pollination possible when pollinators are in short supply.

The tip of the corolla surrounds the anthers and stigma

self-pollinating flowers

It takes a lot of energy for a plant to produce the color and nectar necessary to attract pollinators, and sometimes it is advantageous for a flower to fertilize itself. Self-pollination allows plants that grow in challenging conditions to preserve the traits that help them to succeed in those environments. It may also be the reason that small or sparse populations have survived. For many species, self-fertilization is a useful "backup plan," deployed at times when pollinators are scarce.

A single raceme of a bleeding heart plant may produce 3–15 pendulous flowers

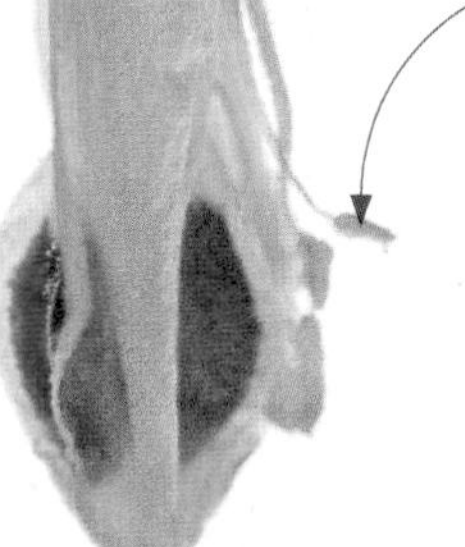

The pink sepals flex back as the flower matures

Anthers are very close to the stigma

The petals of open dog violet flowers are large and attract pollinators

Fallback strategy
Some plants have two types of flowers. If the spring blooms of the dog violet (*Viola riviniana*) are not pollinated, all is not lost. In the fall, it produces more flowers at soil level. These closed (cleistogamous) blooms self-fertilize and produce seeds without the wind or insect activity.

Rosa Centifolia.
Rosier à cent feuilles.
P. J. Redouté
Langlois.

Narcissus x odorus, c. 1800
This exquisite watercolor of a scented daffodil is by Austrian artist Franz Bauer, who was the first resident botanical illustrator at the Royal Botanic Gardens, Kew. He was also the official botanic painter to King George III.

plants in art

blooms for royalty

In the golden age of botanical illustration spanning the late 18th and early 19th centuries, leading artists were favored in the royal courts of Europe and achieved international status for their work. Their watercolors could be reproduced with exquisite accuracy thanks to major advances in printing and copperplate engraving techniques.

Often referred to as the "Raphael of flowers," Belgian artist Pierre-Joseph Redouté published more than 2,000 plates depicting 1,800 plant species during his lifetime. He studied plant anatomy with French aristocrat Charles Louis L'Héritier de Brutelle, and floral art under Gerard van Spaendonck, who was miniature painter to Louis XVI of France. Redouté was first appointed as court artist and tutor to Queen Marie Antoinette. After the French Revolution he was engaged in Empress Josephine's project to create one of the finest gardens in Europe at the Château de Malmaison. Her rose garden included 200 different types of roses, many of which appear in Redouté's three volumes of *Les Roses*—a work that is still used today to identify older varieties.

In 1790, the Austrian artist Franz Bauer was appointed as the first botanical artist in residence at the Royal Gardens at Kew. Bauer was both a scientist and a skilled artist, whose work includes microscopic studies of plant anatomy.

***Jacinthe Double*, 1800**
The work of Flemish artist Gerard van Spaendonck combines the traditional Dutch mastery of flower painting with French sophistication. *Jacinthe Double* is one of 24 plates using stipple engraving that was developed further by Redouté.

***Rosa centifolia*, c. 1824**
A hybrid rose with a clear, sweet fragrance, "the rose with a hundred petals" is featured in a stipple engraving from Pierre-Joseph Redouté's *Les Roses* (left). Redouté perfected the technique of stipple engraving copper plate with fine dots to reproduce the gradations of color seen in his flawless watercolors and allow the "light" of the paper to shine through. The prints were then finished by hand in watercolor.

> "… among the flowers which have received, in the highest degree, the gift of mutability, none can be compared to the Rose …"
>
> CLAUDE ANTOINE THORY, FOREWORD, *LES ROSES*, 1817

TEMPERATURE CHANGES

Many flowers open and close because the fluids in their cells respond to temperature changes, which make the cells expand and contract, and expanding cells create surface pressure that forces petals open. The inner and outer sides of tulip petals differ in temperature by up to 50°F (10°C). As sunlight warms the flower, the surface temperature of the inner petal rises and its cells expand, pushing the flower open. As the temperature drops, cells on the inner surface contract first, pulling the flower closed again.

Opening hours
A lotus (*Nelumbo* sp.) flower lasts for 3–4 days, opening at dawn and closing at dusk. The first day it opens only partially to enable its stigmas to receive pollen, then it closes completely. The next two mornings it opens more fully, releasing scent to attract bees, flies, and beetles.

closing at night

Some flowers open or close in response to external stimuli, such as touch, or changes in light, temperature, or humidity. These factors trigger physical reactions in many species, but flowers that close at night may also have a reproductive agenda. By closing, they are protecting their pollen and sex organs from the elements, and reducing the risk of being eaten or damaged by nocturnal predators—and therefore increasing the chances of attracting pollinators by day.

The petals close rapidly as the light fades and temperatures drop

The petals close so tightly on the first day that the flower resembles a bud

Each flower has only two sepals

Closing at night
When lotus flowers (*Nelumbo* sp.) close at night, chemical changes inside the receptacles generate heat, creating temperatures up to 104°F (40°C) higher in the flower than the outside air. The heat releases scent, which the lotus, lacking nectar, needs in order to attract pollinators when it opens the next day.

Sepals protect the flower bud as it emerges

Glandular hairs cover the sepals and provide extra protection for the flower bud

The ovary, the swollen area at the base of the pistil, develops into a hip after pollination

bud defenses

Buds are usually protected by sepals, but some plants use tiny hairs (trichomes) to strengthen their defenses. These are thought to trap a layer of air around the bud, which insulates it from the elements and regulates temperature and moisture content. To further deter pests, some hairs release chemicals when they are touched.

Anther, which produces pollen

Sepals, covered with trichomes inside and out, fold back when the flower opens, and conceal the developing hip

Glandular hairs also protrude from the edges of the leaf

Protected inside and out
The eglantine rose (*Rosa rubiginosa*) is often found scrambling through hedges or over scrub at woodland edges. Its pests are numerous, so it needs an arsenal of protection for its buds. This defense is provided by trichomes, which protect the leaves and surround the seeds inside the hips (fruit). They are so effective that in the past the hairs inside the hips were extracted and used as "itching powder."

PLANT HAIRS

Made up of one or many cells, plant hairs, or trichomes, grow from the epidermis. Those that secrete protective substances (glandular trichomes), are usually multicellular. The secretions are stored in a glandlike cell at the trichome's tip.

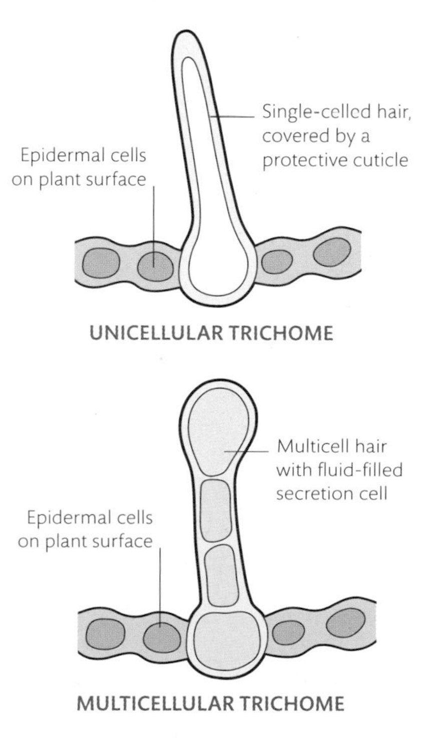

Flower buds turn pink as they develop

Spiky protection
The spines on a teasel flower are actually stiff, sharp bracts that protect its developing buds.

Phased flowering
Protected by its spiny bracts, a single teasel (*Dipsacus fullonum*) inflorescence can produce about 2,000 blooms that begin to open in a ring around the middle. Bands of flowers at the top and bottom mature later, weeks after the blooms in the central ring have died.

White-tipped, upcurved bracts are longer at the top of the flower head than at its base

The central flowers have died, leaving just the prickly bracts

armored **flowers**

When it comes to survival, plants are at a disadvantage—they cannot move away or hide when threatened by predators. Many plants protect their leaves and stems with spines, prickles, or thorns, but the common teasel (*Dipsacus fullonum*) has also developed sharp defenses on its flower heads. This "body armor" allows pollinators to visit the open flowers, while protecting the buds and developing seeds.

Long, prickly bracts curve up and around the flower head to form a protective cage

Anthers and filaments protrude from pale pink to purple tubular corollas

MULTIPURPOSE SPINES

The spiny bracts of some flowering plants, such as burdock (*Arctium* sp.), serve a dual purpose. Not only do they protect the flower head by repelling potential predators, but the hooked tips of the resulting burrs—the inspiration for the hook-and-loop fastener—also latch onto the fur of passing animals. This helps disperse the burdock seeds over a wide area.

Protective bracts end in hooked tips

BURDOCK

colorful bracts

Apart from the spectacular fall displays shown by the leaves of certain trees, color in the plant world is most strongly associated with flowers. Modified protective leaves called bracts can be just as vibrant as flowers, though, and are often mistaken for flower parts, particularly in hot-climate species. Bracts can also function as brightly hued petals, such as in crimson Central American poinsettias, and their colors are irresistible to pollinators.

Mistaken identity
The bracts of many tropical species often overshadow the nondescript flowers they protect. The South American hanging lobster claw plant (*Heliconia rostrata*) features strikingly bright crimson and yellow bracts, which attract hummingbirds to pollinate the tiny flowers they enclose.

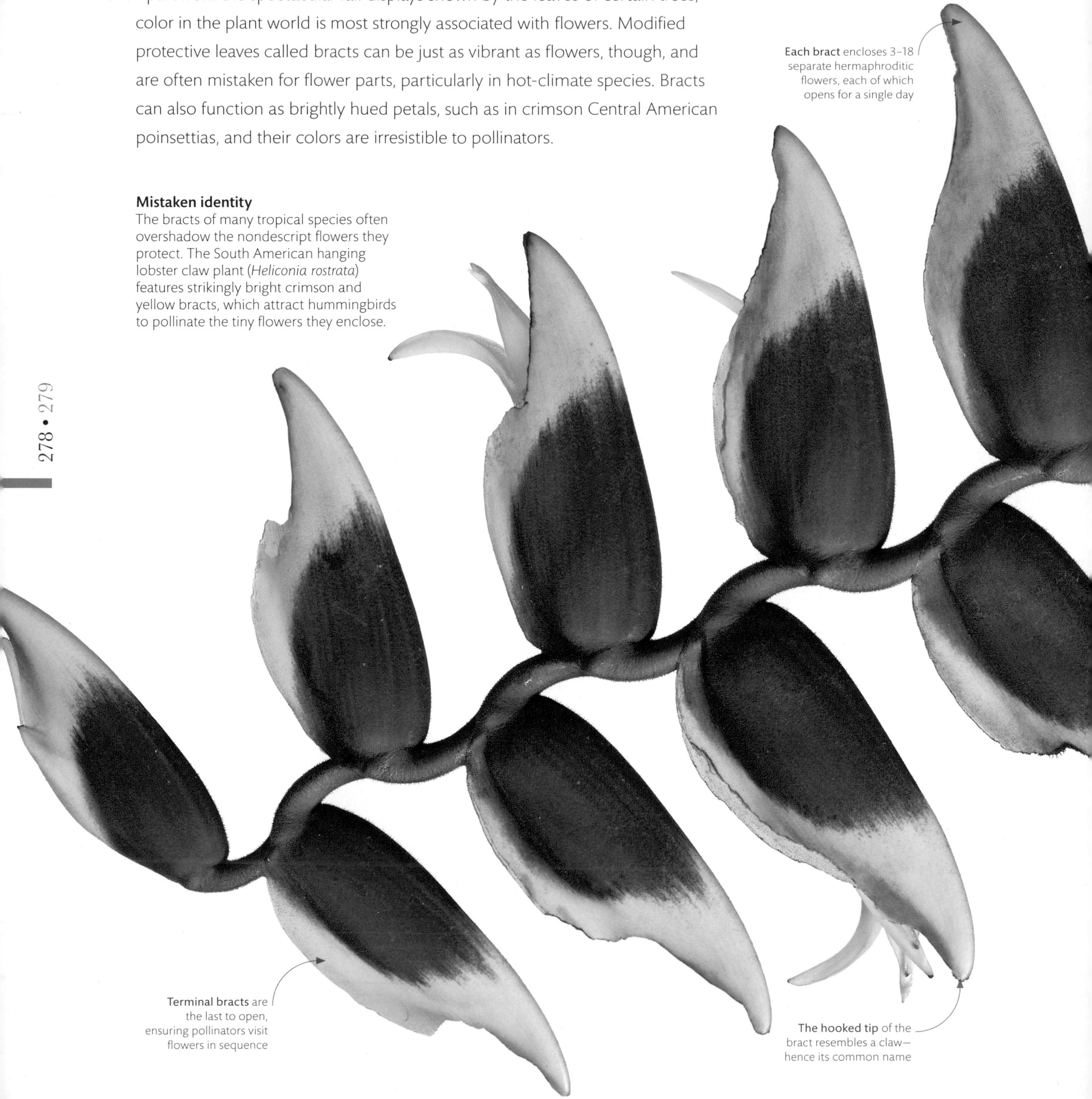

Crimson-colored stems add to vivid inflorescence display

The dominant red color on the top of the bracts attracts pollinators from above

Pouchlike bracts conceal delicate purple flowers with yellow spots; both the bract and flower colors attract insects

A single sepal protrudes from each flower, allowing access to nectar

Red, upturned bracts conceal tubular flowers that appeal to hummingbirds

Lighting up the shade
Tropical species use shape as well as vibrant bracts to stand out in shade. Species such as the Peruvian *Ruellia chartacea* and the Malaysian beehive ginger (*Zingiber spectabile*) flower in the understory, but use very different bract arrangements to attract pollinators.

RUELLIA CHARTACEA

ZINGIBER SPECTABILE

M.V.W.

Dutchman's Pipe
Mary Vaux Walcott contributed watercolor illustrations to the book *North American Pitcher Plants*, published by the Smithsonian Institution in 1935. This artwork features a species of *Aristolochia* commonly known as the Dutchman's Pipe. The name is taken from the shape of the flowers that resemble smoking pipes once common in the Netherlands and northern Germany.

plants in art

american enthusiasts

The expansion of the North American railroads during the 19th century gave adventurers, naturalists, and scientists access to the diverse and unexplored habitats of the vast continent. Avid photographers and artists were drawn to remote areas, such as the Rocky Mountains, to capture images of landscapes and wildlife. Notable among these were intrepid female painters who produced remarkable collections of botanical artworks.

Born into a prosperous Quaker family from Philadelphia, Mary Vaux Walcott (1860–1940) first visited the Canadian Rockies on a family holiday in 1887, and was captivated by the landscape. Subsequently returning during most summer holidays, she reveled in the outdoor life and became a keen mountain climber and amateur naturalist. It was these interests that she would combine with her lifelong passion for painting.

On one visit to the Rockies, Walcott was asked by a botanist to paint a rare blooming plant, and the results led her to pursue botanical illustration. For many years she would traverse the rugged terrains of North America, seeking out significant and new species of wild plants, and creating hundreds of watercolor paintings. Some 400 of these were reproduced in a five-volume book set titled *North American Wildflowers*, published by the Smithsonian Institution between 1925 and 1929. Walcott received much acclaim for the work and her very engaging and botanically accurate images, and was hailed as the "Audubon of Botany."

On some of her expeditions Walcott was joined by her childhood friend Mary Schäffer Warren (1861–1939), who shared her sense of adventure and a talent for painting. The publication *Alpine Flora of the Canadian Rocky Mountains* (1907), inspired by her late naturalist husband, featured many of Warren's striking watercolors of plants and flowers. The works produced by these trailblazing women represented a new age of discovery and revealed to the world the true, and little known, beauty of the flora of North America.

Formal education
Roses on a Wall (1877) is a painting by the popular Philadelphia artist George Cochran Lambdin. Renowned for his formal paintings of flowers, Lambdin is known to have had Mary Schäffer Warren as a pupil. Some historians believe that Mary Vaux Walcott may also have studied with him.

> "... to collect and paint the finest specimens obtainable, and to depict the natural grace and beauty of the plant without conventional design."
>
> MARY VAUX WALCOTT

Reproduction without flowers

Male and female cones may be produced by separate plants, but when both occur on the same tree, they are often found in different parts of the canopy to promote cross-pollination. On the Atlas cedar (*Cedrus atlantica*), male pollen cones appear mainly on the lower branches. High above them, the female cones are more likely to receive pollen blown from a neighboring tree.

Male cedar cones grow to about 3 in (8 cm) long

The soft scales of male cones release clouds of pollen in the fall

Pollen grains collect on needles before being blown to female cones

Seed-bearing cones
The female cones of cedars (*Cedrus atlantica*), may take up to two years to mature. The process of fertilization itself often takes a year to complete, as male pollen-grain tubes push slowly beneath the female cone's scales to deliver sperm to ovules. In the following months, tiny winged seeds develop on the underside of the scales.

Young, green female cones become a woody barrel shape as seeds develop on their scales

Each broad scale releases two winged seeds

cone reproduction

Although they produce pollen and ovules, gymnosperms—an ancient group of plants that includes cycads, ginkgos, and conifers—share little in common with flowering species. They reproduce by means of male and female cones, creating seeds over a much longer period. The term "gymnosperm" literally means "naked seed," a term that refers to the ovules of female cones, which emerge fully exposed and are not surrounded by a protective ovary.

MALE AND FEMALE CONES

In most gymnosperms, male and female cone structure is different. A male cone usually lives for a few days. It is softer-textured, longer, and slimmer than the female cone, with bract scales arranged in a spiral around a central stem; each scale contains a pollen sac on its lower surface. Female cones are wider and more substantial, with spirally arranged ovule-bearing scales. Each scale carries one or more ovules, and these develop into seeds once they have been pollinated.

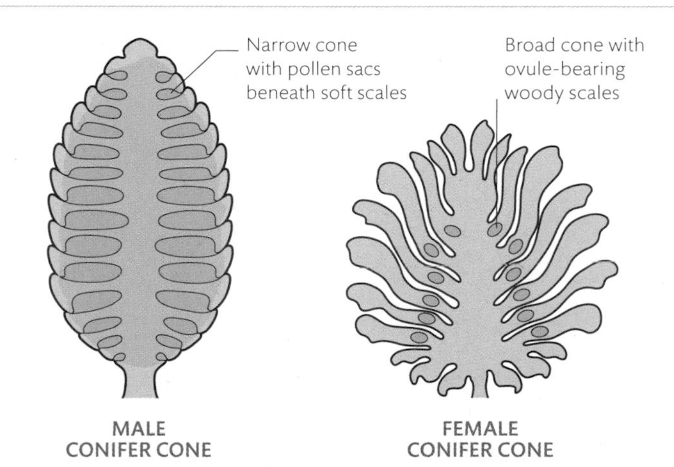

seeds. and fruits

seed. a plant's unit of reproduction, from which another such plant can develop.

fruit. the structure that surrounds a plant's seeds, often sweet, fleshy, and edible.

Enclosed seeds
The seeds of angiosperms form within fruits. Honesty plants have disklike seed capsules, called siliculae, which enclose the developing seeds.

Placental lines show where seeds were attached to the ovary

A central septum divides each silicle into two seed-containing valves

A seed coat, also known as a testa, protects the seed after release

seed structure

Whether they bear cones or fruit, all non-flowering plants (gymnosperms) and flowering ones (angiosperms) reproduce by seed. And while there are differences in how the naked seeds of a conifer develop compared with the enclosed seeds of a flowering plant, the seeds of both have the same basic structure—an outer seed coat, stored nutrients, and a developing embryo.

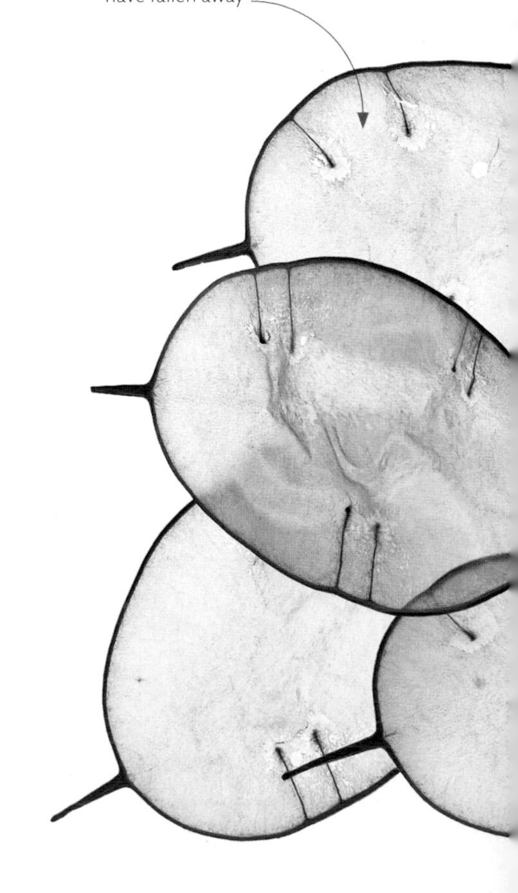

The silvery septums remain on the plant long after the valves have fallen away

INSIDE SEEDS

All seeds have seed leaves (cotyledons)—monocot seeds have one and most other seed plants have two. Some cotyledons provide food for the plant embryo, as does endosperm in a monocot. Both types of seed contain epicotyls: shoots for upper stems and leaves; hypocotyls, which form lower stems; and radicles, which form roots.

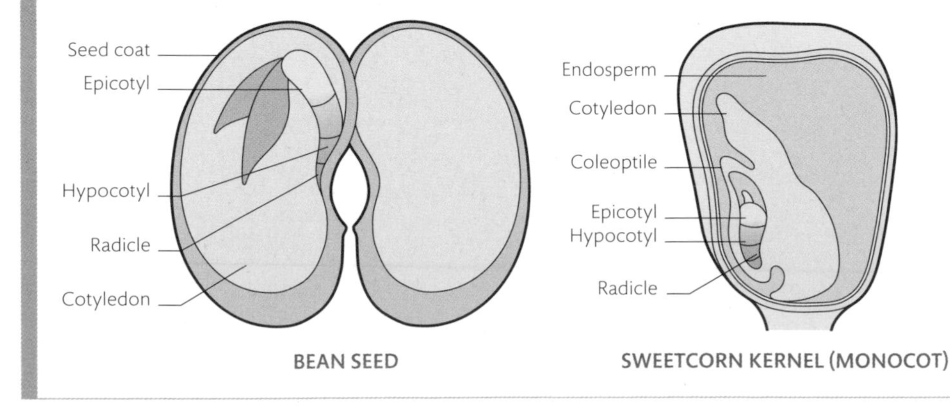

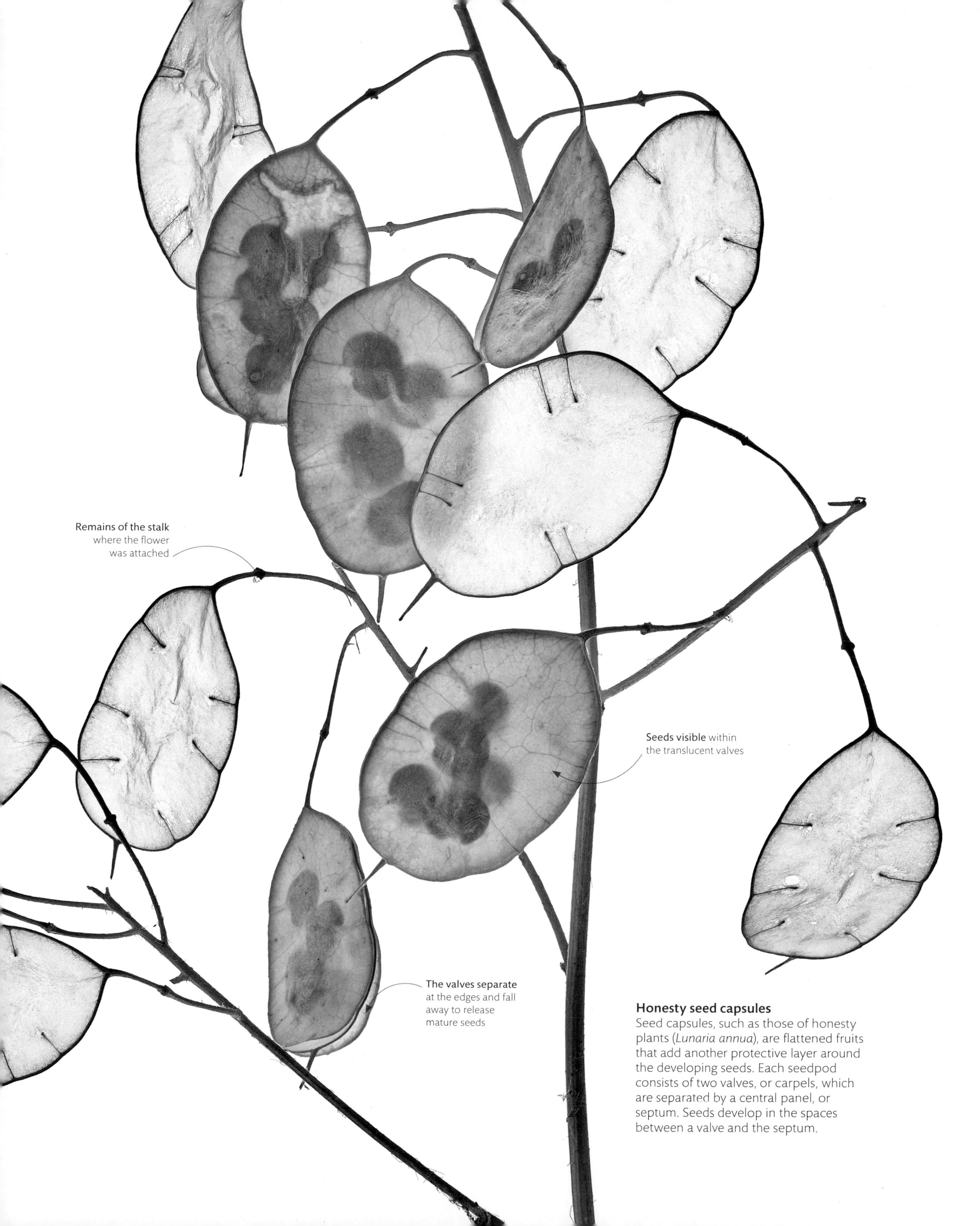

Honesty seed capsules
Seed capsules, such as those of honesty plants (*Lunaria annua*), are flattened fruits that add another protective layer around the developing seeds. Each seedpod consists of two valves, or carpels, which are separated by a central panel, or septum. Seeds develop in the spaces between a valve and the septum.

naked seeds

Developing without surrounding ovaries, the naked seeds of gymnosperms are exposed to the environment. Like those of flowering plants, naked seeds have coats, but they mature within cones instead of fruits. The most familiar are woody conifer cones, whose scales shield seeds as they ripen. Others, such as yews, produce single seeds that grow in fleshy cases.

Woody cones
Seed-bearing conifer cones come in a wide variety of shapes and sizes. Not all cones seem to match the size of the trees that produce them. Those of the giant sequoia are just 2–3 in (5–8 cm) long, although the tree reaches heights of 310 ft (94 m).

Three-pointed bracts, known as "rat tails," protrude from scales

PSEUDOTSUGA MENZIESII

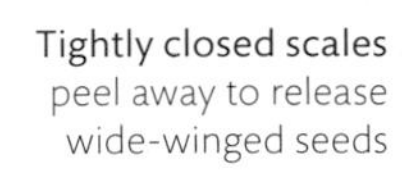

Tightly closed scales peel away to release wide-winged seeds

CEDRUS ATLANTICA

The massive cones are 9–15 in (24–40 cm) long and weigh up to 11 lb (5 kg) when fresh

PINUS COULTERI

Cones live for decades, and only release their seeds if affected by fire, squirrels, or beetles

SEQUOIADENDRON GIGANTEUM

Unusual seed structures

Some gymnosperms produce seed structures that do not look like cones. Yews (*Taxus* spp.) and junipers (*Juniperus* spp.) are conifers, but their seeds mature in fleshy cases called arils. Maidenhair trees (*Ginkgo biloba*) bear male pollen cones, but the seeds develop on stem ends, revealing one seed per "berry" after its fleshy covering decays.

Seed just visible within aril, but still open to elements

The ripe aril enlarges into a brightly colored, fleshy covering around the seed

Immature seed protruding from lighter-green unripe aril

TAXUS BACCATA 'LUTEA'

Ovules develop in pairs, and are exposed at the ends of the stalks

GINKGO BILOBA

A juniper produces tiny cones with merged scales

JUNIPERUS COMMUNIS

Long-term protection

Most of the structures we think of as "cones" are female, often known as "seed cones." They are generally larger and more robust than short-lived male cones. The thick, woody scales that were used to protect developing seeds allow the female cones of many species to remain intact, attached to their parent trees for years after fertilization and seed release.

Scales develop from a central axis called a rachis

A "spike," or umbo, on a scale is a remnant of the cone's first year of growth

SEED DEVELOPMENT
Conifer seeds develop when pollen grains released by male cones fertilize ovules on the scales of female cones. The wind carries pollen to female cones, entering via a small opening called a micropyle. Pollen grains form tubes through which male gametes travel to fuse with female gametes in the ovules. Once fertilized, the ovule develops into an embryo surrounded by a seed coat and protected by scales.

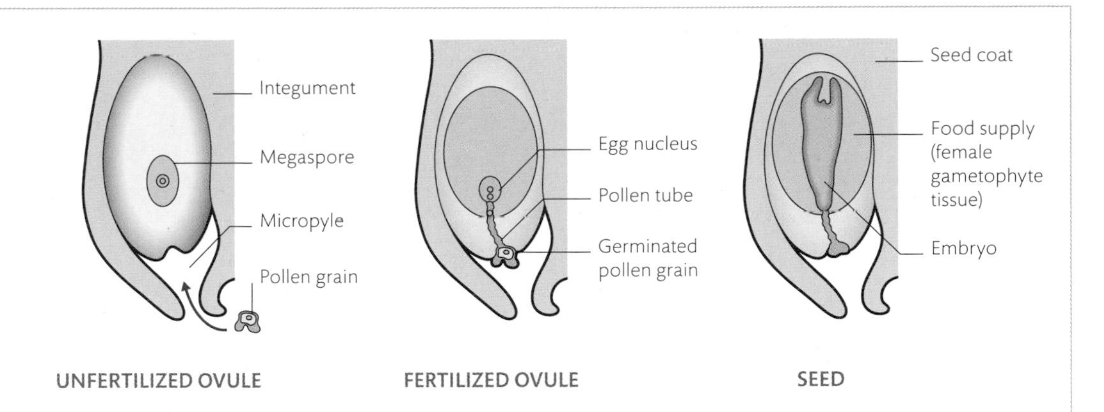

inside seed cones

Gymnosperms have two distinct phases to their life cycles. Both male and female reproductive structures form in cones, where each generates sex cells, or gametes, that are haploid—meaning that the nucleus of each cell carries only one set of chromosomes. Each seed produced by the union of male and female gametes during fertilization is diploid: containing two sets of chromosomes. Many gymnosperm trees are diploid organisms, created by the fusion of two separate haploid cells.

An undetached seed remains lodged between scales

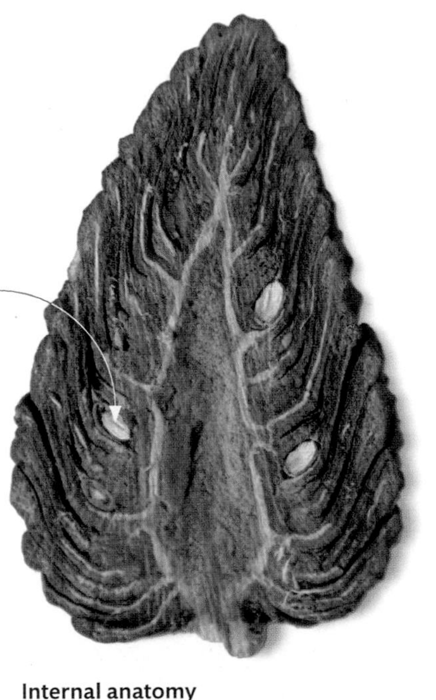

A fertilized ovule has developed into a seed, wedged between scales

Internal anatomy
Cutting through an unopened female cone shows how tightly pressed together the scales are, and how effectively the developing seeds are protected.

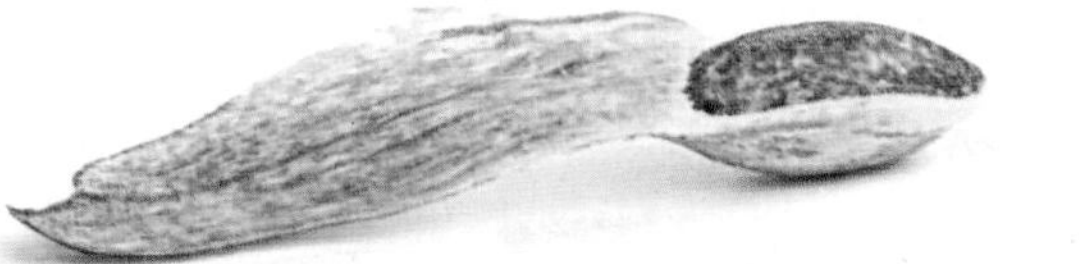

The lantern skeleton remains on the plant months after the seed has dropped

enclosed seeds

Enclosed seeds develop within ovaries, which form fruit coverings that protect developing seeds as they ripen. This extra layer may also act as food to attract animals that aid in seed dispersal. Additional coverings, such as the layers surrounding coconut seeds, can be quite hard, while others are so fragile that, at first glance, they seem to have little purpose at all.

NUT OR SEED?

Botanically, a "nut" is defined as an enclosed seed with a hard outer shell that is indehiscent, meaning that it does not open naturally to release the seed. Although both sweet and horse chestnuts include the word "nut" in their names, neither qualifies as a nut. Both may be protected by formidable spiky coverings, but because the coverings eventually fall apart naturally, what is inside is classed as a seed with a seed coat. Other common misnomers include Brazil nuts and cashews, which both mature in dehiscent pods.

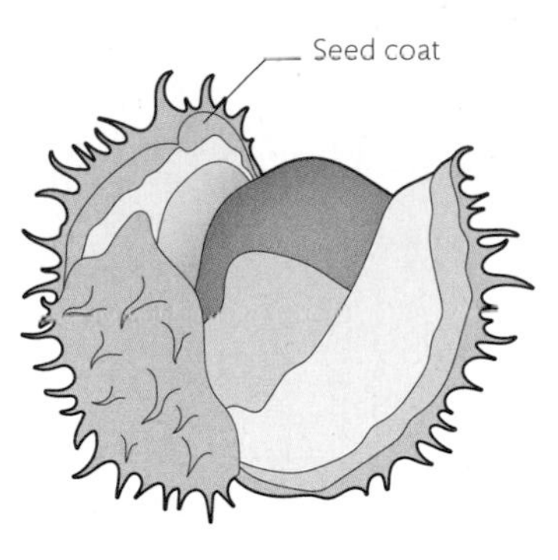

HORSE CHESTNUT

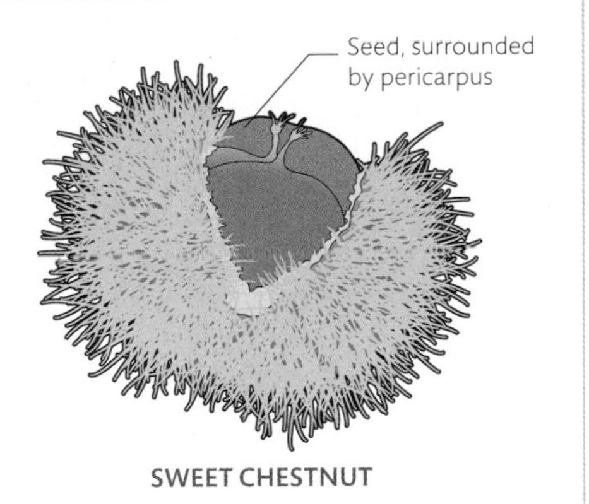

SWEET CHESTNUT

Chinese lanterns

Each of the attractive, paperlike coverings of the Chinese lantern (*Physalis* sp.) are a calyx—the sepals of a flower, which have fused and inflated to envelop a single berrylike fruit. What the "lanterns" lack in durability they make up for in other ways. While the fruits they surround may be edible, the calyx is toxic, and this characteristic, combined with its ability to provide protection against the weather, forms an effective enclosure.

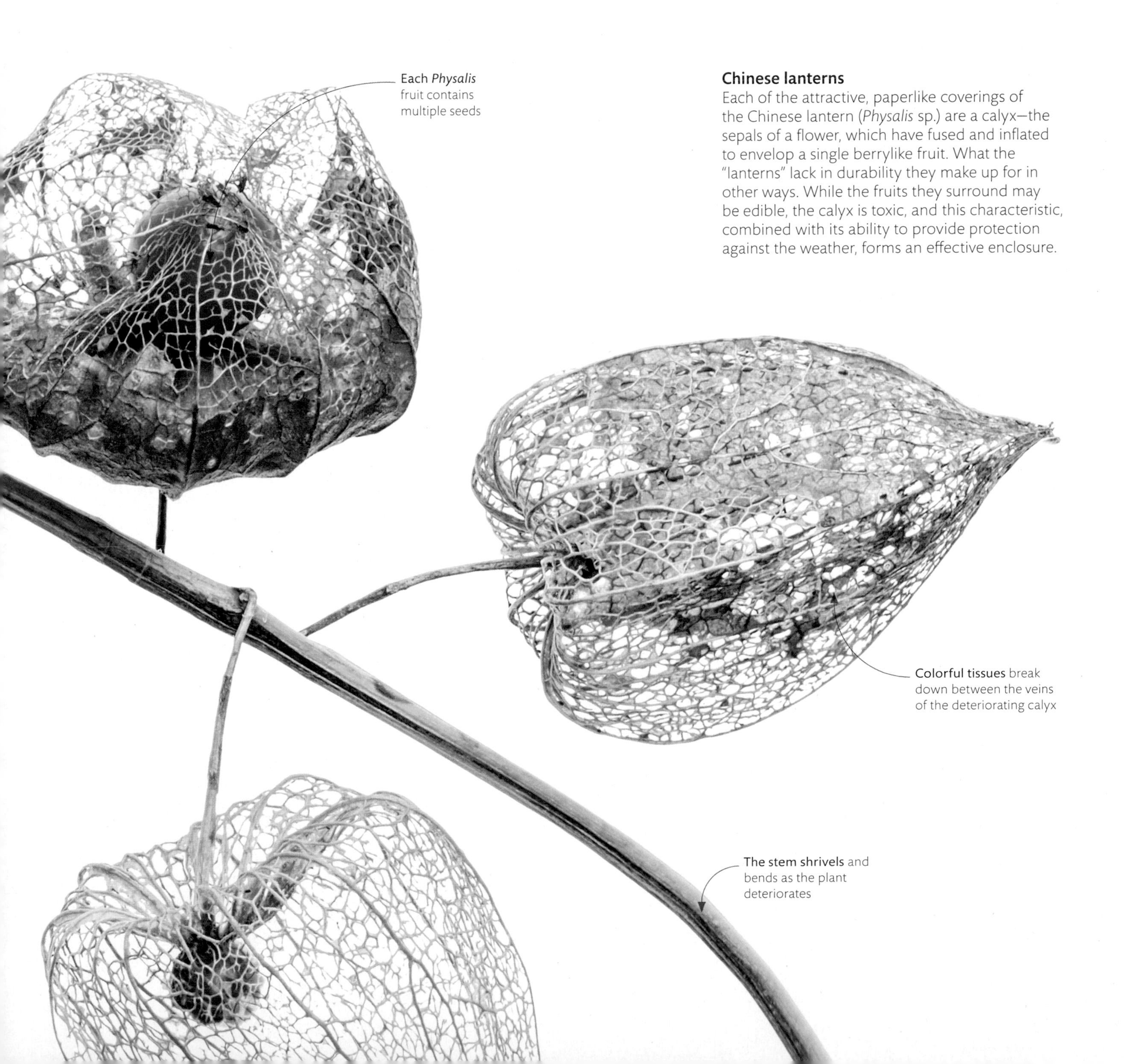

Each *Physalis* fruit contains multiple seeds

Colorful tissues break down between the veins of the deteriorating calyx

The stem shrivels and bends as the plant deteriorates

Mature stems are brown and woody and carry fruits produced from the previous year

Flexible green stems bear fresh new blooms awaiting pollination

types of fruits

After a flower is pollinated, the ovules within the ovary develop into seeds. The ovary wall, or pericarp, forms a protective layer enclosing the seeds, which comprises the fruit. The way in which the pericarp develops determines the type of fruit, with some becoming fleshy and edible, while others are dry and largely inedible. In many fruits, the pericarp is differentiated into three layers: the skin, or epicarp; the flesh, or mesocarp; and the stone, or endocarp.

Flower to fruit
At the heart of each madrone (*Arbutus menziesii*) flower is an ovary. Once fertilized, this will develop into a simple fleshy fruit.

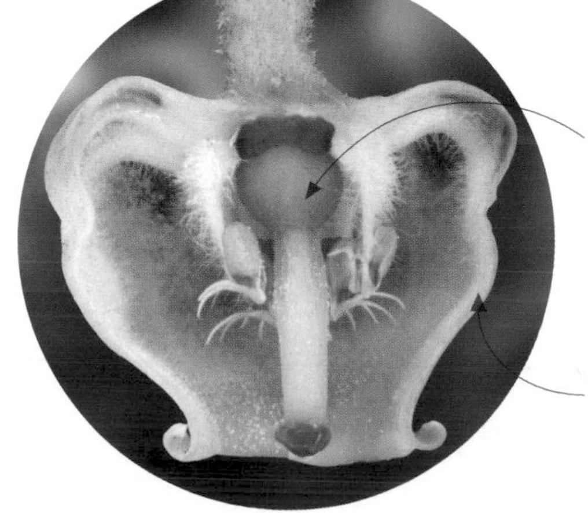

A green ovary sits in the center of the flower, attached to the receptacle

Five petals fuse into an urn-shaped tube that largely encloses the flower parts

Urn-shaped corolla is common in flowers pollinated by bumblebees

Not all flowers develop to maturity; some abort at the bud stage

The color changes from yellow to red when ripe, a signal that the fruits are at their sweetest

Arbutus fruits are popular with birds, who distribute the seeds far from the parent plant

FRUITS AND FLOWER HEADS

Simple fruits come from a single flower with one ovary. Aggregate fruits are also the product of a single flower, but one with several ovaries. Multiple fruits derive from several close-set flowers, while accessory fruits incorporate tissues other than the pericarp.

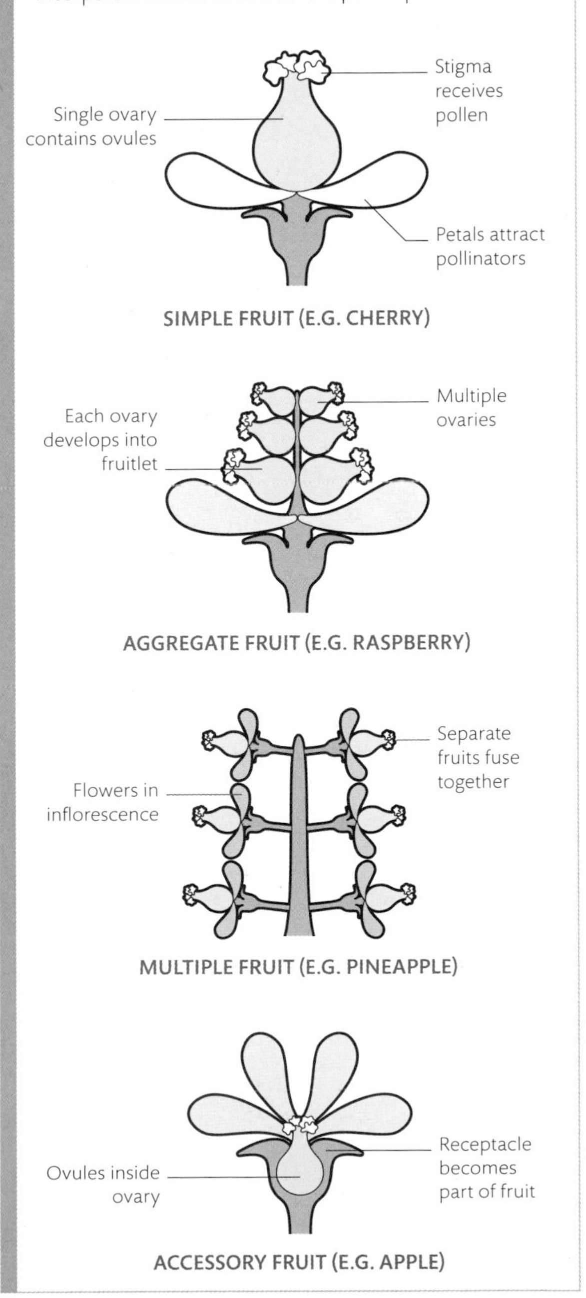

Distinguishing features

Fleshy, strawberry-shaped fruits give the strawberry tree (*Arbutus unedo*) its name, but they are not actually strawberries. Each fruit is formed from the ovary within a single flower and is therefore a simple fruit. By contrast, a strawberry is an accessory fruit with a fleshy receptacle.

Pavement of fruit trees
This mosaic of a pomegranate tree, possibly dating to the reign of Byzantine emperor Heraclius (575–641), was part of a pavement decoration from the Great Palace of Constantinople.

plants in art

ancient gardens

The first gardens were created by the earliest societies of the Middle East, when the need for self-sufficiency led people to enclose plots of land next to their homes. Over time, the practical function of the garden was superseded by people's desire to enhance their surroundings, with the emerging ruling classes using gardens to enjoy their leisure time and reinforce their status.

Glimpses of ancient gardens and plants are found in archaeology, literature, and art across the ancient world.

The first large-scale formal gardens were built by the emperors of ancient Mesopotamia, home to the legendary Hanging Gardens of Babylon. These gardens often combined elaborate irrigation systems and stone landscaping with formal planting of trees and exotic plants acquired on foreign campaigns.

The ancient Egyptians created gardens for both secular and religious purposes, and temples often had gardens in their compounds, where symbolic herbs, vegetables, and plants used in rituals were grown. The Egyptians also cultivated many types of flowers that they used in festive garlands and for medicinal purposes.

Domestic pleasure gardens were rare in ancient Greece. Relatively simple gardens were closely associated with religion, and the trees and plants grown in them were associated with particular deities.

Heavily influenced by Egypt and Persia, garden design and horticultural techniques became highly advanced in ancient Rome. From the town houses of Pompeii to Rome's imperial palaces, gardens were places for relaxation and escape, and often featured art and objects with religious and symbolic meaning.

Timeless garden
A fresco from the Villa of Livia, near Rome, built for the wife of the Roman Emperor Augustus, depicts a garden that is both naturalistic and borne of fantasy. By fruiting and flowering at the same time, the trees and bushes serve to convey the fecund "perpetual spring" of the glorious reign of the Emperor.

> “ If you have a garden and a library, you have everything you need. ”

MARCUS TULLIUS CICERO, *TO VARRO, IN AD FAMILIARES IX, 4*

Thorns help protect the berries from being eaten by predators

Stamens begin to wither as fruitlets emerge after fertilization

Blackberry fruits
Blackberry (*Rubus* sp.) bushes produce long panicles (see p.216), whose branches end in flower buds. The flowers at the end of the stem usually bloom and ripen before the other flowers, so the fruits develop at different times on just one section of a blackberry bush.

HOW BLACKBERRY FRUITS DEVELOP

Each blackberry flower contains many pistils, and each pistil's ovary contains multiple ovules. Every ovule can form a seed surrounded by a single fruitlet, or drupe. Once fertilized, each flower's pistils fuse, forming an aggregate fruit.

flower to **fruit**

When flowers appear in late spring or early summer, they mark the first stage of fruit formation. The next stage occurs when a grain of pollen from a plant of the same species lands on a flower's stigma. This produces a pollen tube that travels through its style. This "tunnel" gives the pollen's nucleus access to the flower's ovule-filled ovary (see pp.184–185), where it fuses with an ovule nucleus, fertilizing it. Fertilization signals the end for the flower, but as its petals wither and fall off, all fertilized ovules transform into seeds, and their surrounding ovaries swell and ripen into fruit.

succulent fruit

dry fruit

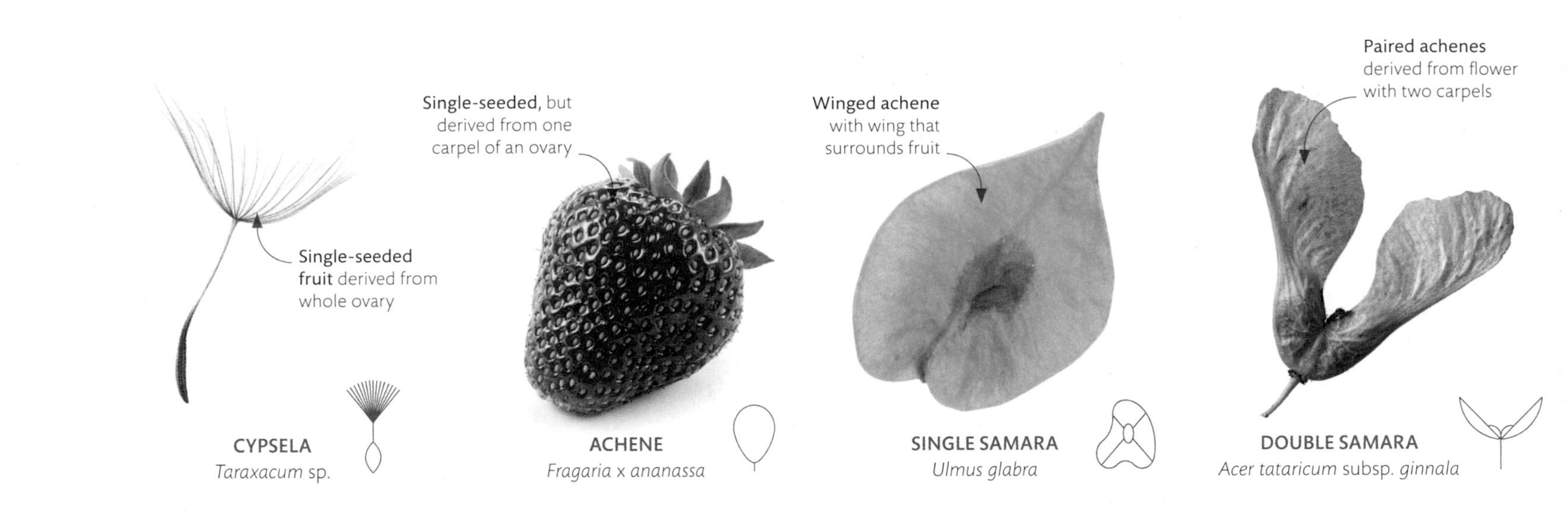

fruit anatomy

Essentially, fruit classification is based on a handful of characteristics, one of the most important of which is texture. Succulent fruits are eaten and dispersed by animals, whereas dry fruits rely on wind, gravity, or animal fur for dispersal. Although inspecting a fruit, and ideally the flower from which it is derived, should make everything clear, botanical classifications can still be surprising. A cucumber, for example, is classified as a berry, while a strawberry is not.

Schizocarp that bursts open to scatter seeds

REGMA
Erodium cicutarium

Seeds enclosed within leathery core
POME
Malus x domestica
Seeds are enclosed within woody pit
DRUPE
Prunus persica
From a single flower with multiple carpels
AGGREGATE
Rubus idaeus
Small fruits fuse together into one unit
MULTIPLE
Maclura pomifera
Hard-walled fruit that does not split open to release the seed
NUT
Corylus colurna
Similar to achenes, but outer coat is fused to the seed
CARYOPSIS
Zea mays
Derived from one carpel and split along two seams
LEGUME
Lathyrus odoratus
May cluster together and each splits along one seam
FOLLICLE
Aquilegia chrysantha
Multiple chambers distinguish capsules from other dry fruits
CAPSULE
Rhodophiala bifida
Capsule develops open pores to shed seeds
POROSE CAPSULE
Papaver somniferum
Often similar to legumes, but derived from two carpels
SILIQUE
Lunaria sp.
Fruit splits into one-seeded pieces (mericarps)
SCHIZOCARP
Trachyspermum ammi

Musa sp.

the banana

The long, slender fruits of a banana plant are actually berries. Store-bought bananas are seedless, but in the wild these fruits are full of seeds hard enough to crack a tooth. There are 68 different banana species, all belonging to the genus *Musa* and hailing from tropical Indomalaya and Australia.

The species in the *Musa* genus differ greatly in height, from the diminutive *Musa velutina*, which rarely grows to be much more than 6 ft (2 m) tall, to the monstrous *Musa ingens*, which regularly reaches 66 ft (20 m). Despite their appearance, bananas are not trees: they do not produce any wood and what looks like a tree trunk is actually the hardened, tightly packed bases of their long, tropical leaves. In fact, bananas are the largest herbaceous plants on the planet.

There are two major flowering forms, upright and drooping. In the upright form, the flowers point toward the sky and are pollinated mainly by birds; in the drooping form, the flowers point toward the ground and are pollinated largely by bats. In both forms, the flowers are produced on a spikelike inflorescence, and each whorl of flowers is protected by large, often showy bracts.

In the wild, bananas only form fruits if they are pollinated. The fruits start out greenish in color but change hue as they ripen. Not all bananas end up yellow: some species produce bright pink fruits. If animals harvest the fruits too early, the banana seeds may not be viable. The color change occurs only when the seeds inside are ready, and it signals to animals that the fruits are fit to eat. Researchers have found that, upon ripening, some bananas fluoresce under ultraviolet (UV) light. This may help animals that can detect UV to find ripe fruits more easily.

Flowers and fruit
Banana flowers are usually yellow or cream in color, and tubular in shape. Birds and bats are the main pollinators of wild bananas. Cultivated bananas produce fruits without pollination, so they are all seedless.

A banana stem (there is only one stem per banana plant) can produce as many as 200 fruits

Banana bunch
Bananas were first cultivated over 7,000 years ago. Today, cultivated bananas all over the world are descended from only two species, *Musa acuminata* and *Musa balbisiana*. Banana farming uses mostly clones of only a few banana varieties, making these plants far more susceptible to disease than wild ones.

Hairs on a flower head signal developing or fertilized spikelets

The dried-out, yellowed inflorescence retains spikelets after the seeds are released

Equipped for dispersal
As blue grama grass seeds ripen, the flower head curves, opening the spikelets and giving the plant its nickname "eyelash grass." Each seed has three bristly awns that are capable of hooking onto fur, clothing, or feathers.

seed distribution

Seeds and spores are the travelers of the botanical world. No matter what type or size they are, seeds have one crucial mission: to transport the genetic material needed to create a new plant. For some species, this is as simple as falling from a parent plant onto fertile ground. For others, it means hitching a ride on whatever is available—wind, water, birds, insects, animals, humans—or even being ingested and deposited many miles away from the plant on which they formed.

The terminal (top) **inflorescence** is in the best position to release seeds on the wind

Each panicle supports 1–3 main flowering branches, usually ending with a terminal inflorescence

Multiple dispersal methods

Launched from stalks up to 12in (30cm) high, seeds of blue grama grass (*Bouteloua gracilis*) have many opportunities to disperse. The seeds catch passing breezes to float several yards away, and some will germinate where they land. They are also eaten by herbivores and grow where they are excreted. Seeds that cling to animal fur or bird feathers travel even farther from their parent plant.

Seeds are expelled from the ends of spikelets when the seedhead matures

Each blue grama inflorescence contains up to 130 spikelets

Takeaway advertising
Brightly colored berries of Tasman flax lilies (*Dianella tasmanica*) are easy to spot from the ground or the air, making them very attractive to birds. By dangling its high-carbohydrate fruits at the ends of long stems, the plant also makes it easy for birds to reach them as they fly past.

Each berry contains five black seeds that pass through birds' digestive systems intact

Vivid colors, such as reds and purples, are known to attract wild birds

Germination boost
Seeds of rowan trees (*Sorbus aucuparia*) dispersed in bird droppings germinate faster than non-digested seeds. This might be due to some chemicals being removed during the digestive process.

A single berry at the end of a stem is easy to pluck

dispersed by diet

Some plant species are much more successful than others at distributing their seeds over wide areas. One of the most effective ways for the plants to accomplish this is by feeding birds. Although many other animals also spread seeds through their droppings, airborne birds can cover a much wider territory, increasing the likelihood of depositing seeds from a recent meal far away from their source.

FORGOTTEN FOOD

Squirrels are among the most famous "wild gardeners" due to their habit of burying, then losing track of, winter nut stashes. By caching food underground for cold-weather meals, these mammals inadvertently plant numerous trees, especially hazels and oaks, whose saplings reveal the lost storage locations the following spring. Buried nuts and acorns have much higher germination rates than those that just fall to the ground from the branch.

Acorn shells protect seed underground

Cupule (cup) formed from bracts at flower base

STRANDZHA OAK (*QUERCUS HARTWISSIANA*)

Green, unripe berries are inconspicuous

TASMAN FLAX LILY (*DIANELLA TASMANICA*)

Red and black for birds

Birds consume a much higher number of red and black fruits and seeds than those of any other color. This is thought to be due partly to birds' well-developed color vision and partly to the nutritional content of these fruits and seeds, or their concentration in certain habitats. Whatever the reason, red seeds, such as those of this saucer magnolia (*Magnolia* x *soulangeana* 'Rustica Rubra'), are high on the avian menu.

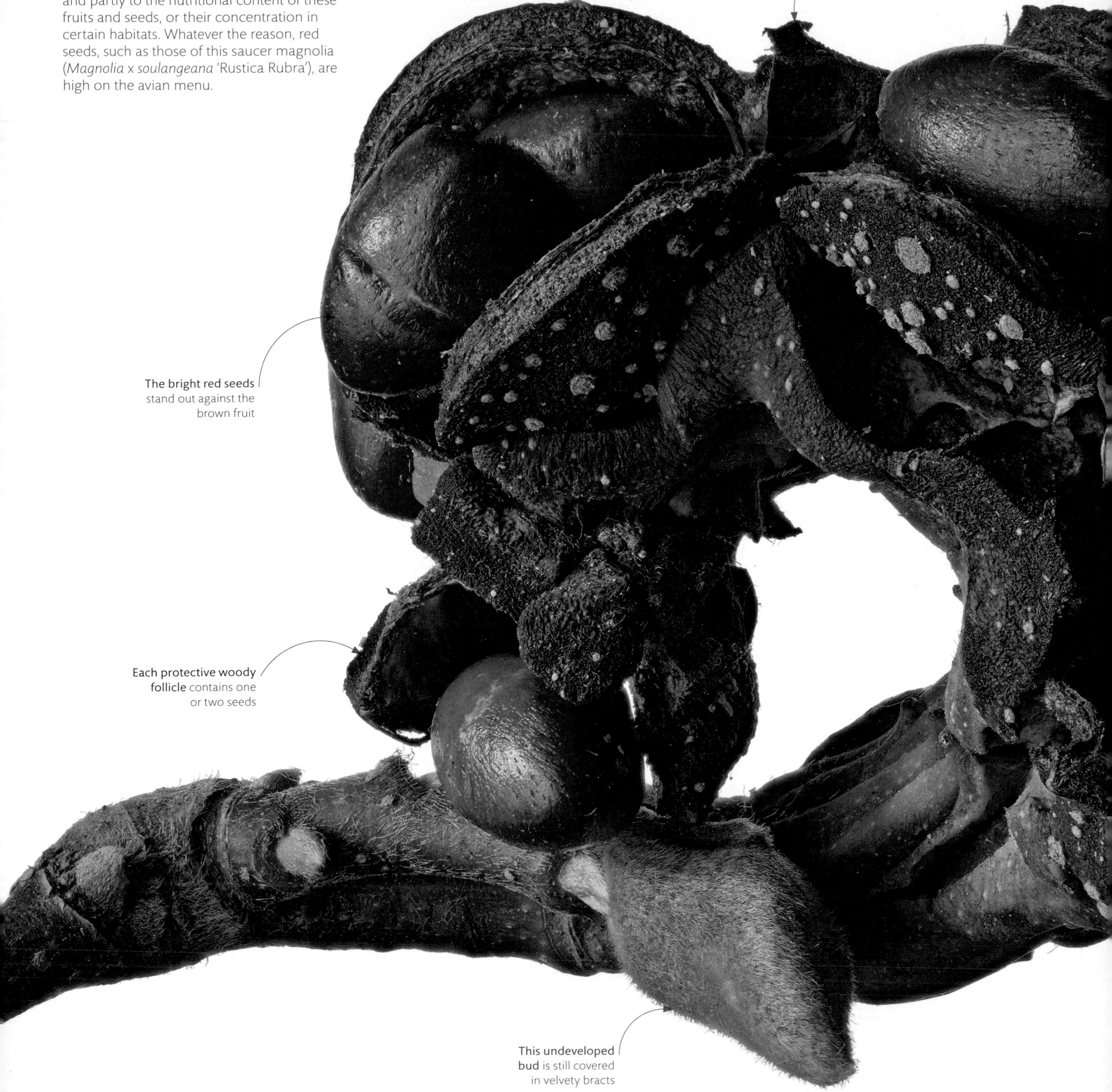

The conelike fruits bend and split open when mature

The bright red seeds stand out against the brown fruit

Each protective woody follicle contains one or two seeds

This undeveloped bud is still covered in velvety bracts

BIRD OF PARADISE
(*STRELITZIA REGINAE*)

The orange arils appeal to monkeys, while the black seeds attract birds

TRAVELER'S PALM
(*RAVENALA MADAGASCARIENSIS*)

Striking blue arils decorate the blue seeds of the traveler's palm, attracting lemurs, which can only see blue and green

Color and seed distribution
Both mammals and birds consume seeds and distribute them in droppings or by hiding them away, but their color preferences vary. Research worldwide shows that birds prefer red and black seeds, while mammals eat mainly orange, yellow, and brown ones. When certain creatures only see or prefer certain shades, plants may have adapted, sometimes simply by adding vibrant arils (hairy coverings) to otherwise drab seeds.

colorful seeds

Like the fruits that bear them, seeds come in a myriad of colors, with protective seed coats ranging from plain black to dazzling reds, oranges, and blues. Although we know that light-colored seeds contain more moisture than dark ones in many species, the reasons why some seeds have such vivid pigments are still unclear. What is certain, however, is that particular colors seem to be preferred by certain animals.

TOXIC RICIN SEEDS

The main function of any seed coat is protection, sometimes protection from being consumed. Toxic seeds contain some of the most deadly substances on earth. Ricin, which is found in castor beans, is so poisonous that just four beans are enough to kill a human. Because the color of castor beans varies from solid white to mottled red and black, many animals and birds die from ingesting it.

CASTOR OIL PLANT
(*RICINUS COMMUNIS*)

Mottled coloring, similar to that of edible lima beans

The caruncle is a spongy outgrowth full of sugars that attracts ants

N.707.
a
b
c
g
f
f
g
d
h
h
e
a. Malus oxymela acida, Saürer Holtzapfel. b. Malus sylvestris fructu rubro minore, Pomme sovage, Holtzapfel. c. Malus sylvestris fructu rotundo viridi, grüne Holtzäpfel. d. Malus Persica flore pleno. e. Malus Persica Sti Laurentii dicta. f. Malus Persica minor, Pesche petit, Pfirsig. g. Malus Persica major molle carne, Pfirsigapfel. h. Malus Persica magna; Boobiter Pfirsig.

art and science

The 18th century is often called the Golden Age of botanical painting, and the illustrations of the botanic artist Georg Dionysius Ehret represent a great meeting of art and science. Ehret matched Carl Linnaeus's groundbreaking approach to the naming and classification of plants and animals with a clear, precise, and beautiful style of plant illustration.

One of the most influential botanical illustrators of all time, German-born Ehret (1708–1770) was the son of a gardener who taught him about nature. Ehret's talent for drawing, eye for detail, and growing knowledge of plants led him to produce botanical artworks that brought him to the attention of some of the world's leading scientists and influential patrons.

Ehret first collaborated with the famous Swedish botanist and taxonomist Carl Linnaeus on *Hortus Cliffortianus* (1738), a catalog of rare plants found on the estate of George Clifford, the governor of the East India Company. Under Linnaeus's direction, Ehret documented every part of the plants in beautiful, yet scientifically accurate and detailed artworks, in a style that become known as the Linnaean style of botanical illustration.

Ehret went on to illustrate most of the important botanical publications of the day, and produced numerous illustrations for collectors and institutions, including the Royal Botanic Gardens at Kew.

Pineapple (*Ananas sativus*)
Ehret's work included pencil, ink, and watercolor studies of plants from around the world, such as this pineapple, which had recently arrived in London to be studied at the Chelsea Physic Garden, one of the oldest botanic gardens in Britain.

Details of *Malus* fruit and flowers
These illustrations of apple and peach flowers and fruit are mezzotint engravings that were colored by hand. Ehret was commissioned by the Dutch apothecary Johann Wilhelm Weinmann to illustrate *Phytanthoza iconographia*, but he completed only half the illustrations due to the low fee paid by Weinmann.

“The genius of Georg Dionysius Ehret was the dominant influence in botanical art during the middle years of the eighteenth century.”

WILFRID BLUNT, *THE ART OF BOTANICAL ILLUSTRATION*, 1950

UNICORN FRUIT

Some of the largest hitchhiking seedheads are produced by plants known as devil's claws, but the spiky pods don't emerge right away. The woody seed capsule of *Ibicella lutea* forms within a large, hornlike fruit, earning it the name of "unicorn plant."

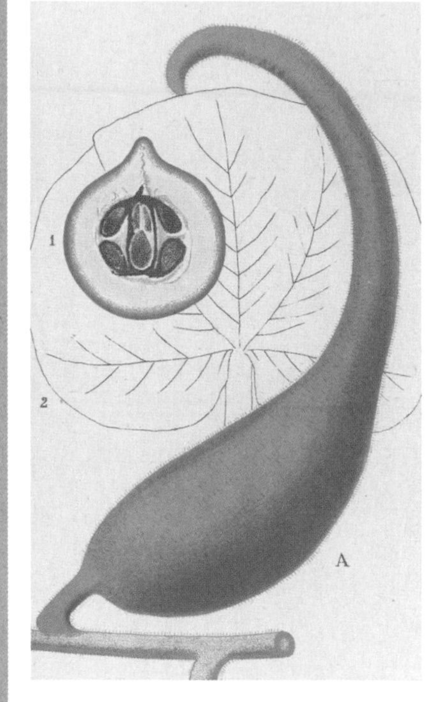

***IBICELLA LUTEA* SEED CAPSULE**

The elongated primary claws curve upward to increase the chances of latching onto animals

Shorter secondary claws sometimes form between the primary claws

Barbs cover all the claws of the seedhead, and hook on to the feet and legs of passing animals

animal couriers

To avoid overcrowding, plants have to disperse seeds over as wide an area as possible. Some plants use wind and water, while others expel their seeds via "exploding" seedheads. Many plants use animals that share their environments to scatter their seeds. "Sticky" seed capsules covered in hooks, barbs, and painful spines latch onto hide and hoof, and only release their seeds when the capsule is rubbed off, crushed, or torn open, often many miles away.

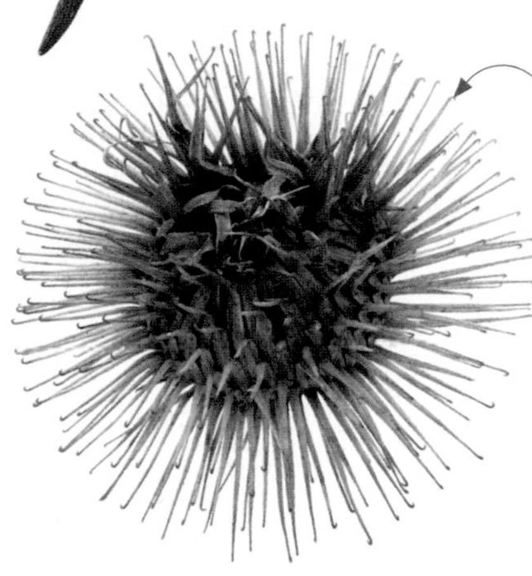

BURDOCK SEEDHEAD
(*ARCTIUM SP.*)

The seedhead is up to 6 in (15 cm) in diameter, with barbed spines

GRAPPLE PLANT SEEDHEAD
(*HARPAGOPHYTUM PROCUMBENS*)

Hitchhiker seeds
Clinging seedheads are different shapes, depending on their environment. The flexible, hooked bracts of small burdock burrs (*Arctium* sp.), native to Europe and Asia, catch on any passing creature. The sharp, woody, spines of the African devil's claw or the grapple plant (*Harpagophytum procumbens*) dig into large animal hooves.

Attachment points for spines that once covered the entire seedhead

Up to 110 seeds emerge from the center of each seed capsule

Clawed seed capsules
The South American devil's claw plant (*Ibicella lutea*) produces about 150 fierce-looking seed capsules. Each one grows up to 8 in (21 cm) long and has two long, curved, barbed "claws" with razor-sharp points. The main body of an unbroken capsule is also covered in spines.

Ideal launchpad
The tree of heaven (*Ailanthus altissima*) is so named because it quickly achieves heights of 80 ft (24 m) or more—perfect for wind-borne seed dispersal. A single tree can launch a million seeds a year—and because of this, it can become an invasive species in non-native lands.

Single-seed samaras break off both individually and in whole clusters

The rigid, seamlike edge stabilizes the samara as it spins

The semitransparent samara wing allows seeds to glide up to 300 ft (90 m) away

The division line of the double samara, where the seeds will separate from each other

The pericarp wall extends and elongates to form a thin, membranelike wing

The pedicel attaches the seed to the tree and separates from the stem when it matures

The pericarp layer encloses a single seed

seeds with **wings**

Putting enough distance between the parent plant and its seeds is vital for trees, which can easily grow too close together. The seeds of maples, sycamores, and other species have evolved winglike extensions that enable them to catch the breeze and travel much farther. Whether they glide, spin, or float, all winged seeds, called samaras, take their chances on the air.

Dual dispersal
Some wind-dispersed species, such as the crimson-leaved Tatar maple (*Acer tataricum*), produce mirror-image samaras. The wings and seeds are part of a single fruit, which began as one unit but then divided into two segments with membranelike wings as it matured. Each half of the samara can produce a new tree.

Veins on the wing form a rippled surface, creating turbulence that aids lift

HELICOPTER FLIGHT

Maple and other helicopter samaras rotate like spinning tops as they fall to the ground. The seeds usually have pitched wings, like helicopter or propeller blades. Their spinning lowers the air pressure over the seeds' upper surface, helping them to descend more slowly.

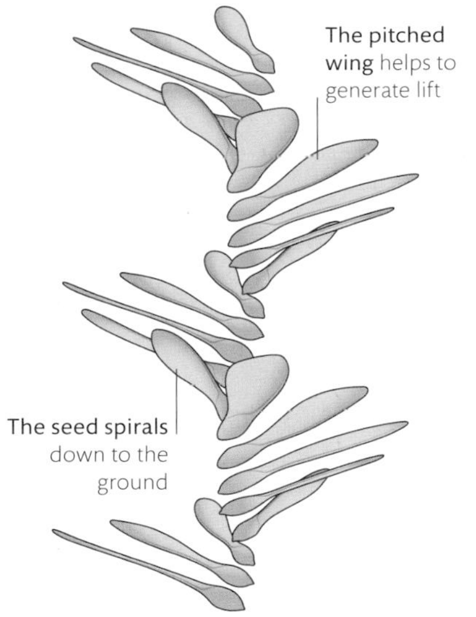

AUTOROTATION OF HELICOPTER SAMARAS

Taraxacum sp.

the dandelion

Named after their jagged leaf edges (from the French *dent-de-lion*, meaning "lion's tooth"), dandelions are loved by children for their puffy seed heads and despised by gardeners for spoiling lawns. Most species are native to Eurasia but have been widely spread by human help; these botanic travelers are now found in almost all of the world's temperate and subtropical regions.

Dandelion is a catchall term for some 60 species belonging to the genus *Taraxacum*; the one most often encountered is the common dandelion (*Taraxacum officinale*). Key to the success of this adaptable plant is its reproductive strategy. Dandelions are among the first plants to bloom in spring, making them an important resource for insects when little else is in flower, and helping to ensure they are pollinated. Having flowered in spring, they will often bloom again in the fall. Crucially, though, the dandelion can produce seeds without being pollinated. When this occurs, the seeds grow into clones of the parent plant (apomixis). With each flower head having up to 170 seeds, and as a single plant is able to produce a total of more than 2,000 seeds, there is a high chance that more than one will grow into an adult plant.

With light, feathery parachutes, dandelion seeds are great aerial travelers. Most land fairly close to the parent plant, but some are blown by the wind or caught on updrafts of warm air and are then transported long distances.

Feathery parachutes
Each dandelion seed is attached by a stalk to a disk of radiating, feathery filaments that form a parachutelike structure.

Early bloomer
A dandelion's flower head is actually a composite of many individual flowers.

The spherical seedhead falls apart once it is ripe, and each single-seeded fruit drops away

The cluster of flowers provides the perfect landing pad for bumblebees and butterflies

Elaborate outer petals help attract insect pollinators to the nectar-rich flowers

Strength in numbers
Scabious flowers are small and are gathered together in pincushionlike flower heads. After the flowers have been pollinated by insects, the petals drop, revealing a spherical head of papery fruits.

seeds with parachutes

Many plants use the wind to distribute their seeds. In open environments, such as meadows and prairies, where trees are few and far between, many plants do this as seeds can travel for long distances. To provide lift, windblown seeds need a sail or, in the case of this scabious, a parachute. Plants with wind-dispersed seeds often hold their flowers up high, so that when the seeds form, they can catch a breeze.

Spiky awns (bristles) catch in plants on the ground, ending the seed's journey away from its parent plant

THE COMPLETE PACKAGE
Each seed contains an embryonic plant complete with a juvenile root (radicle) and shoot (hypocotyl), nestled within one or more cotyledons. These contain nutrients that fuel the plant as it germinates.

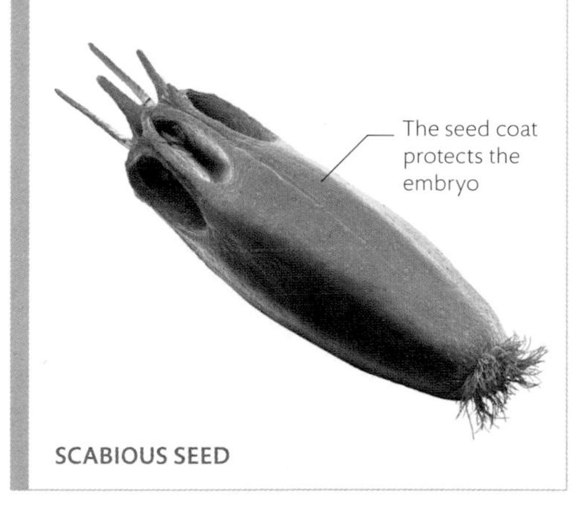

SCABIOUS SEED

Paper moons
The inflorescences of scabious (*Scabiosa stellata*) contain many small flowers, each of which produces a single seed. Every seed has five, spinelike bristles called awns, and is surrounded by papery bracts that form a "parachute." The wind blows the parachute and the awns catch on the ground.

Papery bracts surround each fruit, giving them lift and helping them hitch a ride on a breeze

Silky hairs catch the wind, transporting achenes far from the parent plant

Heading separate ways

Some flowers develop a single fruit, but those of *Clematis integrifolia* produce numerous one-seeded fruits called achenes. This is an advantage because the many fruits spread in various directions, expanding the plant's territory. Each achene has a tail covered in hairs that readily drifts on the wind.

silky seeds

To be transported by the wind, seeds need special adaptations. Some use silky hairs for this purpose, and in cotton (*Gossypium*) and poplar (*Populus*), the fruits are filled with a mass of hairs that are released with the seeds, aiding their dispersal. In other plants, the seed hairs develop into elaborate wings or parachutes. Hairs also help to fasten the seed to the surface on which it eventually lands, where it may be possible to germinate.

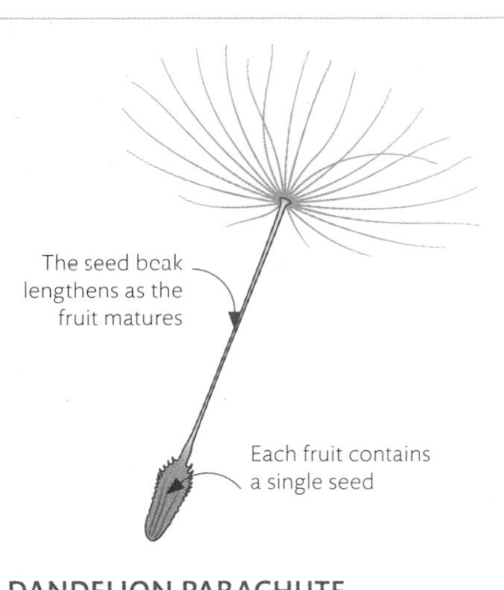

DANDELION PARACHUTE
A dandelion "seed" is actually a single-seeded fruit called a cypsela. Each cypsela has a tuft of hairs that develops from the flower's calyx. A stalk, known as the beak, attaches the hairs to the cypsela, forming the parachute.

A hairy achene tail develops from the flower's old style and stigma

The long stem holds the seedhead aloft, so the wind is more likely to catch the fruits

Numerous achenes remain attached to a central receptacle until they ripen and fall away

A cluster of carpels
At the heart of a clematis flower there are numerous carpels, each of which bears a single seed. These develop into achenes and, at maturity, the cluster of achenes breaks apart.

Asclepias syriaca

common milkweed

Found throughout much of eastern North America, milkweed is probably best known for being the larval food source of the monarch butterfly. At one time, milkweed was grown on a large commercial scale, and its "silk" was harvested as stuffing for pillows, mattresses, and even life preservers.

Milkweed flowers form umbrella-shaped inflorescences. Sweetly scented and ranging from light pink to near purple, individual flowers comprise five reflexed petals and five nectar-filled hoods. Unlike flowers that dust pollinators with copious amounts of pollen, milkweed blooms package their pollen into sticky sacs known as pollinia. The pollinia are held in grooves, called stigmatic slits, on either side of each hood. An insect arriving to drink nectar struggles to grip the flower's smooth surface and may accidentally slip a leg or two into the slits. The pollinia stick to the insect's legs and transfer to another flower when the insect gorges on the nectar of a neighboring milkweed. This pollination strategy favors larger insects; some bees and smaller insects, however, get stuck in the slits or lose limbs in their efforts to escape. The fruits that form after pollination start off as tiny green buds, then swell into large, seed-packed capsules known as follicles. Each flat brown seed in a follicle is adorned by a tuft of silky filaments called a coma.

The sap of milkweed contains toxins to deter mammal herbivores. Despite the toxicity, milkweed provides ample food in the form of leaves and nectar for a range of insects, including the monarch butterfly, that have adapted to its toxic defenses. The fall in monarch numbers is partly linked to the milkweed's decline due to habitat loss and increasing use of herbicides. Changing the fortunes of the plant could revive butterfly numbers.

Silky seeds
New uses for milkweed silk—as thermal insulation for outdoor clothing, acoustic padding in vehicles, and an absorber of oil spills—may herald a renaissance for commercial milkweed farming.

Silky filaments are lightweight, hollow, and have a waxy, waterproof coating

Milkweed seeds are dispersed by wind, borne on their buoyant silk "parachutes"

Milkweed follicle
The seed capsules, or follicles, are 3–4 in (8–10 cm) long and covered with soft prickles and short woolly hairs. Once mature, the follicles split along the side to release the seeds.

HOW SEEDS GERMINATE

Most seeds contain an embryo and a food source to kickstart growth. At germination, roots emerge first, to anchor the seedling, followed by leaves. In most flowering plants, such as beans, a pair of seed leaves (cotyledons) are first to appear; these contain food transferred from the seed. Monocots have only one cotyledon, which may remain within the seed.

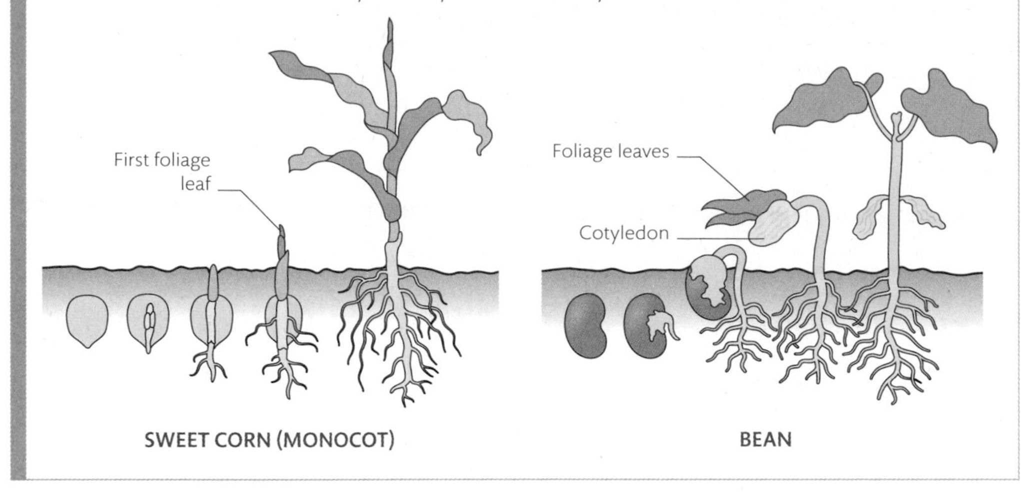

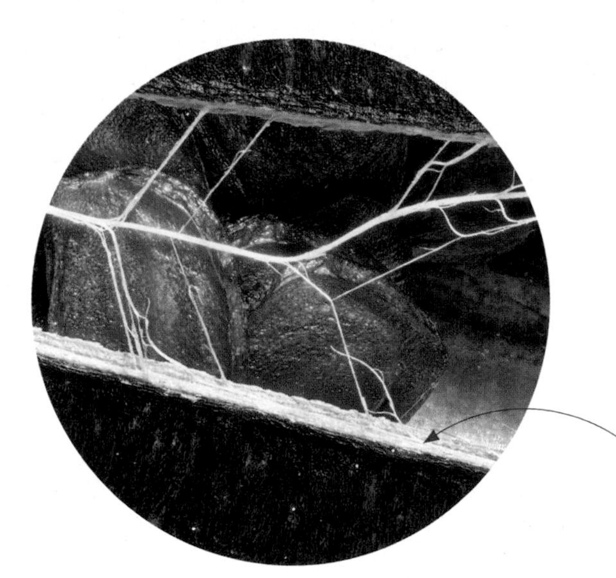

Shaken apart
The papery capsule fruits of the regal lily (*Lilium regale*) contain numerous winged seeds. When the fruits are dry, they split open and shed their seeds—with the help of the strong winter winds that sweep through the precipitous valleys of the lily's natural habitat in western China.

Lily capsules have three chambers, each containing circular, coinlike seeds

pods and capsules

Fleshy fruits are perhaps the best known as they are the ones we eat, but dry fruits are not uncommon. They include pods, capsules, achenes, follicles, and schizocarps, and they may split open to release their seeds (dehiscent) or remain intact to be distributed with the seed inside (indehiscent). Without a juicy coat to attract animals, dry fruits rely on other methods to disperse their seeds. Wind dispersal is widespread, but seeds can also cling to animal fur or fall to earth.

Valves or openings in lily fruit reveal two columns of seeds
The capsules dry out, shrink, and split open in late summer

HYGROSCOPIC MOVEMENT

Seed capsules often explode when they change shape due to moisture loss or desiccation. As the tissues become dry or wet, they contort—a process known as hygroscopic movement. Some plants, such as stork's bill (*Erodium cicutarium*), produce seeds with hygroscopic taillike structures (awns), that twist when wet, and partially bury the seeds. Hygroscopic fruits are usually segmented; when they dry out, the segments change shape, causing the fruits to split or burst open. It is not only dry fruits that pop; some fleshy fruits fill up with water and become pressurized, then split apart suddenly, taking the seeds with them.

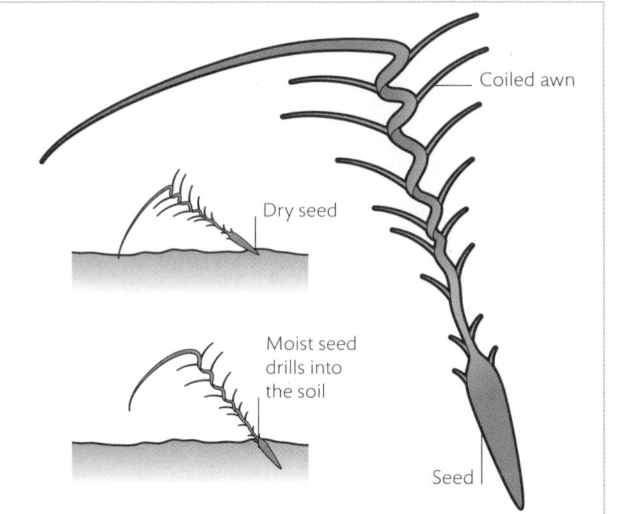

STORK'S BILL (*ERODIUM CICUTARIUM*) SEEDS

Each seed is enclosed within a capsule, called a mericarp

A woody beak holds the five awns together after the fruit splits apart

exploding seedpods

While many dry fruits passively await the wind or an itinerant animal, some plants take charge of their own seed dispersal. Explosive fruits fire seeds away from the parent so that they will land in a less congested patch of habitat. Different plant species have evolved a variety of explosive mechanisms, although most rely on pent-up pressure within the fruit to pop the seeds out.

Seed catapult

Cranesbills like *Geranium sanguineum* are named for their beaklike fruit. Called a regma (see p.300), it contains five seeds arranged around a woody beak. Each seed has its own covering attached to a long awn, and all five awns are fused at the tip of the beak. As the fruit dries, the awns contort, splitting the fruit and flinging the seed away.

Desiccation makes the cell walls within awns distort, curving the awns outward

Willing couriers
An explosive fruit can scatter seeds only a limited distance. Some plants utilize a secondary method of dispersal to expand their range, though. Violet (*Viola*) and broom (*Cytisus*) seeds carry small packages of food called elaiosomes, which encourage ants to pick them up and carry them away.

As an awn dries, it curves outward, pulling the mericarp away from the fruit

Some violet capsules slowly split, then pop seeds out one at a time

PANSY (*VIOLA* SP.)

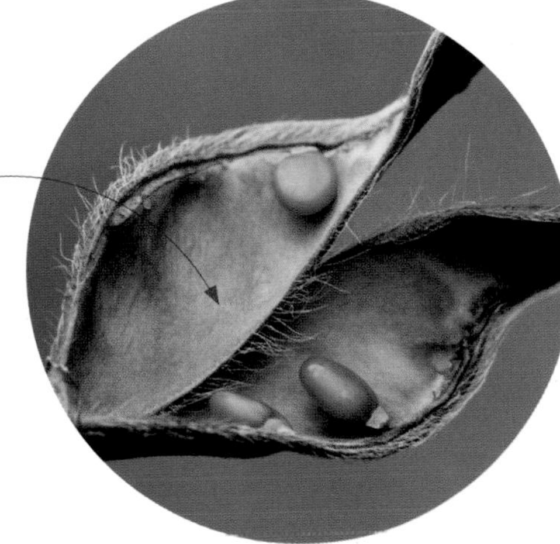

As broom pods dry in the sun, the shady side dries less, making the pod contort, then fly apart

BROOM
(*CYTISUS SCOPARIUS*)

Each regma has five sections, one from each carpel in the flower

Nigella sativa

fennel flower

Nigella sativa, variously known as fennel flower, nigella, black cumin, black caraway, and Roman coriander, has an ancient association with human civilization, having been cultivated for about 3,600 years. The seeds are used as a spice sprinkled on bread and naan, and to produce an herbal oil.

Fennel flower's history of cohabiting with humans makes unraveling its wild origins difficult. Some sources claim that it came from Mediterranean Europe, while others cite Asia or north Africa. Wild fennel flowers are still found in southern Turkey, Syria, and northern Iraq, so it may well have originated in the Middle East.

Up to 2 ft (60 cm) tall, fennel flower is a hardy annual that thrives in a variety of soil types. Despite its common names, it is not related to fennel (*Foeniculum vulgare*), cumin (*Cuminum cyminum*), caraway (*Carum carvi*), or coriander (*Coriandrum sativum*), all members of the carrot family, but is in fact a member of the buttercup family (Ranunculaceae) and a close relative of the popular ornamental love-in-a-mist (*Nigella damascena*). Like other annuals, fennel flower puts all of its energy into flower and seed production. The delicate, blooms are probably what first drew human attention to this plant. The identity of fennel flower's pollinators in the wild is not known, but bees perform the role elsewhere. After pollination, the plant's fruits swell into large, inflated capsules, each with several follicles containing numerous black seeds.

Irresistible to birds, *Nigella sativa*'s tiny, pungent seeds are also widely used as a spice in Indian and Middle Eastern cooking. In the ancient world the seeds and their oil were used to treat a wide range of ailments, and they are still used as an herbal remedy today.

Nigella sativa fruit
The pear-shaped seeds develop in a capsule with up to seven segments (follicles), each of which ends in a long projection formed from the style. The seeds are released when the fruit dries and the follicles rupture.

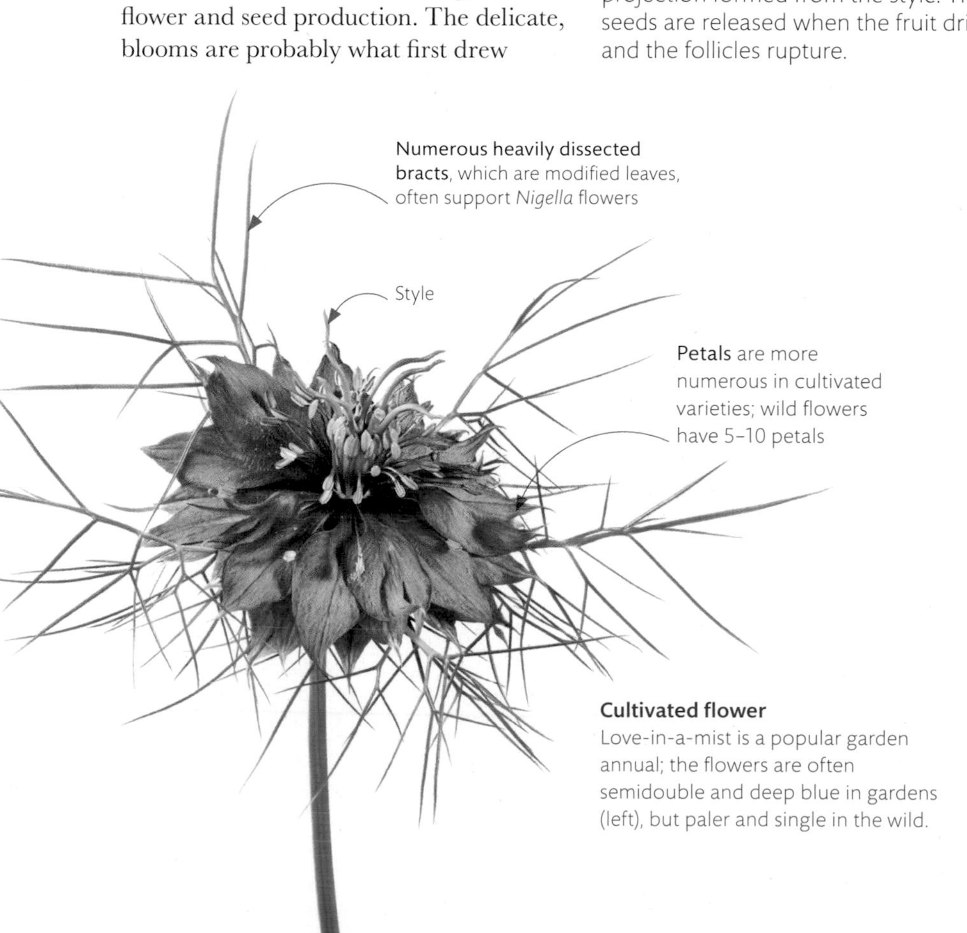

Cultivated flower
Love-in-a-mist is a popular garden annual; the flowers are often semidouble and deep blue in gardens (left), but paler and single in the wild.

Heat treatment
The strange-looking, woody seedheads of some *Banksia* species retain seeds for years, and jettison them only after natural fires or artificial heat treatment makes their follicles snap open. Here the liplike follicles have opened, releasing seeds.

seeds and fire

Ripe seeds usually separate spontaneously from parent plants. Some species, however, particularly those in harsh environments, only release their seed after an extreme environmental event. Fire is a common trigger for conifer forests, but it is also necessary for many other trees and shrubs, such as banksia in Australia. Although wildfires often kill younger trees, the heat makes their conelike fruits open and drop their seeds, months or even years after they have matured.

AIDED BY FIRE

Banksia seeds need fire to trigger their release, but they also thrive in the environment that fire provides. Fire clears ground, eliminating competition from other plants, and fallen seeds sink easily into the soft ash left behind, which helps to protect them from the harsh light.

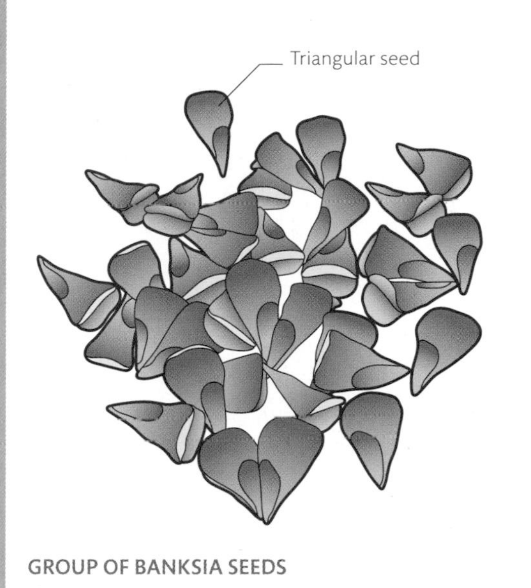

GROUP OF BANKSIA SEEDS

Dense fibers cover the hard inner core of the seedhead

The mass of fibers deters predatory insects after seeds

The outer fibers burn away from the core in a fire

Unopened follicles may indicate that no seed has set

Woody protection
Some banksia seedheads appear to be surrounded by woody "wool," the dried remains of reproductive parts still attached to the seedhead. This woolly barrier helps protect the seeds inside from birds and insects.

The pedicel bends from the weight of the seedhead

Weighty release
Each flower of the water-dwelling sacred lotus (*Nelumbo nucifera*) produces an aggregate fruit, which matures into an extraordinary chambered seedhead that is roughly 3–5 in (7–12 cm) in diameter. As the numerous seeds mature, the seedhead shrinks. Eventually its stem bends under the weight, and the seeds drop into the water.

SEEDS DESIGNED TO FLOAT

Air pockets between the thick, hairy fibers (coir) of a coconut keep it afloat in water. This husk is sandwiched between a protective outer skin (exocarp) and a hard inner shell (endocarp). The coconut "flesh" is food storage tissue (endosperm) that feeds the seed when it germinates, and the "milk" keeps it hydrated. Even after 3,000 miles (4,800 km) in seawater, a coconut can still produce seedlings.

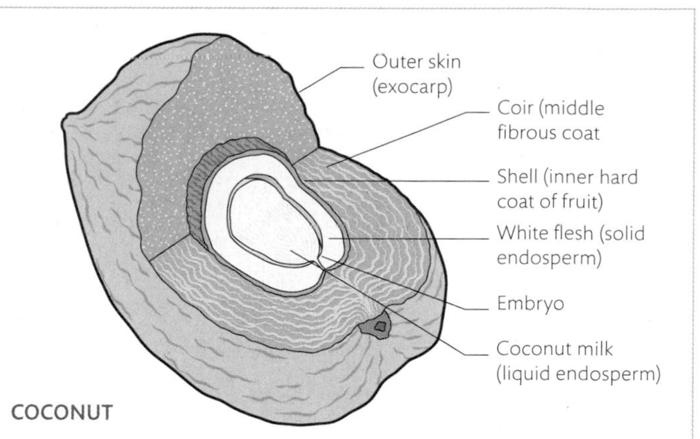

COCONUT

Each seed, or nutlet, is around ⅓ in (1 cm) in diameter

The seed chamber enlarges as the drying seedhead shrinks

water dispersal

Water transports many seeds—a process known as hydrochory. Obviously, wetland plants such as lotuses cast their seed into ponds, rivers, and streams, but the seeds of species such as harebells and silver birch may also be carried in this way. These short freshwater cruises pale in comparison, though, to the epic ocean voyages of tropical seeds such as coconuts.

Protective coating
Seeds need robust seed coats if they are to survive prolonged contact with water. Lotus seed coats are rock-hard, which makes them almost impermeable to water and helps to prevent the seeds from degrading.

The chambered seedhead changes color and turns brown as it dries out

Peanut-sized lotus seeds have germinated when more than 1,000 years old

Exotic fruit
This chromolithograph shows the parts of the papaya plant (*Carica papaya*). It was published in Berthe Hoola van Nooten's *Fleurs, Fruits et Feuillages de l'île de Java* (1863–1864).

plants in art

painting the world

The golden age of botany in the 18th and 19th centuries was characterized by the exploits of explorers and illustrators engaged in a global quest for specimens. Women were largely excluded from scientific discovery, but some intrepid individuals managed to set off on expeditions nonetheless, in search of new plants that they recorded in exquisite paintings.

Nutmeg foliage, flowers, and fruit
North reveals the growth habit of a common culinary spice in this painting made during her stay in the foothills of Jamaica's Blue Mountains during her first major expedition in 1871–1872. The flowers, foliage, and fruit of the nutmeg (*Myristica fragrans*) are shown with a hummingbird (*Mellisuga minima*) and butterfly (*Papilio polydamas*).

Marianne North (1830–1890) was a remarkable Victorian biologist and artist who, in 1871 at the age of 41, began to travel the world, recording flora in her paintings. Her gallery of 832 landscapes and plant, bird, and animal portraits at Kew Gardens in London offered the Victorian public an insight into the natural habitats of exotic specimens before the advent of color photography. North's family connections and wealth made it possible for her to spend 13 years traveling across all continents, but the courage and energy were all her own.

Around the same time, Dutch-born Berthe Hoola van Nooten (1817–1892) was left penniless in Batavia (Jakarta) after her husband's death, and survived by selling her paintings of Java's flora as chromolithographs. Her publication, *The Fruits and Flowers of Java*, was supported by the Queen of the Netherlands.

A century later, Margaret Mee (1909–1988) began 30 years of study and painting in the Amazon rainforest. She recorded several new specimens, some of which were named after her, and painted plants against the rain forest backdrop.

> “I had long dreamed of going to some tropical country to paint its peculiar vegetation on the spot in natural abundant luxuriance.”
>
> MARIANNE NORTH, *RECOLLECTIONS OF A HAPPY LIFE*, 1892

Natural setting
The emerging fruits of the towering ackee (*Blighia sapida*) were painted by North in a natural setting in Jamaica. This West African plant was brought to Jamaica by Captain William Bligh and is named after him.

Bulbils grow in the junction between leaf and rachis

New plantlet forming from bulblet, with leaves partially unrolled

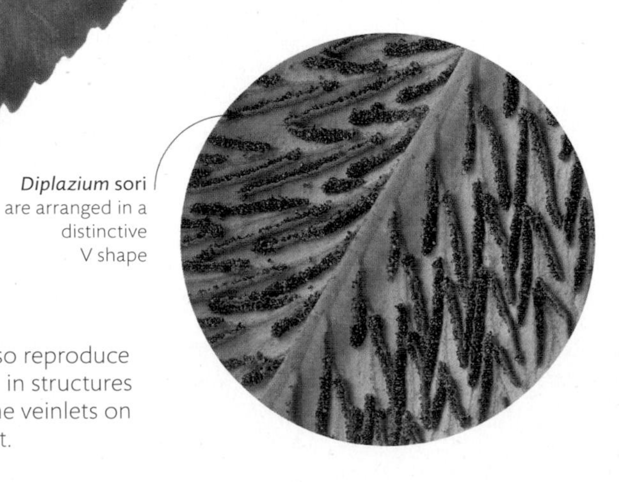

Diplazium sori are arranged in a distinctive V shape

Backup plan
Diplazium proliferum can also reproduce from spores. These are held in structures called sori, situated along the veinlets on the underside of each leaflet.

natural clones

Some plants have evolved more than one way of creating new generations. All ferns, for example, reproduce by spores, but many also make clones of themselves by means of bulblets—tiny secondary bulbs that grow where leaves meet the main stem, or rachis, of a frond. The bulblets produce new plants when they fall off the parent fern or form roots where a frond droops and touches the soil. These new plants are clones of the parent.

Unfurling young frond, or fiddlehead, of plantlet

Mother ferns
Ferns that produce bulblets are referred to as "mother ferns." The African species *Diplazium proliferum* is renowned for producing plantlets along the length of its yard-long fronds, which makes it easy to propagate. As the plantlets develop, the leaves of the main stem wither and turn brown.

The rachis, or main stem

CLONAL COLONIES

Cloning is a successful strategy, but some species never sever the ties between parent and clones. The Pando aspen grove in Utah contains some 47,000 genetically identical trees, and all share a root system that is around 80,000 years old. This not only makes Pando a clonal colony—in effect, a single plant—but also one of the oldest living organisms in the world.

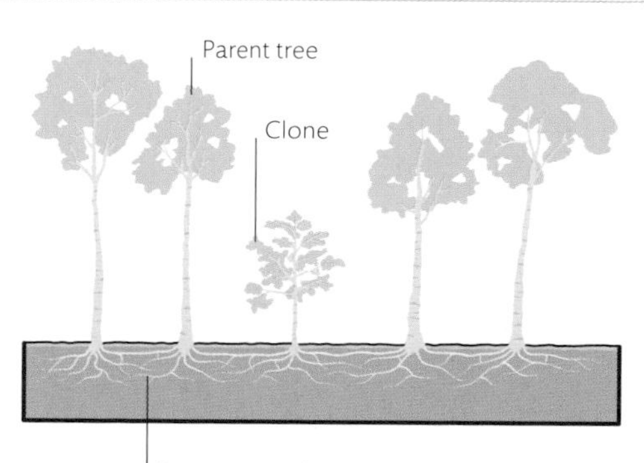

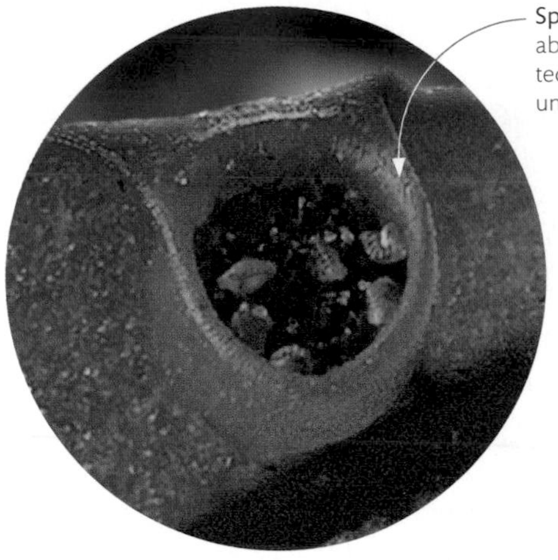

Sporangia are visible from above, although they are technically formed on the underside of the leaf

Cups of spores
The sporangia of *Lecanopteris carnosa* are formed in deep "cups" along the underside of the leaf. The cups fold back onto the upper surface of the leaf.

Each sorus is approximately 1/10 in (2.5 mm) across, and contains many sporangia

spore to **fern**

A wind-dispersed spore that lands in a suitable habitat will germinate into a tiny plantlet. Unlike a seed, the plantlet that grows from the spore—a gametophyte—has only one set of chromosomes. It makes gametes (sperm and eggs) that combine and grow into the more complex plant (sporophyte), which has two sets of chromosomes and is recognizable as a fern.

Pinnate leaves form sori on the tips of many of their lobes

ALTERNATION OF GENERATIONS

Fern gametophytes produce sperm, which travel through the water in their damp habitats to reach and fertilize the egg cell. The sporophyte develops from the fertilized egg (zygote). This alternation of generations, between gametophyte and sporophyte, occurs in flowering plants, too, but flowering plant gametophytes (embryo sac and pollen grains) are very small and dependent on the sporophyte.

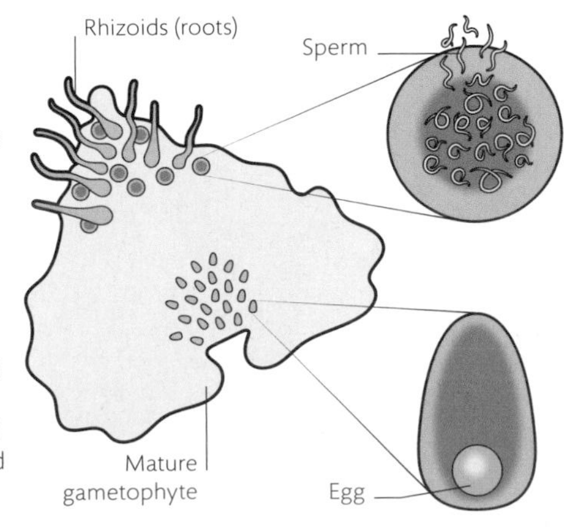

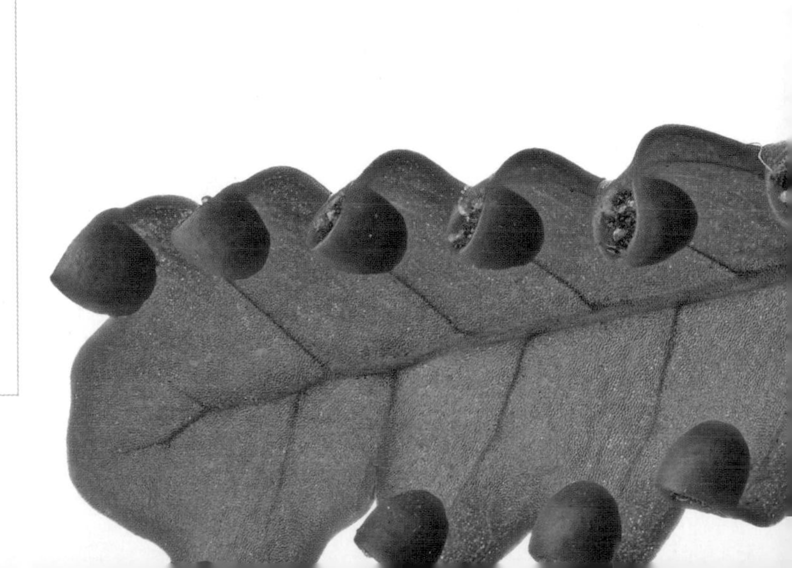

The sporangia are held horizontally above the ground

The leaves reach up to 3 ft (90 cm) in length

Life in the canopy

The ant fern (*Lecanopteris carnosa*) grows on rainforest trees in Indonesia. It produces groups of four spores (tetrads) connected by filaments. These act like a parachute, helping to spread the spores to new trees by wind. Once germinated, the gametophytes are more likely to mate with each other than self-fertilize, improving the genetic diversity of the resulting sporophytes.

Spores may also be dispersed by ants

Tangled tetrads of spores will be dispersed by the wind

The cups holding sporangia do not usually photosynthesize

Pinnae (leaflets) near the tip of the frond form sori; those nearer the base are often sterile

Polytrichum sp.

haircap mosses

Seemingly insignificant and yet highly successful, tiny mosses have been around since at least the Permian Period, some 300 million years ago, and they can still be found growing on every continent today. Some of the most commonly encountered are the haircap mosses of the *Polytrichum* genus.

Haircap mosses have a two-phase life cycle known as alternation of generations. The gametophyte phase is the green, leafy part of the moss. The sporophyte phase is the hairlike growth that can sometimes be seen spouting from the moss stem—hence the common name of haircap. At the tip of the hair is a capsule holding spores. These two phases are genetically distinct. The gametophyte produces the sex cells (sperm and eggs) and has only one set of chromosomes. Once the gametophyte is fertilized, the spore-bearing sporophyte, which has two sets of chromosomes, grows out of the top of the moss. When dispersed spores land in a favorable spot, they grow into new gametophytes.

Mosses evolved long before plants developed vascular tissues for carrying water and nutrients, and most mosses must remain in contact with water for a portion of their lives to survive. However, thanks to a form of convergent evolution, haircap mosses possess rudimentary vascular tissues, allowing them to grow taller and survive longer periods of dehydration. As a result, haircap mosses have conquered many habitats that are too dry for most of their relatives.

Being so hardy, *Polytrichum* mosses play a key role in ecosystem regeneration, and are often the first to colonize barren soil. Patches of haircap moss prevent erosion, and their ability to retain water and reduce temperatures makes them good sites for other plants to germinate.

Hairlike sporophytes
Depending on humidity, the capsules open and close to release their spores when conditions are favorable for dispersal. The sporophyte is entirely dependent on the photosynthetic gametophyte for water and nutrients.

Haircap moss leaves trap water between their cells, helping them hold onto water in dry environments

Green gametophytes
Individual stems are either male or female. Females produce eggs and the males produce sperm, which swim through moisture in the moss clump to fertilize the female plants.

plant families

family. a taxonomic category, or group, of related plant genera.

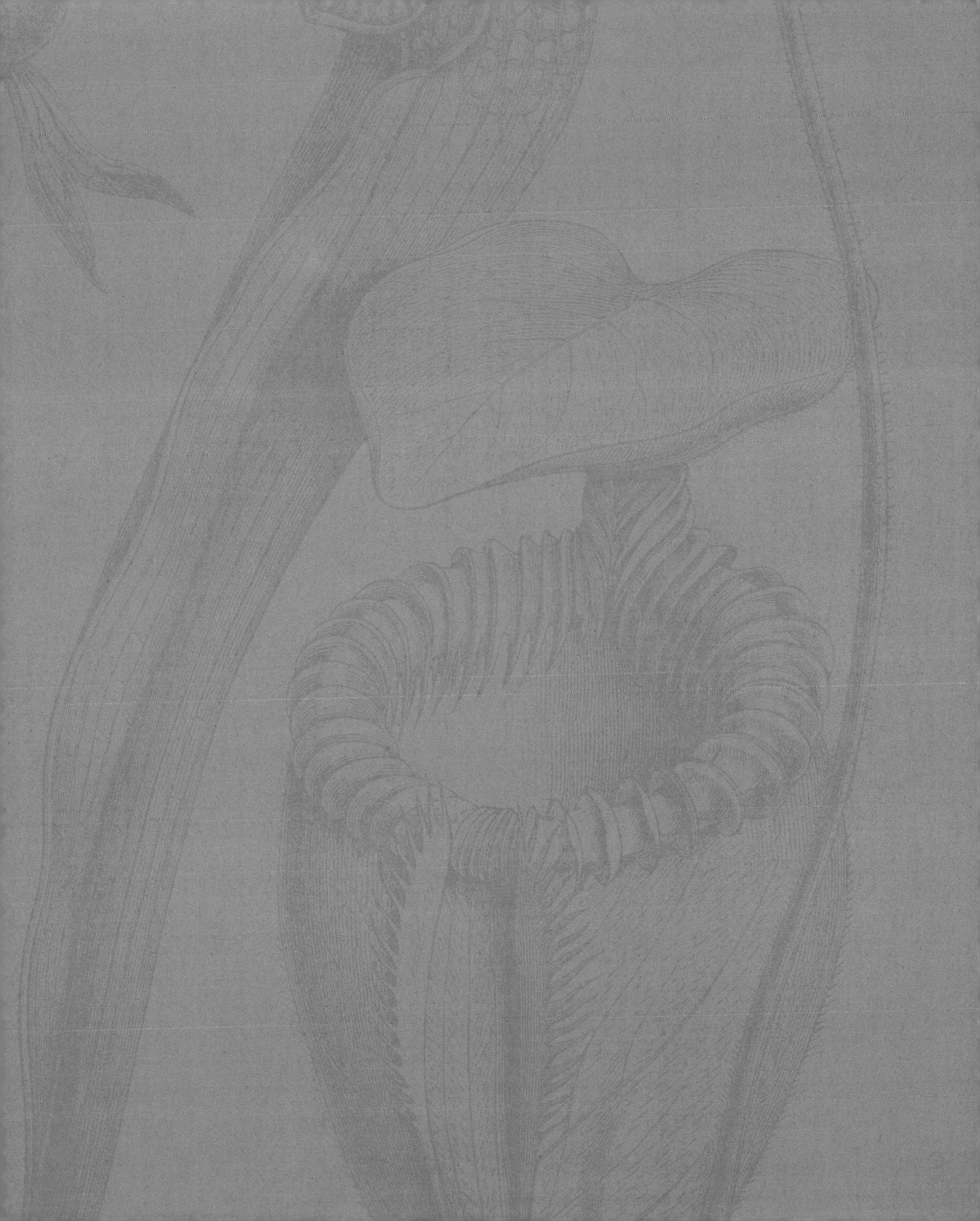

catalog of plant families

Plants within a family share certain characteristics, but can diverge widely in other respects. For instance, a family may contain trees and perennial plants that have a similar flower structure or leaf arrangement. It is, however, often only at the molecular level that close relationships are revealed.

Many plant families are widespread and are found all over the world. The fossil record tells how plants emerged and were carried across the globe as continents broke up and realigned, evolving and adapting to variations in the climate. The genus *Magnolia* has representatives in Asia and the Americas, but the two groups are very different in appearance and behavior despite their close kinship.

This catalog gives an overview of more than 70 significant plant families in the angiosperm, gymnosperm, pteridophyte, and bryophyte plant groups, arranged alphabetically within their orders. Each is illustrated with a botanical drawing of a typical genus, and the accompanying photographs give an indication of the diversity of plants within each family.

Shared characteristics
Each family is informally referred to by the common name of a familiar plant that broadly represents its characteristics. Ranunculaceae, for example, are known as the buttercup family and many plants within it bear cup-shaped flowers. In the onion family (Amaryllidaceae), budding flower clusters are often protected with a spathelike bract.

LEEK (*ALLIUM AMPELOPRASUM*)

calla-lily family

FAMILY Araceae

ORDER Alismatales

This family has virtually worldwide distribution due to the inclusion of the duckweed subfamily (Lemnoideae), which are extremely common water plants. The greatest diversity is found in tropical regions. All members of the family contain toxic calcium oxalate crystals, and edible species, such as the widely grown taro (*Colocasia esculenta*), need to be cooked carefully to remove them. Many produce their small flowers in a spike surrounded by a spathe, which is often a different color than the leaves. They are followed by sometimes fleshy berries. Fruits of the popular houseplant Swiss cheese plant (*Monstera deliciosa*) have a sweet flavor.

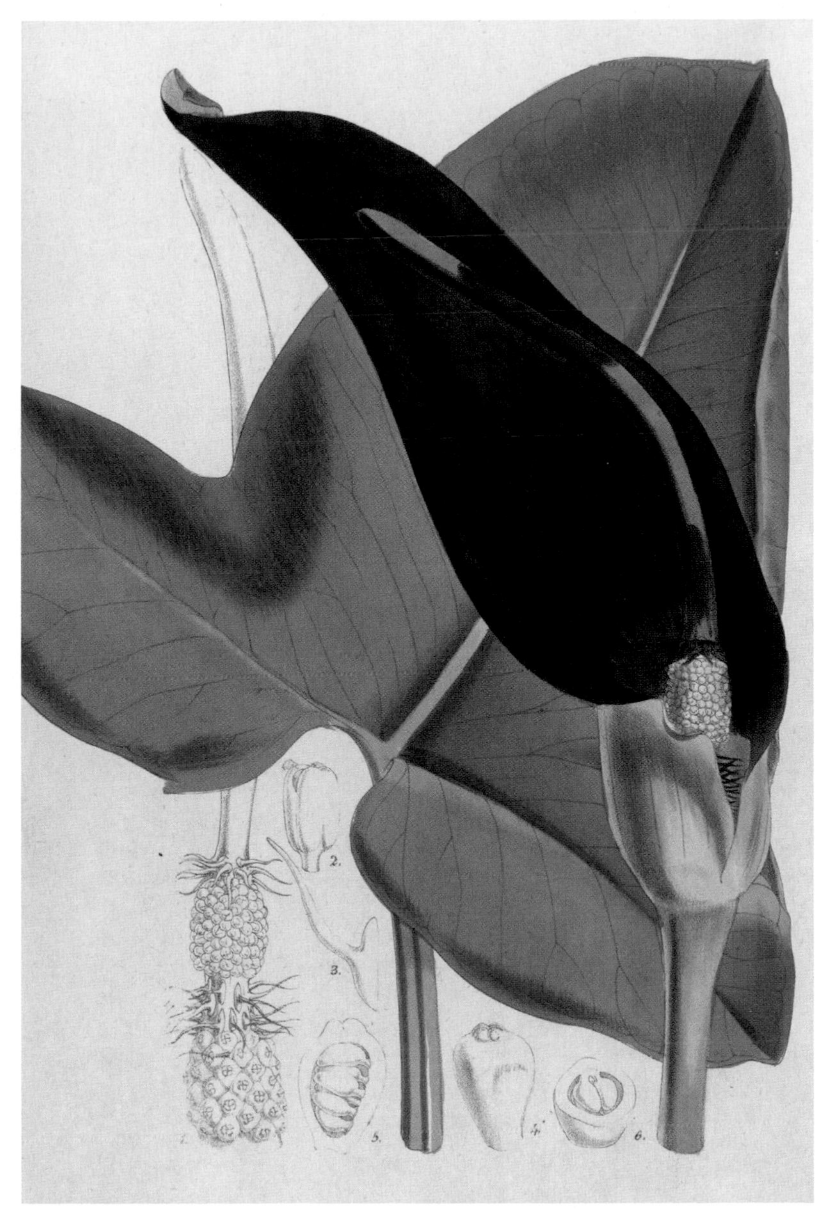

Arum palaestinum
Drawing by W.H. Fitch, *Curtis's Botanical Magazine*, 1865

Amorphophallus titanum

Anthurium andraeanum

Zantedeschia aethiopica

Aglaonema sp.

carrot family

FAMILY Apiaceae (Umbelliferae)

ORDER Apiales

Comprised of mainly perennials and rarely trees or shrubs, the carrot family is distributed virtually worldwide. The leaves are usually compound, and the plants produce flowers in umbels. This family includes many well-known vegetables, herbs, spices, and ornamental plants. Carrots (*Daucus carota* subsp. *sativus*) and parsnips (*Pastinaca sativa*) are important commercial crops, but many others have edible tubers (often starchy), leaves, and seeds, and have been important food crops. Celery (*Apium graveolens* var. *dulce*) is grown for its stems. Sea holly (*Eryngium maritimum*), common in coastal districts in Europe and naturalized in the US, was once believed to be an aphrodisiac.

Sanicula europaea (above, left); *Apium graveolens* (right and below, left)
Drawing by F. Losch, *Kräuterbuch, unsere Heilpflanzen in Wort und Bild,* 1905

Petroselinum crispum var. *neapolitanum*

Anthriscus sylvestris

Anethum graveolens

Eryngium giganteum

holly family

FAMILY Aquifoliaceae

ORDER Aquifoliales

This family is found throughout tropical regions, but has several important species in temperate zones. Consisting of the single genus *Ilex*, it contains often unisexual trees and shrubs, both evergreen and deciduous, that have simple leaves with sometimes toothed or spiny margins. Fruits are red, brown, or black berries, each containing several hard seeds. Fruiting stems of holly (*Ilex* spp.) are often used in floral arrangements and in festive wreaths. Several hollies are widely grown as garden plants, especially for dense, prickly hedges, while in some areas young leaves have been harvested and dried for brewing tea.

Ilex latifolia
Drawing by W.H. Fitch, *Curtis's Botanical Magazine*, 1866

Ilex mucronata

Ilex paraguariensis

Ilex aquifolium

Areca catechu
Drawing from *Köhler's Medizinal-Pflanzen*, 1890

palm family

FAMILY Arecaceae

ORDER Arecales

Found throughout the tropics, with some also occurring in warm-temperate regions, palms are trees or shrubs, and some climbers, with usually unbranching woody stems. Leaves usually spiral out from stem tips. Some genera, such as *Calamus*, are spiny. Palms are among the most useful of all plant families, providing food as well as materials for making shelters and tools. The coconut palm (*Cocos nucifera*) is one of the world's major crops, producing edible coconuts, oils for use in food production and the cosmetics industry, and, as a by-product, fibers (coir) for matting and rope-making. Fruits of the date palm (*Phoenix dactylifera*) have a high sugar content.

Cocos nucifera

Trachycarpus fortunei

Phoenix dactylifera

onion family

FAMILY Amaryllidaceae

ORDER Asparagales

This large family, distributed worldwide, is comprised of bulbs and perennials with fleshy roots. Flowers, usually formed of six tepals, are often tubular and can be carried singly or, more usually, in clusters, which are sometimes geometrically precise. Of its three subfamilies, Allioideae, which includes the edible and ornamental onions (*Allium*) and their relatives, was formerly a family in its own right—Alliaceae. Many of these have been cultivated for thousands of years. Amaryllidoideae, which can be toxic, includes such popular plants as *Clivia*, *Narcissus*, and *Nerine*, familiar both in gardens and as houseplants. Agapanthoideae consists of a single genus, *Agapanthus*, native to South Africa.

Hippeastrum pardinum (syn. *Amaryllis pardina*)
Drawing by W.H. Fitch, *Curtis's Botanical Magazine*, 1867

Narcissus pseudonarcissus hybrid

Clivia miniata

Amaryllis belladonna

Allium 'Purple Sensation'

asparagus family

FAMILY Asparagaceae

ORDER Asparagales

This is now a very diverse family of mainly perennial herbs, many bulbous or succulent, with a few trees, shrubs, or climbers. It is found in nearly all parts of the world. Its seven subfamilies have at times been classified as separate families. They include treelike yuccas, whose early ancestors have been found fossilized; agaves that form rosettes of fleshy, pointed leaves at or near ground level; and the familiar hyacinth. Some are familiar foodstuffs, including the spring vegetable asparagus and camas (*Camassia*), traditionally eaten by native North Americans. Popular garden plants include *Aspidistra*, lily of the valley (*Convallaria*), and *Hosta*. Sisal (*Agave sisalana*) is economically valuable for its fibers.

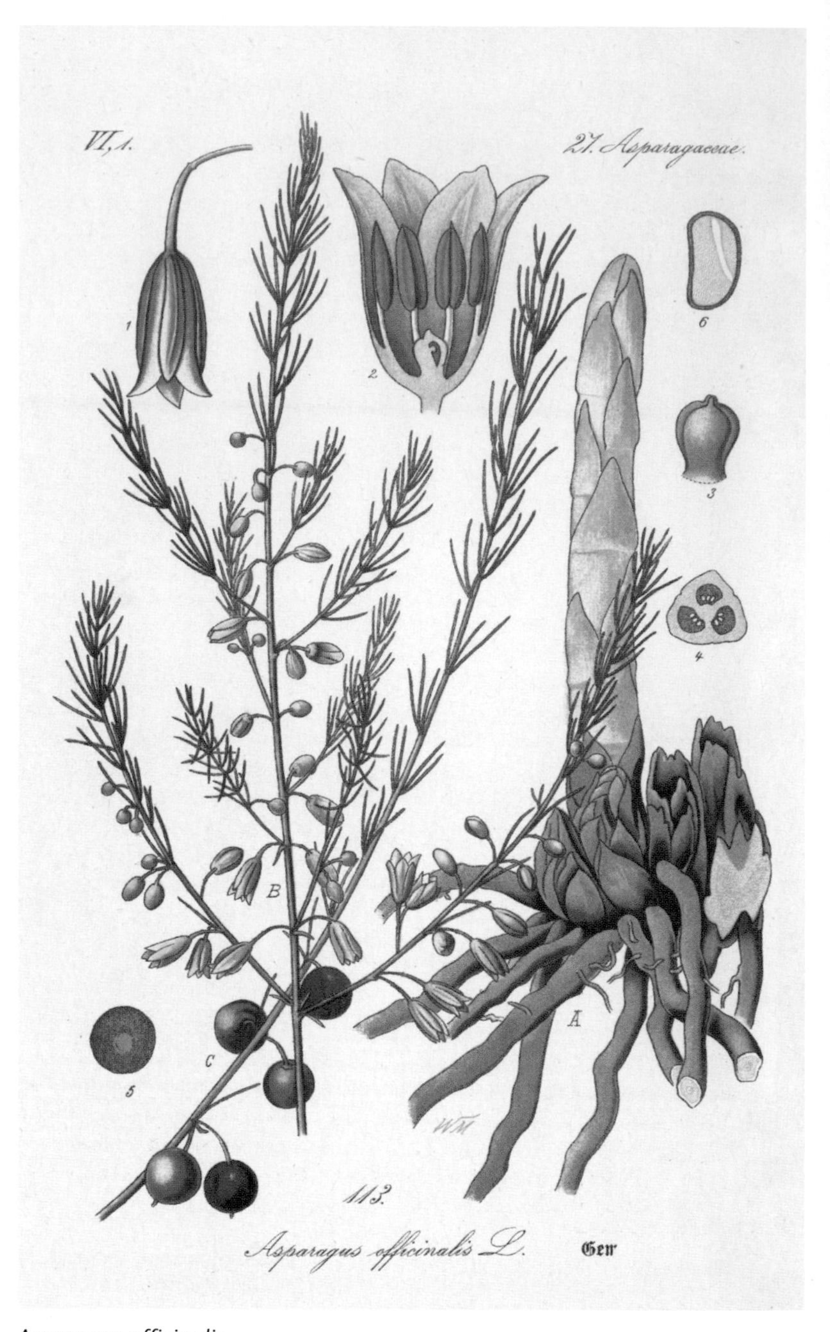

Asparagus officinalis
Drawing by O.W. Thomé, *Flora von Deutschland, Österreich und der Schweiz*, 1885

Hosta ventricosa

Agave parryi

Cordyline australis

Hyacinthus orientalis

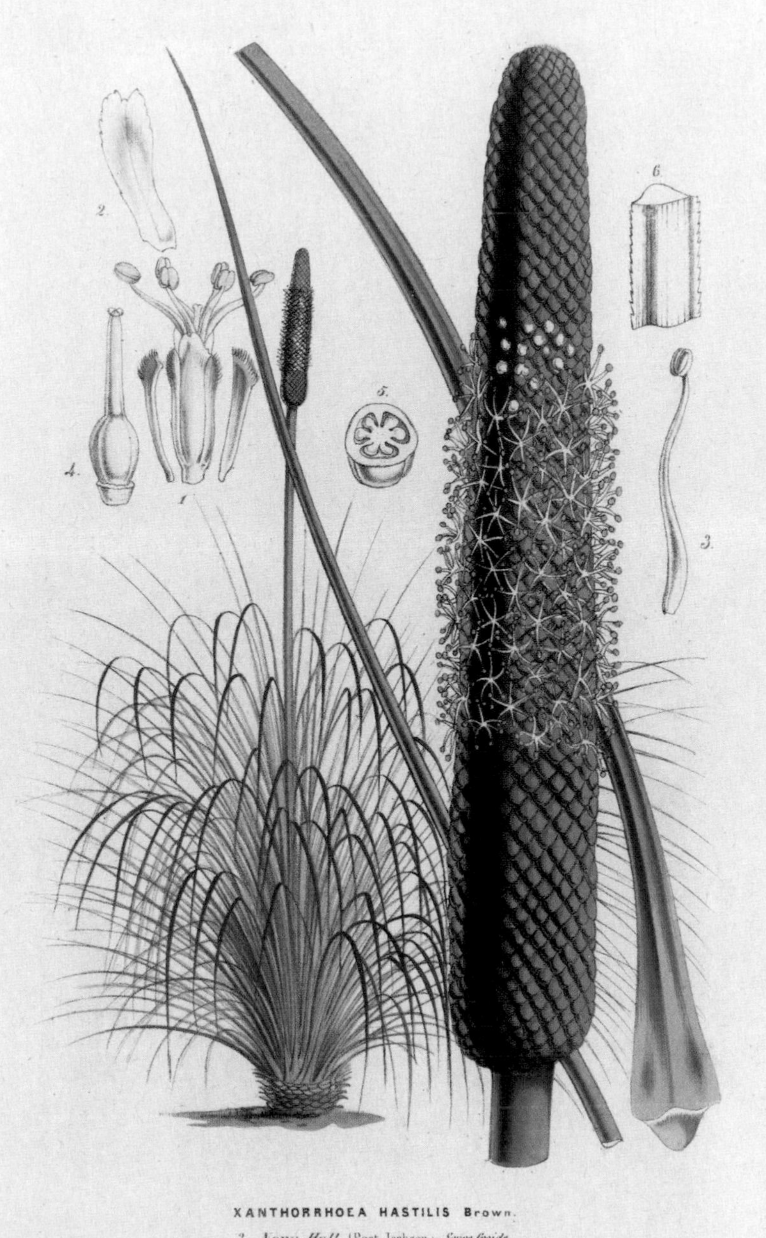

Xanthorrhoea resinosa (syn. *X. hastilis*)
Engraving by L. van Houtte, *Flore des serres et des jardins de l'Europe*, 1853

daylily family

FAMILY **Asphodelaceae**

ORDER Asparagales

Comprised of three subfamilies (Asphodeloideae, Xanthorrhoeoideae, and Hemerocallidoideae), which some authorities would prefer to separate, this diverse family contains perennials, shrubs, climbers, and trees. Flowers, which are sometimes symmetrical in only one plane, are made up of six tepals that may be fused at the base. The succulent *Aloe vera* is widely used for its sap, both in cosmetics and in medicine. Daylily (*Hemerocallis*) is one of several popular garden plants, with a huge number of hybrids, many developed in the US. Some have scented flowers. New Zealand flax (*Phormium tenax*) was formerly grown commercially for its fibers but is now more familiar as an ornamental plant.

Haworthia tessellata

Aloe vera

Kniphofia sp.

Hemerocallis dumortieri

iris family

FAMILY **Iridaceae**

ORDER Asparagales

The iris family may have its origins in the southernmost part of the globe, from where it has spread virtually worldwide. Molecular evidence has led to a reallocation of several genera within the family. *Acidanthera*, formerly a separate genus, now belongs to *Gladiolus*. Plants are mainly perennials, generally with rhizomes, bulbs, or corms, and a few are shrubby. Many are popular as ornamentals, with *Iris*, *Freesia*, and *Gladiolus* being particularly important in the cut-flower industry. *Crocus sativus* is grown commercially for its stigmas that produce the spice saffron. Starch derived from orris root (*Iris pallida*) was used medicinally and as a fixative for perfumes.

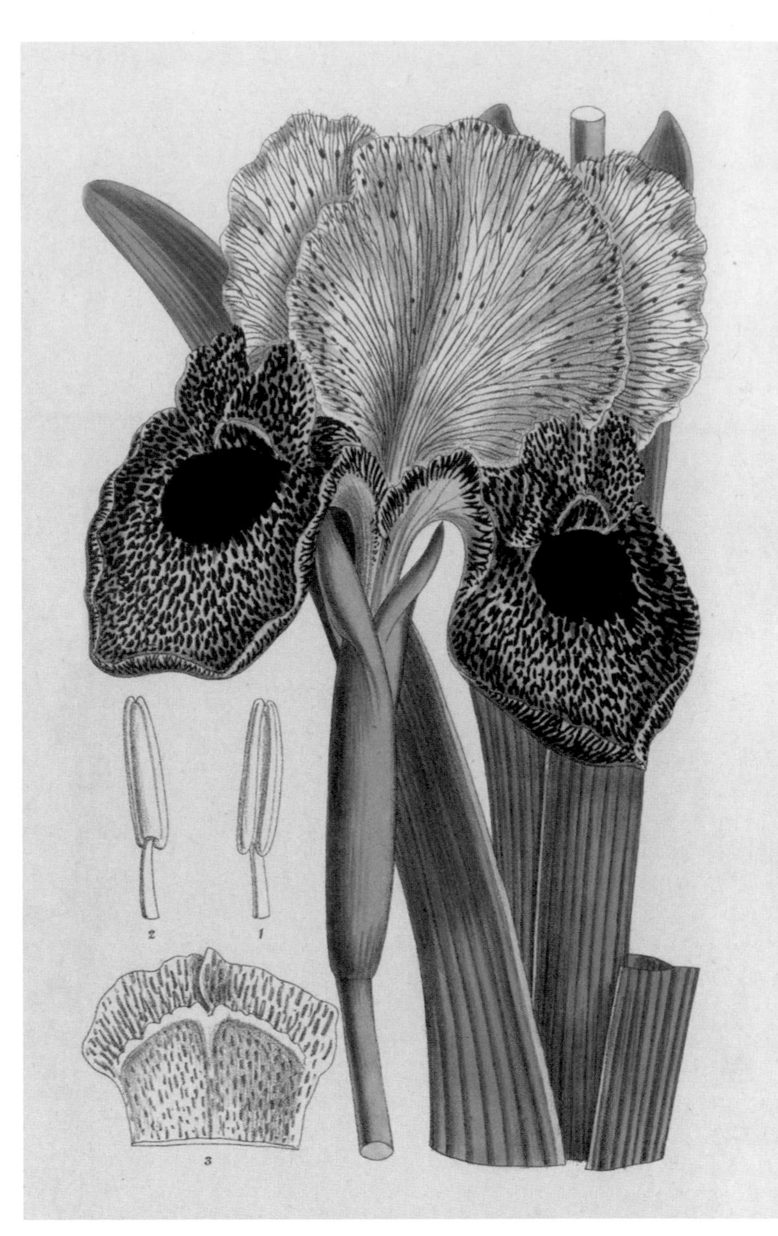

Iris bismarckiana
Drawing by M. Smith, *Curtis's Botanical Magazine*, 1904

Iris sibirica

Crocus goulimyi

Gladiolus murielae

orchid family

FAMILY **Orchidaceae**

ORDER Asparagales

Orchids form the largest family of vascular plants. They are found worldwide but are most diverse in tropical regions. Typically perennials, they are terrestrial or epiphytic, with a few climbers, and have stems that can be swollen into pseudobulbs. Flowers, often large and showy, are comprised of three sepals and three unequal petals. Bold markings guide pollinators toward a landing platform or pouch. A complicated pollination mechanism means that cross-pollination is common, and many artificial genera have been produced for the houseplant industry alongside the popular *Phalaenopsis* and *Cymbidium*. Vanilla flavoring is derived from species of *Vanilla*. The seeds of almost all orchids contain no food, and rely on symbiosis with fungi for early growth.

Orchis mascula
Engraving from J. Kops et al., *Flora Batava*, 1885

Anacamptis pyramidalis

Aranthera 'Anne Black'

Cymbidium aloifolium

Phalaenopsis stuartiana

daisy family

FAMILY Asteraceae

ORDER Asterales

Easily recognizable by their flower structure, members of the daisy family are found in all environments nearly worldwide, but are particularly abundant in desert regions. Each "flower" is in fact a composite head, made up of either tiny disk flowers or petallike ray flowers, or, in many cases, both. Asteraceae is one of the largest plant families, including shrubs, climbers, and trees, as well as the large body of annuals and perennials, such as *Aster* and *Chrysanthemum*, that are common garden plants. Edible members include lettuce (*Lactuca sativa*) and globe artichokes (*Cynara cardunculus* var. *scolymus*). Wormwood (*Artemisia absinthium*) was used to flavor the alcoholic drink absinthe.

Dieteria canescens (syn. *Diplopappus incanus*)
Drawing by S.A. Drake, *Edwards's Botanical Register*, 1835

Cynara cardunculus

Helianthus annuus

Aster alpinus

Taraxacum officinale

Campanula pallida (syn. *C. colorata*)
Drawing by W.H. Fitch, *Curtis's Botanical Magazine*, 1851

bellflower family

FAMILY Campanulaceae

ORDER Asterales

While its distribution is global, most members of this family of both woody and soft-stemmed plants appear in temperate areas. Flowers are nearly always cup-shaped, with the (usually) five petals fused, but in some cases, petals can be almost separate and the flower can open virtually flat. Stems usually contain a milky white sap. Most species of *Campanula* (Latin for "little bell") produce edible leaves, although the genus is most widely grown as a garden plant in cool climates. *Lobelia* is also popular for summer bedding. Many cultivated forms of both genera have blue flowers; a few other genera have edible roots or fruits.

Phyteuma orbiculare

Codonopsis clematidea

Platycodon grandiflorus

Adenophora remotiflora

forget-me-not family

FAMILY Boraginaceae

ORDER Boraginales

The order Boraginales comprises this single family of annuals, perennials, trees, shrubs, and climbers that occur almost globally. One desert species, *Pholisma sonorae*, from California is a parasite with edible tubers. Plants often have rough, hairy leaves and flowers usually have five petals. Many contain alkaloids and have been used widely in traditional medicine. Some have edible fruits or roots. Poultices for treating wounds were often made from the common garden weed comfrey (*Symphytum officinale*). Borage flowers (*Borago officinalis*) have a cucumberlike taste and are used to garnish drinks. The plant is grown commercially for its seed oil, which is used in herbal medicine. *Alkanna tinctoria* (alkanet) is one of several plants in this family that is used as a dye.

Borago officinalis
Drawing by A. Masclef, *Atlas des plantes de France*, 1893

Cordia sebestena

Myosotis arvensis

Alkanna tinctoria

Symphytum officinale

cabbage family

FAMILY Brassicaceae

ORDER Brassicales

Found nearly worldwide, though absent from some tropical regions, the cabbage family is comprised of mainly soft-stemmed and shrubby plants, and a few small trees and climbers. They produce pungent mustard oils (glucosinolates) and many are familiar foodstuffs and flavorings. Flowers are comprised of two pairs of petals arranged in a cross. Many of the edible species have been developed as commercial crops. There are many *Brassica oleracea* cultivars, including kale, cabbage, and broccoli. Other species have edible roots (including swedes and turnips), while oilseed rape (*B. napus* and *B. rapa*) is grown to provide animal feed, vegetable oil, and biofuel. Alongside several popular ornamentals, *Isatis tinctoria* was grown to make the blue dye woad.

Brassica alba
Engraving by J.J. Haid from J.W. Weinmann, *Phytanthoza Iconographia*, 1735–1745

Matthiola fruticulosa

Brassica napus

Brassica oleracea

spinach family

FAMILY Amaranthaceae

ORDER Caryophyllales

Occurring worldwide, but showing most diversity in warm-temperate and subtropical regions, members of the spinach family are mainly soft-stemmed plants and shrubs. Flowers are produced in cymes, dense heads, or spikes. The most important species economically is beet (*Beta vulgaris* subsp. *vulgaris*), with sugar beet (*Beta vulgaris* var. *altissima*) also important as an alternative to sugarcane. South American quinoa (*Chenopodium quinoa*) has become an important health food, while samphire (*Salicornia* species) is collected from salt marshes and marketed as a luxury vegetable. The family also includes spinach and other plants grown for their iron-rich leaves. Forms of *Amaranthus* and *Celosia* are popular ornamentals.

Amaranthus blitum
Engraving from J. Kops et al., *Flora Batava*, 1846

Gomphrena globosa

Amaranthus caudatus

Spinacia oleracea

cactus family

FAMILY Cactaceae

ORDER Caryophyllales

Disocactus flagelliformis (syn. *Cactus flagelliformis*)
Drawing by G.D. Ehret, *Plantae selectae*, 1752

Natives of the Americas, with the exception of *Rhipsalis baccifera*, which is found in tropical Africa and South Asia, cacti are mainly trees, shrubs, and climbers of distinctive appearance. Some are epiphytes. Their succulent stems, sometimes clumping, are globular or columnar, and erect or scrambling. Stems can be segmented, with the segments sometimes flattened. While they can be smooth, they are often ribbed and have spines that can be short or long, and needlelike or fine, like hairs. Flowers are often large and striking. Some, such as the prickly pear (*Opuntia ficus-indica*), produce edible fruits. Others, such as *Lophophora williamsii* (peyote), are the source of hallucinogens. Many are popular houseplants. In some warm regions, they have escaped cultivation and have become naturalized.

Opuntia microdasys

Carnegiea gigantea

Echinopsis candicans

Mammillaria senilis

carnation family

FAMILY Caryophyllaceae

ORDER Caryophyllales

Largely consisting of annuals and perennials, this family is one of the most widely distributed, showing greatest diversity in temperate zones. Some can survive harsh conditions in Antarctica and high altitudes in the Himalayas. Flowers usually have five petals. The carnation, *Dianthus caryophyllus*, has long been cultivated for its scent and to make soap. Hybridizing between this species and others has produced many plants suitable for gardens and the cut-flower industry. *Silene viscaria* is one of several plants with sticky leaves that allow it to trap insects that might otherwise cause it harm. It is possible, although it has not been demonstrated, that the plant is able to derive nutrients from the insects.

Dianthus juniperinus (syn. *Dianthus arboreus*)
Drawing by F. Bauer, from J. Sibthorp and J.E. Smith, *Flora Graeca*, 1825

Saponaria ocymoides

Stellaria holostea

Silene alpestris

thrift family

FAMILY Plumbaginaceae
ORDER Caryophyllales

This family is found mainly in the coastal districts of Europe and North America, although it occurs throughout the world. Closely related to the Polygonaceae, these perennials, shrubs, and climbers are divided into two larger subfamilies, with the mangrove *Aegialitis* the sole occupant of a third. Flowers have five petals, usually with a central nectar disk, and leaves are arranged spirally. Some are grown as ornamentals. *Ceratostigma* is common in gardens, while species of *Plumbago* are popular garden shrubs in frost-free areas. Statice (*Limonium sinuatum* and other species) comes in many colors and the flowers retain their color well when cut and dried.

Ceratostigma plumbaginoides
Drawing from *Addisonia*, 1920

Goniolimon tataricum

Plumbago auriculata

Armeria maritima

Limonium spectabile

Polygonum vacciniifolium
Engraving by L. van Houtte, *Flore des serres et des jardins de l'Europe*, 1853

knotweed family

FAMILY Polygonaceae

ORDER Caryophyllales

Although spread virtually worldwide, most members of this family are found in temperate regions in the northern hemisphere. Plants include both soft- and woody-stemmed species. The (usually) bisexual flowers are small and carried in spikes, racemes, panicles, or heads. One of its three subfamilies, Symmerioideae, is comprised of a single species, *Symmeria paniculata*, from West Africa. The family includes several food crops and ornamentals. Buckwheat (*Fagopyrum esculentum*) is an important commercial crop that has been grown since antiquity. Species of *Rheum* include both rhubarb—whose stems are cooked with sugar to make them palatable—and ornamental plants. Japanese knotweed (*Fallopia japonica*) is invasive and considered a pernicious weed in many areas.

Rheum rhabarbarum

Persicaria bistorta

Rumex sanguineus

dogwood family

FAMILY Cornaceae

ORDER Cornales

Found mostly in temperate regions of the northern hemisphere, but also occurring in parts of the tropics and the southern hemisphere, the dogwood family is comprised of evergreen and deciduous woody plants and rhizomatous perennials. Molecular analysis has reduced what was once a larger family to two genera, *Cornus* and its close relative *Alangium*. Usually they have opposite leaves. Flowers, produced in heads, sometimes have colorful bracts and can be followed by edible fruits. The fruit of the Cornelian cherry (*Cornus mas*) has a higher vitamin C content than an orange. Several species are popular garden trees and, because of their high tannin content, are also of value in medicine. Regular coppicing or pollarding can provide whippy stems for basket making.

Cornus capitata (syn. *Benthamia fragifera*)
Drawing by W.H. Fitch, *Curtis's Botanical Magazine*, 1852

Cornus alba

Cornus mas

Cornus florida

Begonia beddomei
Drawing by M. Smith, *Curtis's Botanical Magazine*, 1884

begonia family

FAMILY Begoniaceae
ORDER Cucurbitales

Widespread throughout the tropics, with a few species occurring in temperate zones, this family is comprised of two genera, *Begonia*, and the much smaller *Hillebrandia*, found only in Hawaii. Mainly perennial and succulent, occasionally epiphytic, but with a few tending to be more shrublike, plants sometimes form tubers. Flowers are unisexual and both sexes can appear within the same inflorescence. While some species have edible leaves, begonias are grown mostly as ornamentals, and most plants sold commercially are complex hybrids. Semperflorens begonias, with brightly colored flowers, are used for bedding. Tuberous begonias have larger flowers, and males are often double and very showy. Rex begonias are grown for their attractively marked leaves, often as houseplants.

Begonia 'Escargot'

Begonia imperialis

Begonia rex

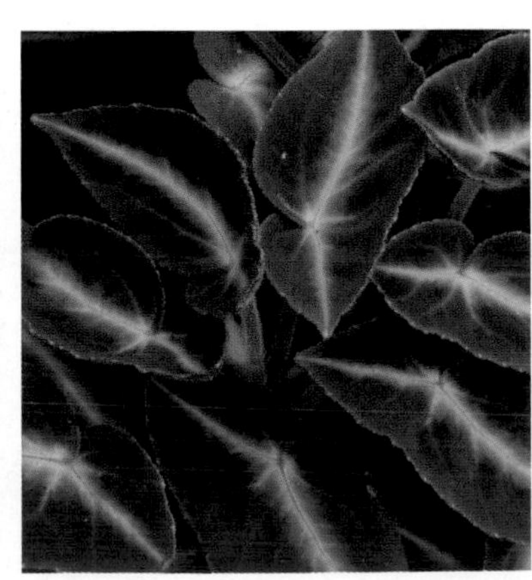

Begonia listada

cucumber family

FAMILY Cucurbitaceae

ORDER Cucurbitales

Mainly comprised of climbers, both soft-stemmed and woody, this family is found most often in tropical and subtropical areas. The fruits are usually berries containing one to many flattened seeds. Molecular analysis suggests an Asian origin. Many of these plants are of economic value, not only for their edible fruits but for the oil that can be extracted from the seeds and used in cooking and salad dressings. Cucumbers (*Cucumis sativus*) have been cultivated since antiquity and are eaten either raw or pickled. Squashes (*Cucurbita* species) are cooked as a vegetable. Their thick skins can be used as bowls, flasks, and even musical instruments. Loofah sponges are made from *Luffa cylindrica*.

Cucurbita digitata
Drawing from *Revue horticole*, 1863

Momordica cochinchinensis

Citrullus lanatus

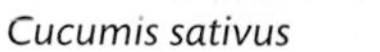

Cucumis sativus

Lagenaria siceraria

elder family

FAMILY Adoxaceae

ORDER Dipsacales

Formerly exclusively consisting of perennials, this family, distributed mainly across the northern hemisphere, has been expanded on the basis of molecular evidence to include some shrubs. They have opposite leaves, and flowers are followed by fleshy fruits. Shrubby elder (*Sambucus*) and *Viburnum* species are valuable garden plants, often with fragrant flowers and edible, though tart, fruits. The sweetly scented flowers of *Sambucus nigra* are often used to flavor cordials. The shrub itself was thought to have magical properties: the scent of the leaves repels flies and was used to ward off evil. Legend has it that the biblical figure Judas hanged himself from an elder and from then on the fruit had a bitter taste.

Adoxa moschatellina
Engraving from G.C. Oeder et al., *Flora Danica*, 1761–1883

Viburnum tinus

Viburnum plicatum

Viburnum opulus

Sambucus nigra

heather family

FAMILY Ericaceae

ORDER Ericales

Found virtually worldwide, this diverse family is comprised of woody and soft-stemmed plants, both evergreen and deciduous. Many have an association with soil-borne mycorrhizal fungi. Flowers are usually bell- or cup-shaped, and are sometimes followed by fleshy berries. Heathers (*Calluna* and *Erica*) often colonize tracts of ground, functioning as a food source for grazing mammals and pollinating insects. Shoots of *Calluna* were used to make brooms and dyes, as well as to brew beer. Some genera bear edible fruits, including blueberries and cranberries (forms of *Vaccinium*), which are important commercial crops. Many are popular garden plants, especially *Rhododendron*, of which a vast number of hybrids have been developed.

Erica chamissonis
Drawing by W.H. Fitch, *Curtis's Botanical Magazine*, 1874

Erica carnea

Rhododendron catawbiense

Cyanococcus sp.

Kalmia latifolia

Polemonium caeruleum
Engraving by C.A.M. Lindman, *Bilder ur Nordens Flora*, 1922–1926

jacob's-ladder family

FAMILY Polemoniaceae

ORDER Ericales

Mainly comprised of annuals and perennials, with a few woody species, this family is found in the northern hemisphere, extending southward in the Americas down the Andes toward the tip of South America. Flowers are often saucer- or funnel-shaped, usually with five petals. Native Americans used several genera as soap and for their supposed medicinal benefits. Qantu (*Cantua buxifolia*) was sacred to the Incas and is the national flower of Peru. In gardens, *Phlox* is particularly important. In addition to the vast number of border plants derived from *Phlox paniculata*, many dwarf species, including *P. subulata*, are popular with alpine garden enthusiasts. *Cobaea scandens* (cup-and-saucer vine) is often grown as an annual climber.

Cobaea scandens

Cantua buxifolia

Phlox divaricata

primrose family

FAMILY Primulaceae

ORDER Ericales

Absent from the world's major deserts, this family is otherwise cosmopolitan, comprised of both soft-stemmed and woody plants including climbers and mangroves. Leaves are usually arranged spirally and flowers generally have five petals, often fused into a bell- or urn-shaped tube. Many genera contain species that are important in horticulture, particularly *Cyclamen* and *Primula*, extensive breeding of which has resulted in a range of plants for the garden and potted plant culture. A few plants have medicinal uses: false black pepper (*Embelia ribes*) is used to combat flatulence and tapeworm, while the fruits of *Maesa* are bactericidal and used to treat intestinal parasites. The Maya used *Jacquinia* flowers to decorate temples.

Primula capitata
Drawing by M. Smith, *Curtis's Botanical Magazine*, 1887

Lysimachia punctata

Soldanella alpina

Cyclamen persicum

Primula vulgaris

birch family

FAMILY Betulaceae

ORDER Fagales

Spread across the northern hemisphere, with some found in mountainous tropical regions, this family is comprised of deciduous trees and shrubs. They have alternate leaves, usually with toothed margins. Male flowers are carried in pendent catkins; females are carried in short catkins that can be erect. Birch (*Betula* species) are common forest trees and many are also valued as ornamentals, not only in gardens but as street trees and in urban landscaping. The wood of the birch and alder (*Alnus*) is often used to make furniture. Hornbeam (*Carpinus betulus*) is popular for hedging. Hazelnuts (*Corylus avellana* and *C. colurna*) and filberts (*C. maxima*) are important crops.

Betula lenta
Drawing from *Köhler's Medizinal-Pflanzen*, 1890

Ostrya virginiana

Corylus sp.

Betula papyrifera

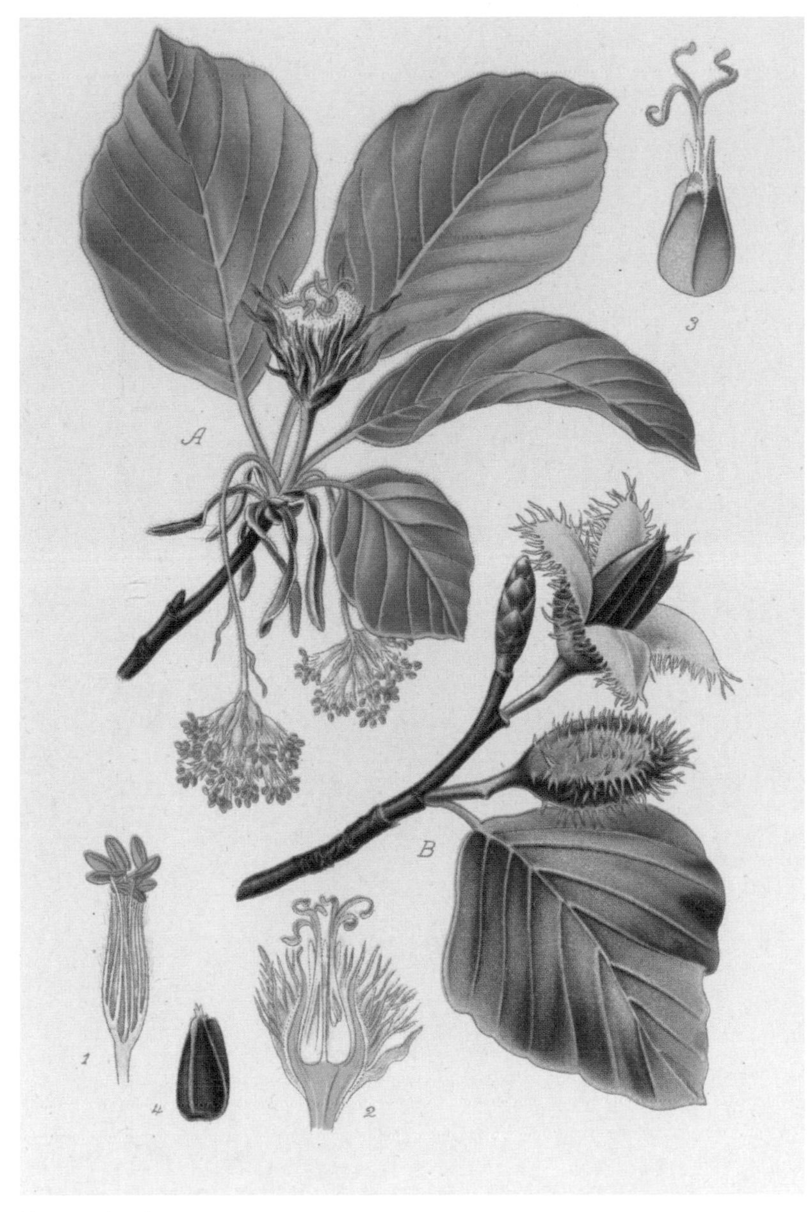

Fagus sylvatica
Drawing by A. Masclef, *Atlas des plantes de France*, 1893

beech family

FAMILY Fagaceae
ORDER Fagales

Mainly comprised of deciduous or evergreen trees, with a few shrubs, this family is found in the temperate northern hemisphere and occasionally in the tropics. The leaves are usually alternate and often covered with hairs. Fruits are nuts attached to hardened bracts that are either cuplike, as in acorns, or enclose the fruit, as in chestnuts. The sweet chestnut (*Castanea sativa*) was introduced into Northern Europe by the Romans and is often found in forests, sometimes forming an association with edible truffles. Other *Castanea* species also have edible fruits. Oak acorns (from *Quercus* species) have served as a famine food and as pig fodder. Oak and beech (*Fagus*) are important lumber trees.

Quercus robur

Fagus sylvatica

Lithocarpus dealbatus

Castanea sativa

walnut family

FAMILY Juglandaceae

ORDER Fagales

Deciduous and evergreen trees and shrubs make up this family, which is distributed across the northern hemisphere and parts of the southern hemisphere. The leaves are usually alternate and pinnate, with leaflets that are often aromatic. The flowers are generally carried in pendent panicles, followed by often edible nuts. *Juglans regia* has long been cultivated for its fruits (walnuts) and the oil that can be extracted from them. The husks also yield a dye. *Carya illinoinensis*, the state tree of Texas, produces pecans, which have similar uses. Many species are grown for their wood, which has an attractive grain and is extremely durable.

Juglans regia
Drawing from *La Belgique horticole, journal des jardins et des vergers*, 1853

Juglans sp.

Carya ovata

Pterocarya fraxinifolia

periwinkle family

FAMILY Apocynaceae

ORDER Gentianales

This diverse family, widespread from the Arctic to the tropics, is comprised of both woody and soft-stemmed plants, including some succulents that look like cacti. Cut stems often produce a clear, latexlike sap that can be used to make rubber. Flowers have five petals and are often scented, sometimes unpleasantly. Several species have edible fruits, although the seeds can be poisonous, while others have edible roots or tubers. Many contain alkaloids that have medicinal applications, although others are extremely toxic. Java indigo (*Marsdenia tinctoria*) is one of several used as a source of blue dye. The frangipani (*Plumeria*) is a popular ornamental shrub or large tree in the tropics.

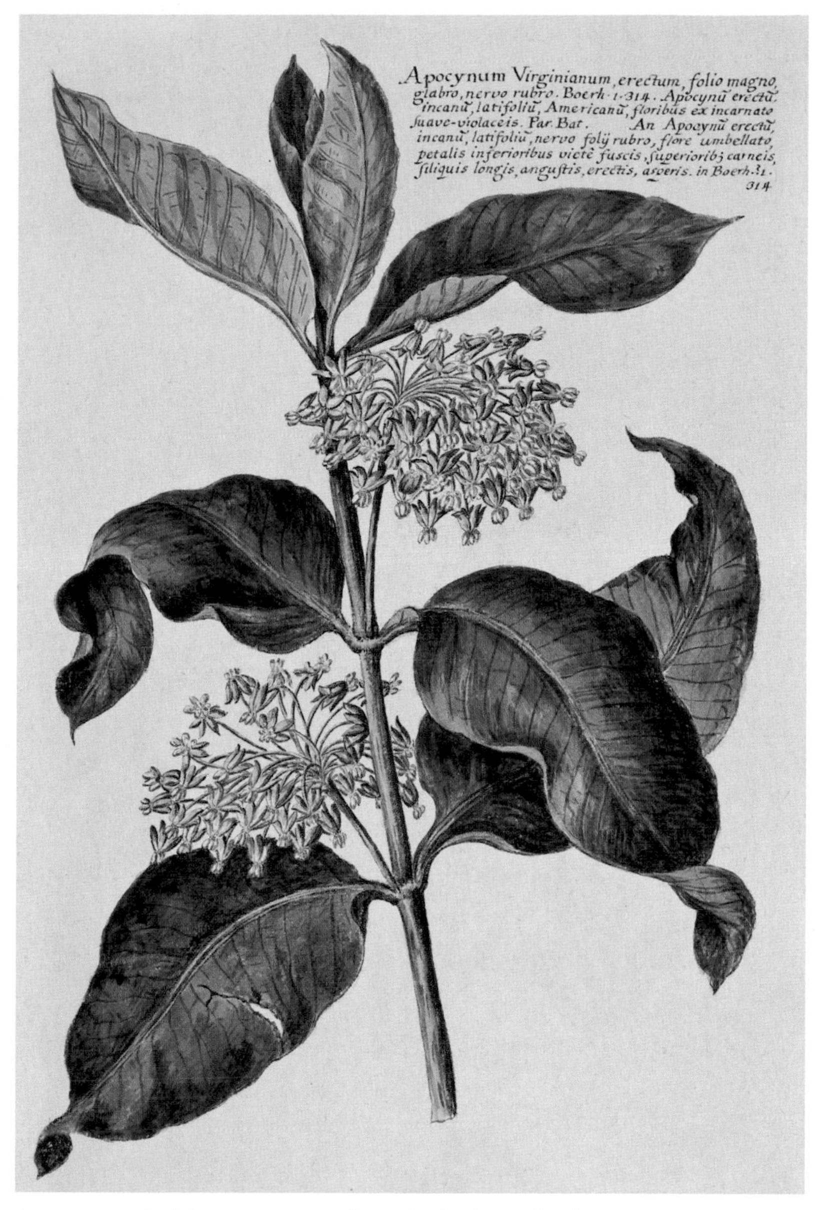

Apocynum virginianum erectum (now *Asclepias syriaca*)
Drawing from G.B. Morandi, *Hortulus Botanicus Pictus Sive Collectio Plantarum*, 1748

Stapelia gigantea

Hoya carnosa

Plumeria obtusa

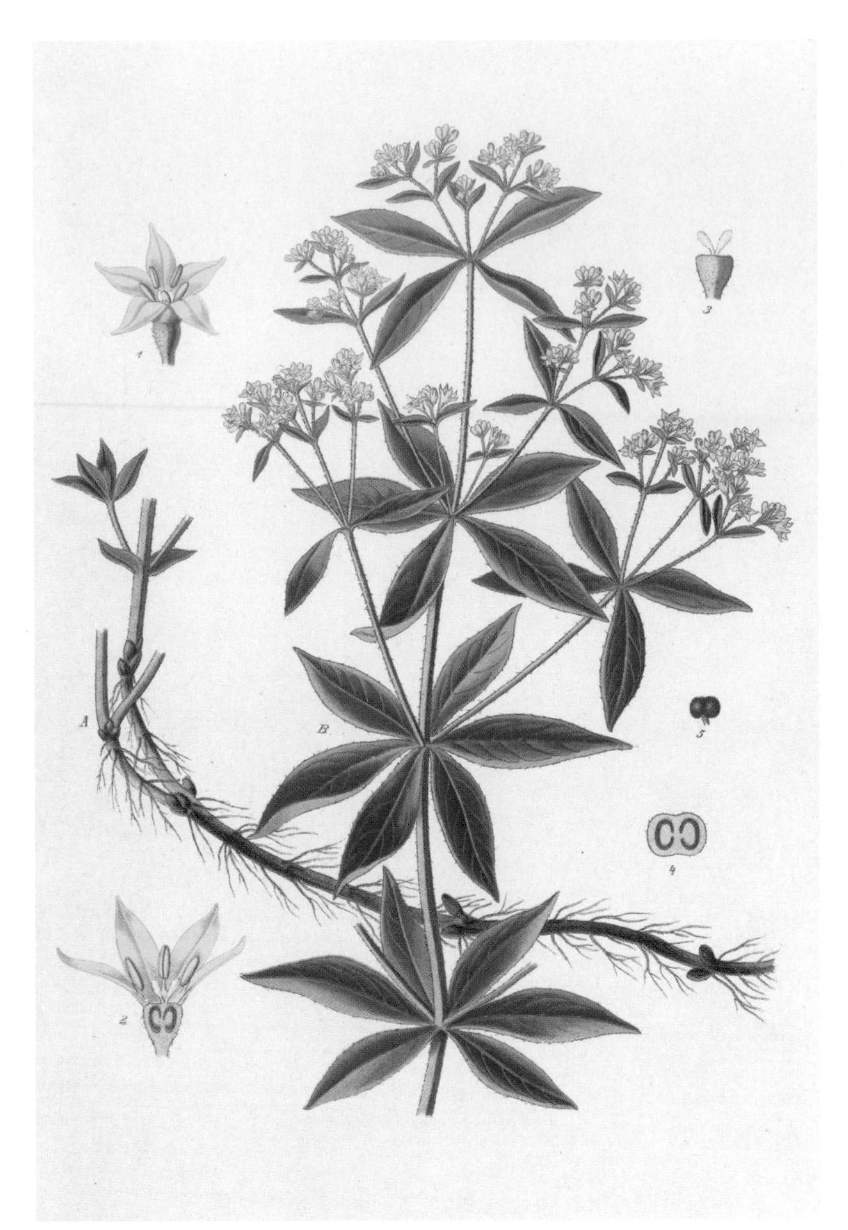

Rubia tinctorum
Drawing from *Köhler's Medizinal-Pflanzen*, 1890

coffee family

FAMILY Rubiaceae
ORDER Gentianales

This family of both woody and soft-stemmed plants is absent only in polar regions and permanently dry deserts, and is particularly diverse in tropical rainforests. They have simple leaves that are usually opposite or apparently whorled. Bracts beneath the flowers are sometimes large and brightly colored. Coffee (*Coffea*) is one of the world's major commodities. Various species are cultivated for their seeds (beans), depending on the local climate and altitudes. Quinine is found in the bark of several species of *Cinchona* and is used medicinally and to flavor Indian tonic water. *Gardenia augusta* is a popular potted plant, with fragrant white flowers.

Coffea arabica

Morinda citrifolia

Ixora coccinea

Gardenia jasminoides

crane's-bill family

FAMILY Geraniaceae

ORDER Geraniales

Five genera comprise this family of annuals, perennials, and shrubs, found mainly in temperate areas and at high altitudes in the tropics. Plants are usually hairy and can be fragrant, and some are succulent or cormose. Flowers usually have five petals. Southern African *Pelargonium* shows great diversity and extensive hybridizing among five of the species (*P. inquinans*, *P. zonale*, *P. cucullatum*, *P. grandiflorum*, and *P. peltatum*) has produced thousands of cultivars that are popular potted plants worldwide. Oils extracted from other species, particularly *Pelargonium graveolens*, are used in perfumes and as food flavorings. Species and cultivars of *Geranium* and *Erodium* are common garden plants.

Geranium lucidum (left) and *G. robertianum* (right)
Engraving by C.A.M. Lindman, *Bilder ur Nordens Flora*, 1922–1926

Erodium cicutarium

Pelargonium quercifolium

Geranium sanguineum

bear's-breeches family

FAMILY Acanthaceae

ORDER Lamiales

Mainly perennials, shrubs, and climbers, with a few evergreen trees and mangroves, members of this family are found largely in the tropics, with a few genera in temperate regions. Flowers, which often have large, showy bracts, are made up of five petals forming a two-lipped corolla that is sometimes curved and scalloped. Many are popular in gardens and as houseplants, including the lollipop plant (*Pachystachys lutea*) and black-eyed Susan (*Thunbergia alata*). The lobed leaves of *Acanthus mollis* and *Acanthus spinosus* are a familiar motif in architecture, first used by the ancient Greeks to adorn the capitals of Corinthian columns, then by the Romans. It can now be seen worldwide.

Acanthus spinosus
Painting by H.S. Holtzbecker, *Gottorfer Codex*, 1649–1659

Ruellia tuberosa

Thunbergia grandiflora

Acanthus mollis

Justicia brandegeeana

trumpetvine family

FAMILY Bignoniaceae

ORDER Lamiales

Found throughout the tropics and with some members in warm-temperate areas, this family is comprised mainly of evergreen trees, shrubs, and perennials, as well as climbers that are twining or use leaf tendrils to attach to their host. They usually have opposite leaves. The flowers are made up of five petals fused into a two-lipped corolla with two upper and three lower lobes, creating a characteristic trumpet shape. The fruits are sometimes large and woody. Those of the calabash tree (*Crescentia cujete*) are used to make bowls and musical instruments. Some family members have medicinal qualities, occasionally hallucinogenic or supposedly aphrodisiac. *Catalpa*, *Jacaranda*, and *Campsis* are among the genera grown in gardens for their beautiful flowers.

Bignonia callistegioides (syn. *Clytostoma callistegioides*)
Drawing by M.E. Eaton, *Addisonia*, 1932

Tabebuia rosea

Catalpa bignonioides

Tecoma stans

Jacaranda mimosifolia

gloxinia family

FAMILY Gesneriaceae

ORDER Lamiales

Found in the tropics, with a few examples in temperate zones, this is a family of perennials, shrubs, small trees, and climbers. Sometimes epiphytic or rock-dwelling, they usually have succulent or fleshy leaves and stems covered with dense hairs. Flowers have four or five petals and are often two-lipped, varying in shape. The lower lip sometimes forms an elongated pouch, as in *Calceolaria*. Others have tubular, funnel-, bell-, or cup-shaped flowers, or flowers that open flat. This family includes several plants that are popular in gardens or for indoor display, including slipperworts (*Calceolaria* hybrids), gloxinias (forms of *Sinningia speciosa*), African violets (*Saintpaulia* hybrids), and Cape primroses (*Streptocarpus* hybrids).

Smithiantha zebrina
Drawing by W.H. Fitch, *Curtis's Botanical Magazine*, 1842

Saintpaulia 'Bright Eyes'

Episcia dianthiflora

Microchirita lavandulacea

mint family

FAMILY Lamiaceae

ORDER Lamiales

Distributed worldwide, this is a family of annuals, biennials, and perennials, with a few woody plants. Young stems are often square in cross-section and leaves are usually opposite. Many have glandular hairs with aromatic oils. Flowers have petals fused into a two-lipped corolla, sometimes with large, petal-like bracts. Many are familiar culinary herbs such as mint (*Mentha* hybrids), basil (*Ocimum basilicum*), and oregano (*Origanum vulgare*). The scented oils of lavender (*Lavandula* species) and bergamot (*Monarda*) are used in cosmetics. The genus *Salvia* includes not only the culinary sage (*Salvia officinalis*) but a wide range of garden ornamentals. Teak (*Tectona grandis*) is a tropical tree grown for its wood.

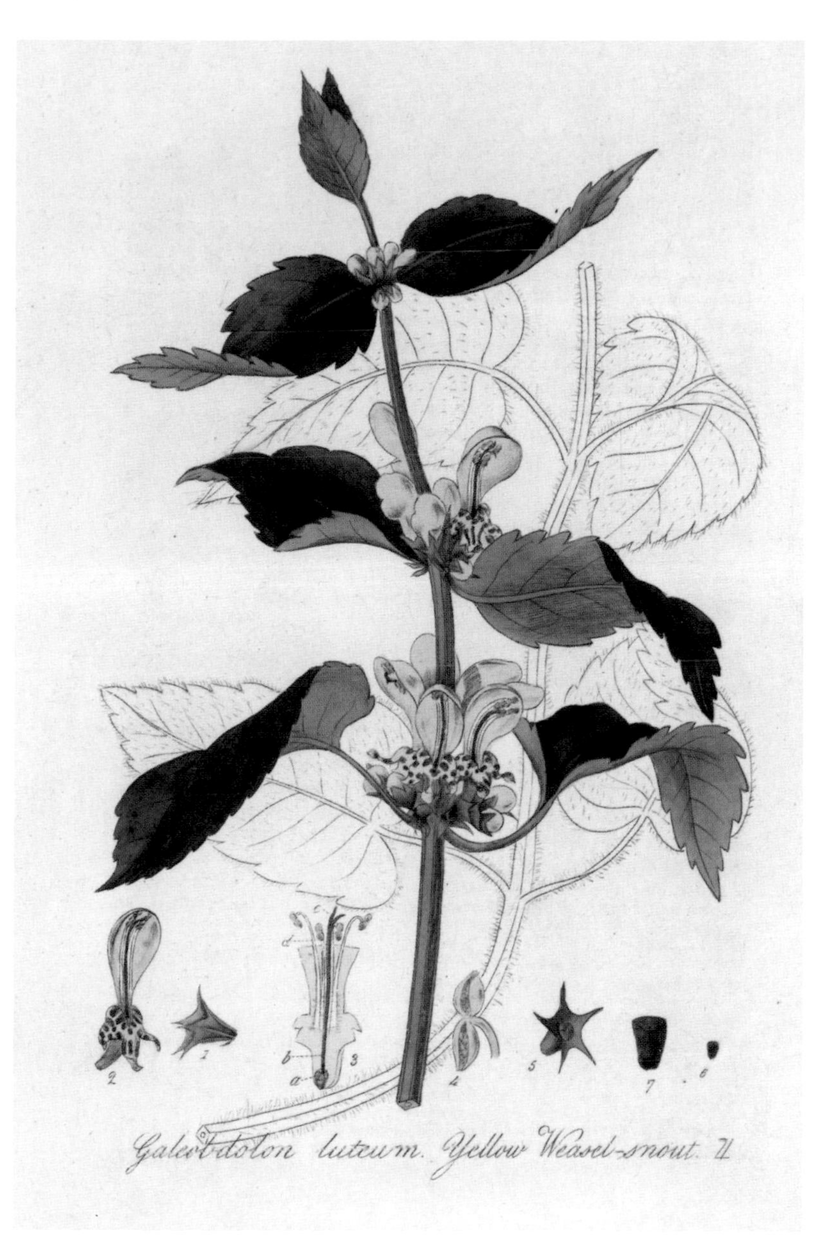

Lamium galeobdolon (syn. *Galeobdolon luteum*)
Drawing from W. Baxter, *British phaenogamous botany*, 1834–1843

Lavandula angustifolia

Mentha suaveolens

Scutellaria baicalensis

Rosmarinus officinalis

olive family

FAMILY Oleaceae

ORDER Lamiales

This family, comprised of woody plants including climbers, is found virtually worldwide. They usually have opposite leaves and flowers made up of four petals, although those of *Jasminum* are sometimes more numerous. Because of their oil-rich fruits, olives (*Olea europea*) are an important crop, and have long been cultivated in temperate areas. Fruits harvested for eating are first cured to reduce their bitterness. A white dove with an olive branch is a universal symbol of peace. *Forsythia* and lilac (*Syringa*) are widely grown in gardens. *Jasminum* species are also popular; those with scented flowers are grown commercially for use in perfumes and to flavor tea. Ash (*Fraxinus excelsior*) is an important lumber tree.

Olea europaea
Drawing by P.J. Redouté, *Traité des arbres et arbustes*, 1812

Jasminum sambac

Syringa reticulata

Osmanthus x burkwoodii

Fraxinus sieboldiana

speedwell family

FAMILY Plantaginaceae

ORDER Lamiales

Found virtually worldwide, including arid regions, this is a variable family of woody and soft-stemmed plants and submerged aquatics. Flowers are cup-, bell-, or funnel-shaped or are two-lipped, with two upper and three lower lobes. Some plants are important in medicine. Foxglove (*Digitalis*), though toxic, is a source of digitalin, which is used to treat heart problems. *Plantago afra* has a laxative effect and can case constipation. Popular ornamentals include snapdragon (*Antirrhinum*), toadflax (*Linaria*), and *Penstemon*, of which there are many hybrids. *Philcoxia*, from Brazil, is carnivorous, with subterranean stems and leaves that sit on or just below the soil surface and are equipped with glands that trap and digest nematodes.

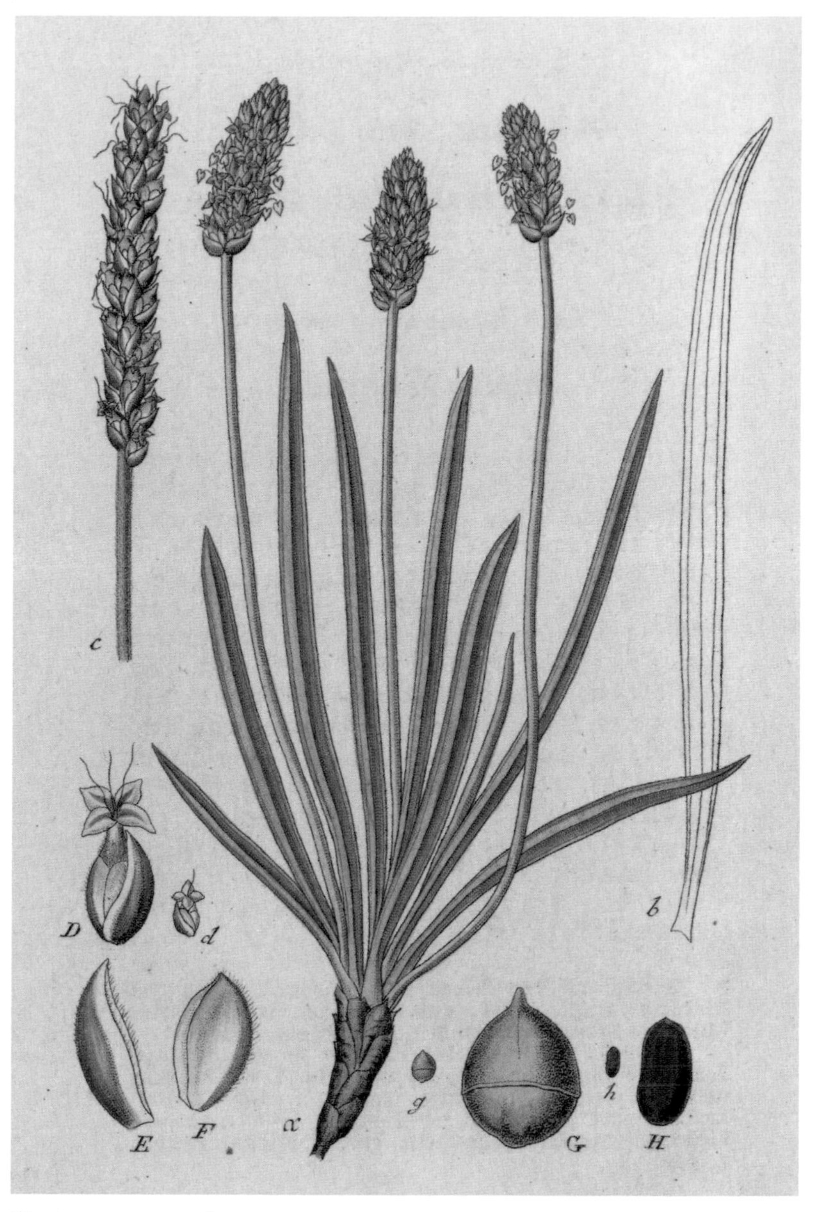

Plantago neumannii
Engraving from J. and J.W. Sturm, *Deutschlands Flora*, 1841–1843

Veronica spicata

Globularia cordifolia

Plantago asiatica

Digitalis obscura

figwort family

FAMILY Scrophulariaceae

ORDER Lamiales

Found worldwide, except in polar regions and some deserts, this family has expanded following DNA sequencing and includes soft-stemmed plants and shrubs—deciduous and evergreen—with only a few trees and climbers. Flowers have usually five petals, fused into a tube-, bell-, or trumpet-shaped or two-lipped corolla. Butterfly bush (*Buddleja*), of the former Buddlejaceae, is one of several important garden plants, others including mullein (*Verbascum*) and Cape figwort (*Phygelius*). In many areas, *Buddleja davidii* escaped cultivation and has become a weed, seeding in walls and cracks in the pavement as well as on waste ground. The African *Buddleja salviifolia* can become treelike and is often planted for its wood.

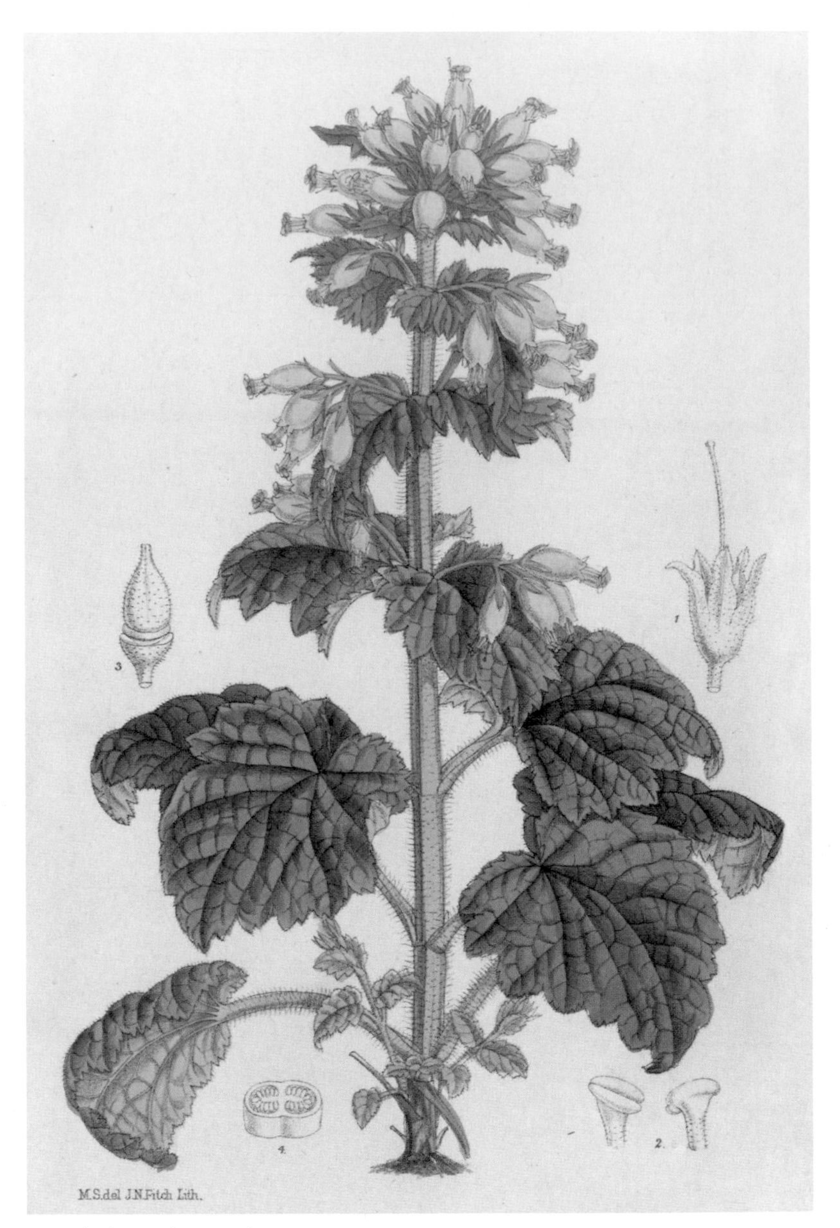

Scrophularia chrysantha
Drawing by M. Smith, *Curtis's Botanical Magazine*, 1882

Phygelius capensis *Buddleja davidii* *Verbascum nigrum* *Nemesia caerulea*

bay-laurel family

FAMILY Lauraceae

ORDER Laurales

Widely distributed, this family consists of trees, shrubs, and *Cassytha*, a genus of parasitic climbers. "Laurisilva forests" are found in certain subtropical regions, including the Canary Islands. Their leaves are simple and flowers are often produced in panicles. One of its most important members is the avocado (*Persea americana*), widely grown in warm regions for its tasty fruits and oil used in cosmetics and soaps. The bark of several *Cinnamomum* species provides the spice cinnamon. Familiar as hedging in the Mediterranean, bay (*Laurus nobilis*) was a symbol of victory in ancient Greece and Rome, where stems would be woven into crowns or wreaths. Leaves are used as a flavoring for casseroles and soaps.

Laurus nobilis
Drawing by W. Müller, *Köhler's Medizinal-Pflanzen*, 1887

Litsea japonica

Persea americana

Cinnamomum camphora

Lilium auratum
Drawing by W.H. Fitch, *Curtis's Botanical Magazine*, 1862

lily family

FAMILY **Liliaceae**

ORDER Liliales

Found mainly in temperate regions of the northern hemisphere, this is a family of perennial plants that form underground bulbs or rhizomes. Flowers, comprising six tepals, are often carried singly and can be marked with blotches, spots, or streaks. *Fritillaria meleagris* is one of several with checkered markings. Bulbs have long been used as a foodstuff, but now the family is of greater importance to the flower bulb industry, particularly in the Netherlands. In addition to *Fritillaria*, *Lilium*, and *Tulipa*, *Cardiocrinum*, *Erythronium*, and *Tricyrtis* are also widely grown. The white flowers of *Lilium candidum* have long symbolized purity and appear in iconography associated with the Virgin Mary and, later, Elizabeth I of England.

Tulipa gesneriana

Erythronium oregonum

Lilium longiflorum

Fritillaria meleagris

tuliptree family

FAMILY Magnoliaceae

ORDER Magnoliales

This family consists of two genera of trees and shrubs, found widely separated in the Americas and in Asia, in both temperate and tropical regions. *Liriodendron* was formerly widespread throughout the northern hemisphere. Flowers are produced singly on stalks, either at the tips of branches or on side shoots, and are most frequently pollinated by beetles. They are followed by conelike fruits. While both *Liriodendron* and *Magnolia* are used for their lumber, most are grown as garden ornamentals, with magnolias valued for their spectacular, often fragrant flowers. Oils from *Magnolia champaca* are used to make perfume. The tulip tree (*Liriodendron tulipifera*) is the state tree of Kentucky, Tennessee, and Indiana.

Magnolia hodgsonii (syn. *Talauma hodgsonii*)
Drawing by W.H. Fitch, *Illustrations of Himalayan plants*, 1855

Liriodendron tulipifera

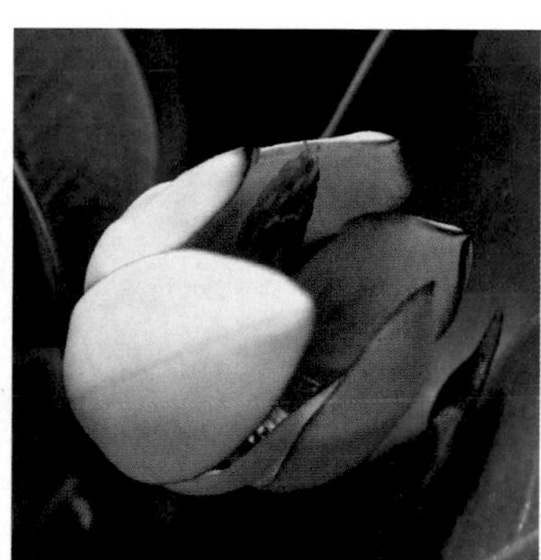

Magnolia figo

Magnolia denudata

spurge family

FAMILY Euphorbiaceae

ORDER Malpighiales

These woody and soft-stemmed plants are found worldwide except for polar regions. Sometimes cactuslike, they have stems that can contain a milky sap. Bracts are often fused into a cuplike structure and resemble flowers, while the true flowers are grouped into a cyathium. Plants provide food, oil, rubber, and lumber. Roots of cassava (*Manihot esculenta*) are an important source of starch in tropical regions. Seeds of sachi inchi (*Plukenetia volubilis*) yield an oil high in omega-3. As a source of rubber, *Hevea brasiliensis* was a major commercial crop in the 19th century. Several *Euphorbia* species are common in gardens, and some succulent types, such as crown of thorns (*E. milii*) and the poinsettia (*E. pulcherrima*), are popular houseplants.

Euphorbia peplus
Engraving from J. Sowerby *et al.*, *English botany, or, Coloured figures of British plants*, 1868

Euphorbia pulcherrima

Manihot esculenta

Acalypha hispida

Codiaeum variegatum

willow family

FAMILY Salicaceae

ORDER Malpighiales

Found in most parts of the world but absent from the Antarctic and some deserts, this family is comprised of trees and shrubs, usually with alternate leaves. Flowers are often produced in catkins. *Salix arctica* is the most northerly distributed vascular plant (alongside *Papaver radicatum*). Some tropical genera have edible fruits. A source of salicin, the bark of white willow (*Salix alba*) was used as a painkiller in Europc, where other species were regularly coppiced or pollarded to provide material for basketry, fences, and fish traps. Retaining its original rootstock, a colony of *Populus tremuloides* in Utah is estimated to be up to 80,000 years old, placing it among the oldest living organisms on the planet.

Salix viminalis
Drawing by A. Masclef, *Atlas des plantes de France*, 1893

Salix lanata

Populus nigra

Azara serrata

myrtle family

FAMILY Myrtaceae

ORDER Myrtales

Dominant in many Australian and Brazilian plant communities, the evergreen trees and shrubs that make up the members of this family are found throughout the tropics and in some warm-temperate regions. Most produce oils. Young leaves can be smaller than mature leaves and differently shaped. Flowers, followed by dry or fleshy fruits, usually have numerous stamens. Guavas (*Psidium guajava*) are among several species that are eaten. *Eucalyptus* species are grown for their lumber and essential oils that are used medicinally, in confectionary, and in cosmetics. Other genera provide spices such as cloves and allspice; tea-tree oil is extracted from *Melaleuca alternifolia*. Sacred in classical mythology, myrtle (*Myrtus communis*) has been used traditionally in wedding bouquets; it is also common in gardens.

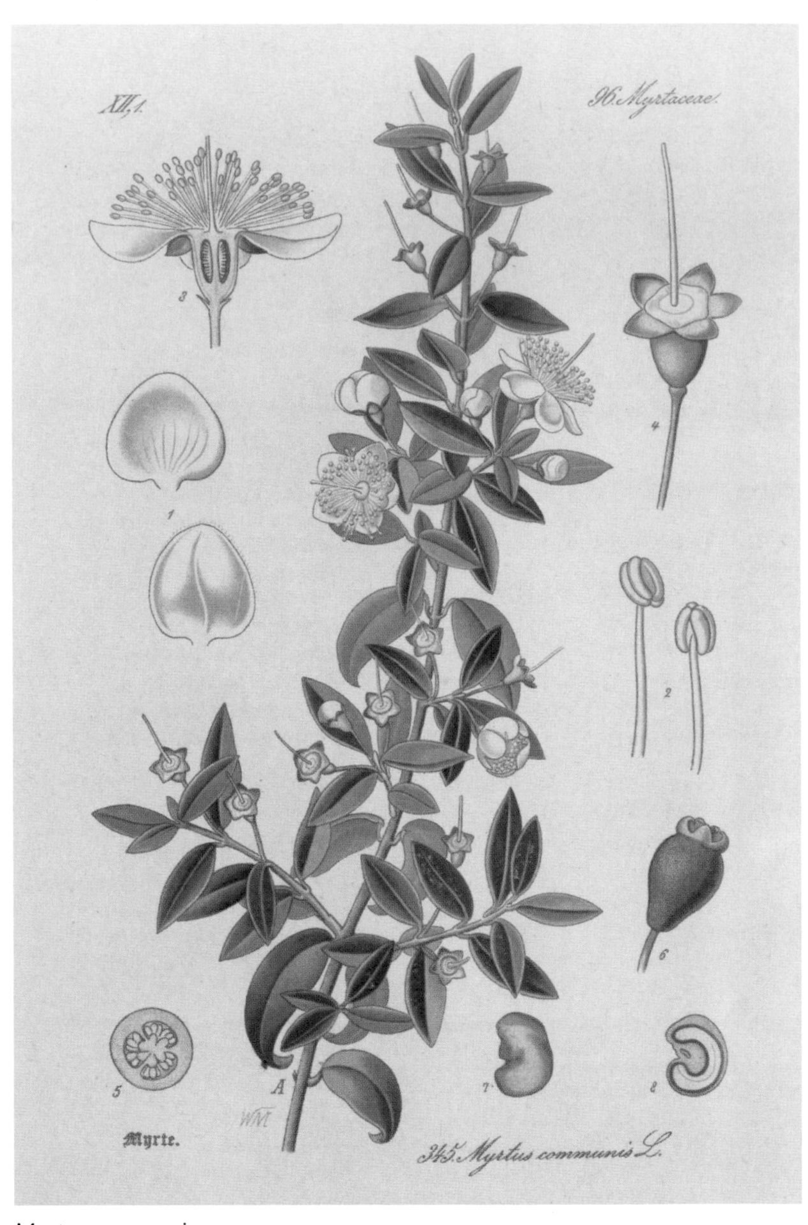

Myrtus communis
Drawing by O.W. Thomé, *Flora von Deutschland, Österreich und der Schweiz*, 1885

Callistemon sp.

Psidium guajava

Eucalyptus sp.

Melaleuca nesophila

fuchsia family

FAMILY **Onagraceae**

ORDER **Myrtales**

Comprising soft-stemmed and woody plants—mainly shrubs, with only a few trees and climbers—this family is found nearly worldwide, including in California deserts. Plants have simple leaves, and flowers are followed by fruits that carry numerous seeds. Evening primrose (*Oenothera*) is grown not only as an ornamental but also for the oil that can be extracted from its seeds and used in medicines. *Clarkia* and *Godetia* are also common in gardens, but the most important is *Fuchsia*, of which a vast number of hybrids have been produced. Both flowers and fruits of fuchsia are sometimes eaten. *Fuchsia magellanica* is often used as hedging; in many regions, it has become naturalized.

Fuchsia magellanica (syn. *Fuchsia gracilis*)
Drawing by E.E. Gleadall, *The beauties of flora*, 1839

Epilobium angustifolium

Oenothera lindheimeri

Oenothera biennis

water lily family

FAMILY Nymphaeaceae

ORDER Nymphaeales

This family of aquatic annuals and perennials is widespread, but is not found in frozen or dry regions, New Zealand, or southern parts of Australia and South America. They can have rhizomes or tubers, and the simple leaves are submerged or floating or held above the water surface. The solitary flowers are often showy and can have many petals; they are followed by spongy berries containing many large seeds. Of the five genera, *Nymphaea*, with its many hybrids, is important in gardens. Some species have edible rhizomes and seeds, and flowers of *N. caerulea* from the Nile are used as narcotics in traditional ceremonies. *Victoria amazonica* produces the largest undivided leaves of any plant; their veining influenced the construction of the Crystal Palace in Victorian London.

Nymphaea elegans
Drawing by W.H. Fitch, *Curtis's Botanical Magazine*, 1851

Nuphar lutea

Nymphaea 'Escarboucle'

Victoria amazonica

Euryale ferox

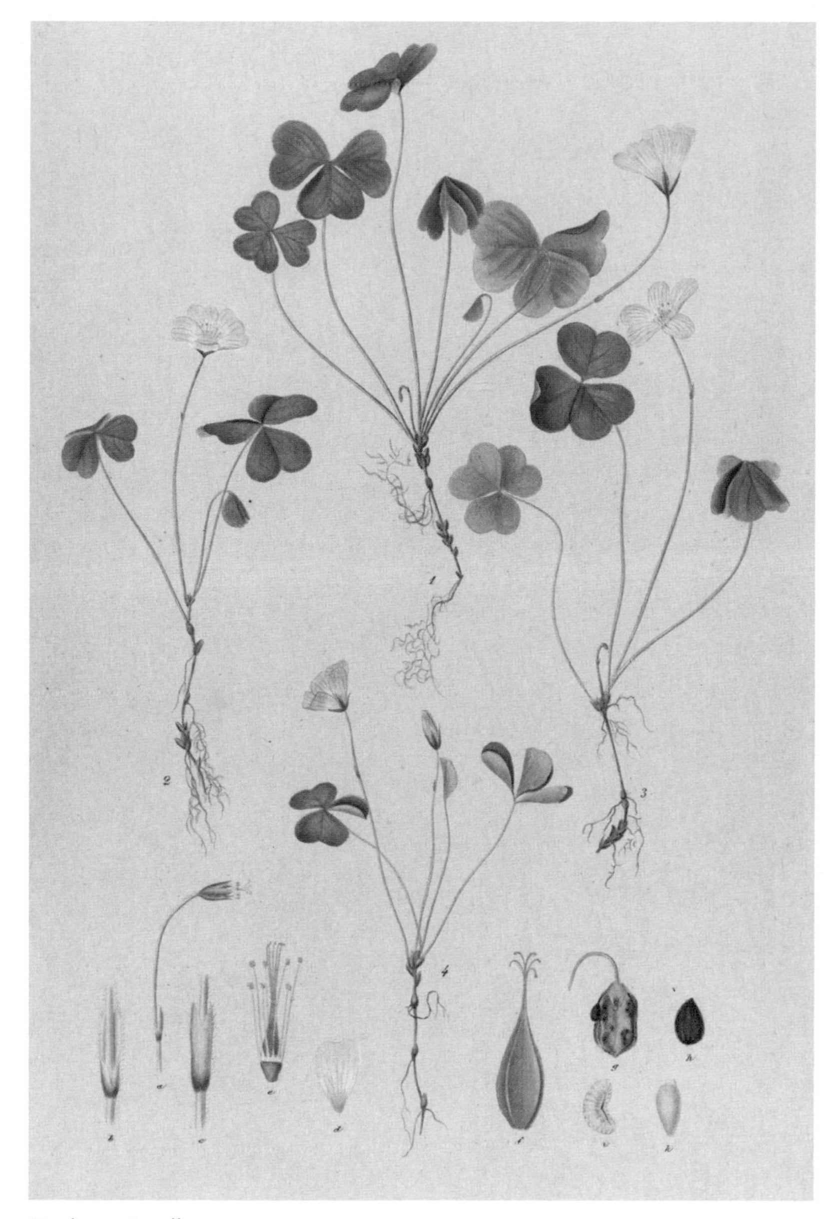

Oxalis acetosella
Engraving from G.C. Oeder et al., *Flora Danica*, 1761–1883

wood-sorrel family

FAMILY Oxalidaceae

ORDER Oxalidales

Found globally, apart from in the great deserts, members of this family are sometimes succulent, and include both soft-stemmed (mainly perennial) and woody plants. They are particularly diverse in tropical and subtropical regions. Flowers have five petals, and many species form bulbs, tubers, or thickened rhizomes. *Oxalis tuberosa* is grown in South America as an alternative to potatoes. Leaves of other species are also edible. Although most species are used as garden plants, the family also includes some invasive weeds. The starfruit or carambola (*Averrhoa carambola*) is grown in the tropics; the ribbed fruits are exported while still unripe and are usually sliced as a star-shaped garnish in fruit salads. Ripe fruits are sweeter but do not travel well.

Oxalis triangularis

Averrhoa carambola

Biophytum zenkeri

Oxalis adenophylla

pineapple family

FAMILY Bromeliaceae

ORDER Poales

These rosette-forming plants (bromeliads), which can be terrestrial or epiphytic, have spirally arranged, sometimes toothed, leaves and stems that can become woody. They are found in tropical and warm-temperate parts of the Americas, with a single species (*Pitcairnia feliciana*) in West Africa. Flowers are carried at the ends of stems and often have large, colorful bracts; they are pollinated predominantly by hummingbirds. The most commercially important member is the pineapple (*Ananas comosus*), whose juicy fruits are composed of berries that coalesce with the flowers' fleshy bracts. The leaves are sometimes harvested for their fibers. Many other bromeliads are popular as houseplants, particularly *Aechmea* and *Tillandsia*, and the smaller species known as air plants.

Ananas comosus (syn. *Bromelia ananas*)
Engraving by J.J. Plenck, *Icones Plantarum*, 1807

Tillandsia dyeriana

Aechmea fasciata

Guzmania lingulata

Tillandsia stricta

sedge family

FAMILY Cyperaceae
ORDER Poales

Sister to the rushes (Juncaceae), this family of annuals, perennials, shrubs, and lianas is often found in wet areas worldwide. The perennials usually have rhizomes, and stems are often triangular in cross section. The flowers do not have petals. Use of papyrus (*Cyperus papyrus*) to make a paperlike material dates to at least the 3rd millennium BCE, when it was naturalized around the Nile in Egypt. Kept in dry conditions, early documents have survived virtually unblemished. The species is also used for handicrafts in parts of Africa. *Cyperus esculentus* is grown for its edible tubers (tigernuts).

Cyperus rotundus (syn. *Cyperus comosus*)
Drawing by F. Bauer, from J. Sibthorp and J.E. Smith, *Flora Graeca*, 1806

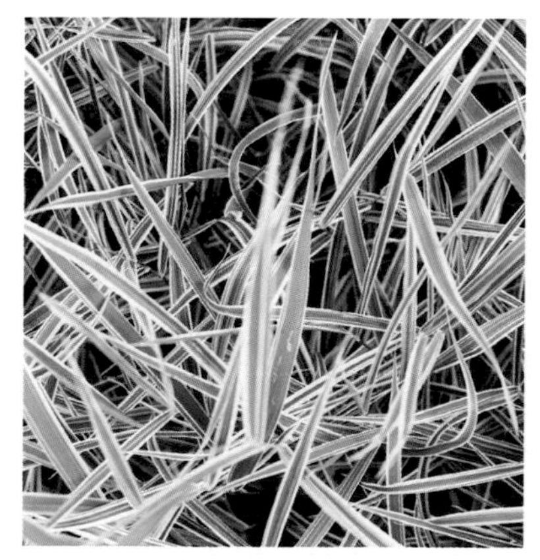

Carex morrowii 'Variegata'

Scirpus cyperinus

Eriophorum angustifolium

Cyperus papyrus

grass family

FAMILY Poaceae

ORDER Poales

This large family comprises annuals, perennials, and woody bamboos, found globally in every habitat, sometimes colonizing great tracts to create grasslands. They have jointed stems, with generally hollow internodes and leaves arranged alternately, sheathing the internode at the base. Flowers are produced in spikelets. Grasses are arguably the most important economic plants, providing cereal crops, grazing and fodder for livestock, and biofuels. Wheat (*Triticum*) is the principal cereal in temperate climates, used to make bread, flour, and foodstuffs such as pasta and couscous that are suitable for storing. Other species are components of alcoholic drinks. Sugarcane (*Saccharum* species) is grown throughout the tropics. Some bamboos are grown for their fibers that are used in construction.

Poa abbreviata
Engraving from G.C. Oeder et al., *Flora Danica*, 1761–1883

Triticum aestivum

Phyllostachys aureosulcata

Zea mays

Cenchrus longisetus (syn. *Pennisetum villosum*)

poppy family

FAMILY Papaveraceae
ORDER Ranunculales

Comprised of mainly annuals and herbaceous perennials, with some woody genera, this family is widespread in the northern hemisphere, with a few members in the southern hemisphere. Plants often have rhizomes or tubers. Stems, with leaves arranged spirally, are usually erect and exude a watery sap when cut. As the source of drugs such as codeine, heroin, and morphine, the opium poppy (*Papaver somniferum*) has been illicitly traded, leading to unrest and even wars. Seeds of this species and *Papaver rhoeas* are commonly used in baking. Other species, such as bleeding heart (*Lamprocapnos*), Californian poppy (*Eschscholzia*), plume poppy (*Macleaya*), blue poppy (*Meconopsis*), and tree poppy (*Romneya*), are familiar as ornamentals in gardens.

Papaver pilosum
Drawing by W.H. Fitch, *Curtis's Botanical Magazine*, 1853

Lamprocapnos spectabilis

Papaver orientale

Eschscholzia californica

Corydalis flexuosa

Ranunculus repens
Engraving by C.A.M. Lindman, *Bilder ur Nordens Flora*, 1922–1926

buttercup family

FAMILY Ranunculaceae

ORDER Ranunculales

Found in temperate regions, with a few species in the tropics, this family is made up mainly of perennials, annuals, climbers, and small shrubs; some are aquatic. Their leaves are usually alternate, and the flowers generally have numerous stamens in spirals or whorls. *Clematis*, of which there are many garden forms, and *Ranunculus* are among the most widespread genera of flowering plants. Though many others are toxic, *Nigella sativa* produces seeds commonly used in baking, and goldenseal (*Hydrastis canadensis*), a natural antibiotic, has been used as a medicine by Native Americans. Monkshood (*Aconitum*), *Anemone*, *Delphinium*, hellebore (*Helleborus*), and pasqueflower (*Pulsatilla*) are all popular garden plants.

Anemone sp.

Delphinium elatum

Aconitum napellus

Ranunculus repens

mulberry family

FAMILY Moraceae

ORDER Rosales

Comprised of both woody and soft-stemmed plants, some epiphytic, this family is widely distributed, though absent from parts of North America, Eurasia, and the southern hemisphere. Stems usually contain a milky sap. Leaves are usually alternate and fruits are drupes (occasionally achenes). Many species of fig (*Ficus*) are pollinated by specialized wasps. *Ficus carica* has long been valued for its fig fruits and is naturalized in the Mediterranean region. Another important crop is the breadfruit (*Artocarpus altilis*), grown in parts of the tropics, especially Jamaica. Mulberries (*Morus* sp.) are also grown for their fruits in temperate regions. The white mulberry (*Morus alba*) is cultivated as a food plant for farmed silk moth caterpillars (*Bombyx mori*).

Morus nigra
Drawing by O.W. Thomé, *Flora von Deutschland, Österreich und der Schweiz*, 1885

Morus alba

Ficus carica

Artocarpus heterophyllus

rose family

FAMILY Rosaceae

ORDER Rosales

This family of woody and soft-stemmed plants is found worldwide, with the exception of frozen regions and some dry areas. The shrubs and trees often have spiny branches. Usually, their leaves are alternate and flowers have five petals, followed by variable fruits, many of economic worth. Drupe-bearers include peaches, apricots, cherries, plums, and almonds (*Prunus*), while pomes include apples (*Malus*) and pears (*Pyrus*). Almond seeds (*Prunus dulcis*) are the most commonly eaten nut, and yield an oil suitable for use in cooking and making cosmetics. Strawberries (*Fragaria* x *ananassa*) are complex hybrids, the fruits being achenes held on the outside of a fleshy receptacle. In gardens, roses (*Rosa*), with thousands of cultivars, *Alchemilla*, *Cotoneaster*, *Photinia*, and *Sanguisorba* are widely grown.

Rosa pendulina (syn. *Rosa alpina*)
Drawing by H. Thiselton-Dyer, *Curtis's Botanical Magazine*, 1883

Malus sylvestris

Prunus dulcis

Fragaria vesca

Rosa glauca

nettle family

FAMILY Urticaceae

ORDER Rosales

Found globally, apart from in dry or frozen regions, this family comprises both woody and soft-stemmed plants, including lianas and epiphytes. Some have aerial or stilt roots. Leaves are arranged spirally, and some plants carry stinging hairs that can embed themselves in the skin, causing inflammation. Fruits are achenes. The common nettle (*Urtica dioica*), an invasive weed in some areas, can be eaten as a vegetable and, along with other species, has also been used to provide fiber for clothing. A few other members of the family have edible fruits or tubers. In Central and South America, species of *Cecropia* are used by fire ants for nesting.

Urtica dioica
Drawing by O.W. Thomé, *Flora von Deutschland, Österreich und der Schweiz*, 1885

Pilea peperomioides

Pilea involucrata

Cecropia peltata

cashew family

FAMILY Anacardiaceae
ORDER Sapindales

Native to tropical regions, with a few in temperate zones, members of this family are trees, shrubs, and climbers, with resinous bark and sap that turns black on contact with the air. Their leaves are usually alternate, and fruits are mainly drupes. Cashews (*Anacardium occidentale*) are grown widely in the tropics for their nuts, from which oil is also extracted. The fruits have a swollen pedicel—cashew apple—that is edible but eaten only locally. Other important crops include mangos (*Mangifera indica*) and pistachios (*Pistacia vera*). Many species cause contact dermatitis, one of the most troublesome being poison ivy (*Toxicodendron radicans* and *T. rydbergii*), widely found in North America.

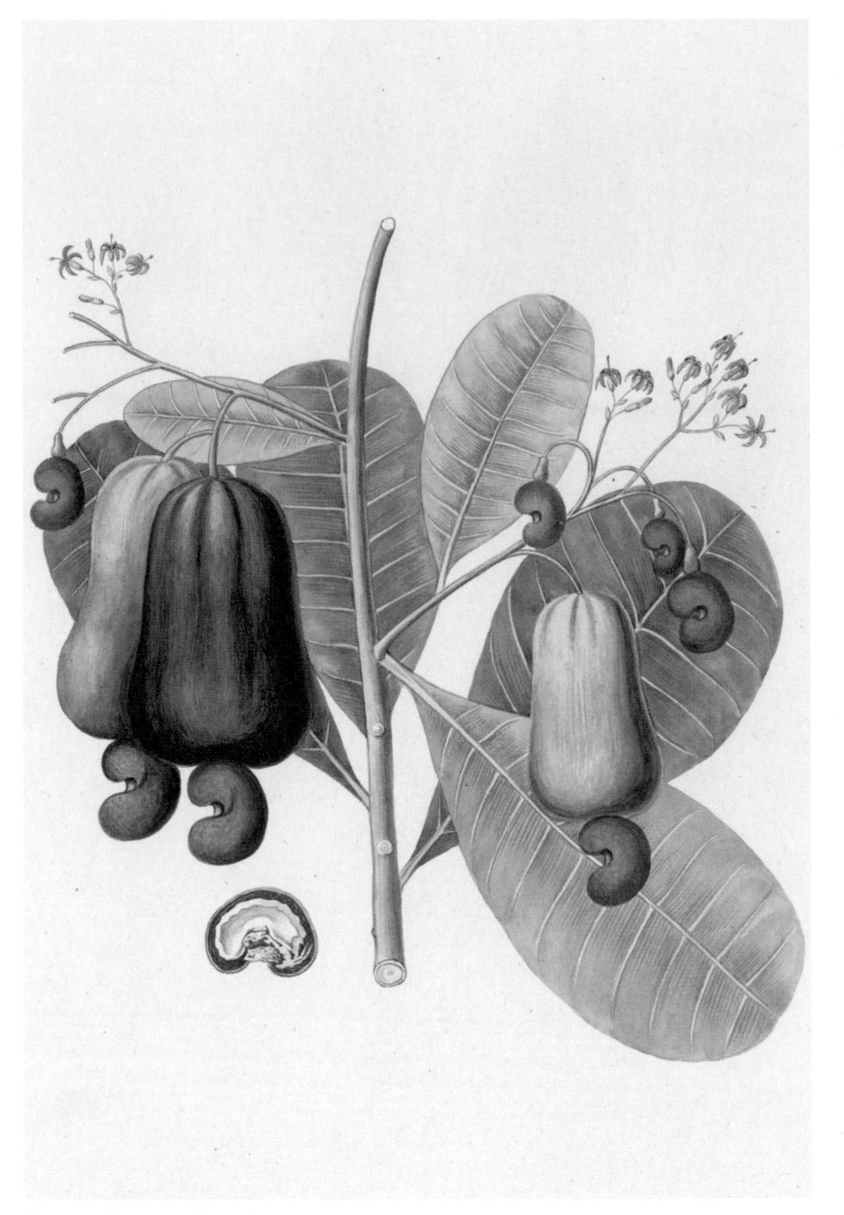

Anacardium occidentale
Drawing by N. J. Jacquin, *Selectarum stirpium Americanarum historia*, 1780–1781

Mangifera indica

Pistacia vera

Schinus molle

Rhus typhina

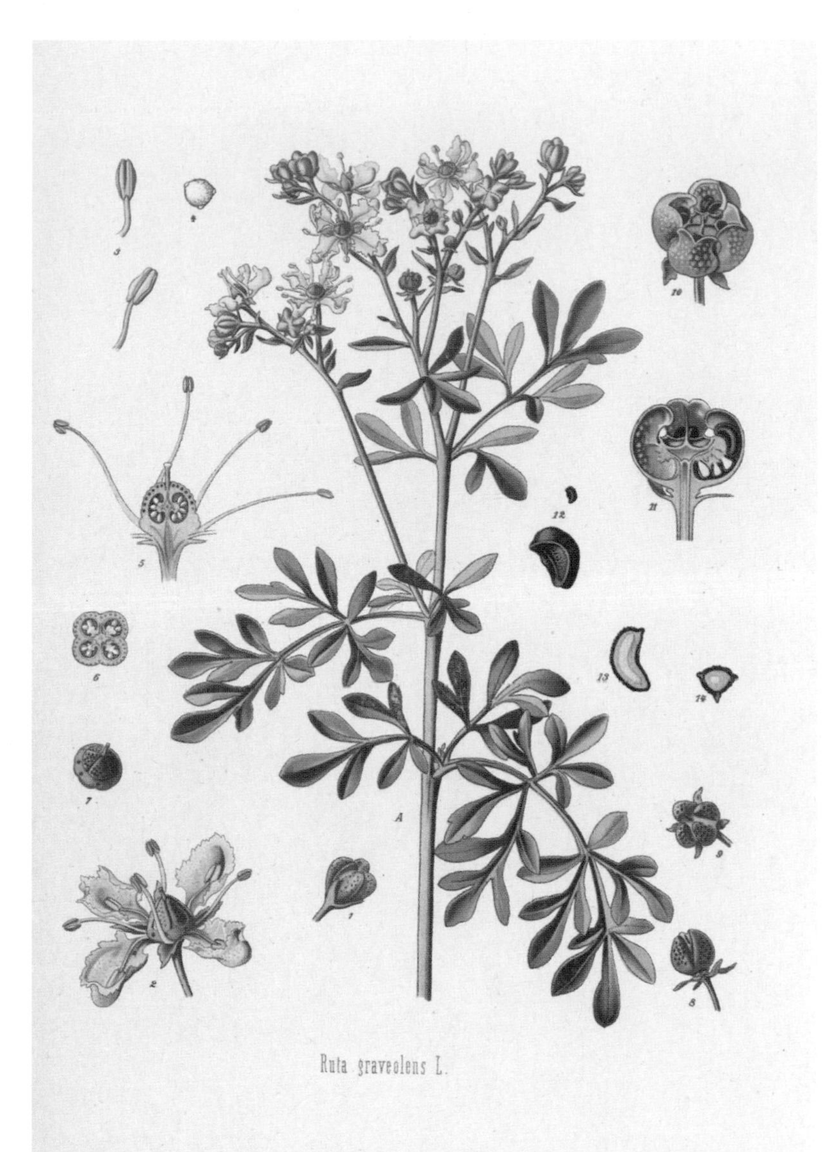

Ruta graveolens
Drawing by W. Müller, *Köhler's Medizinal-Pflanzen*, 1887

citrus family

FAMILY Rutaceae

ORDER Sapindales

Found nearly globally but mainly in the tropics and subtropics, these usually aromatic, woody or soft-stemmed plants can have spiny or scandent stems. Flowers have three, four, or five petals and are followed by variable fruits. In some cases, the fruit has a leathery skin covered with oil glands over a spongy layer that protects a juicy pulp containing the seeds. *Citrus* is the most important genus. Lemons, limes, and oranges (among others) are essential crops in warm climates; the oil is used in perfumery and cosmetics. Most are complex hybrids whose lineage is difficult to unravel. Rue (*Ruta graveolens*) is often planted in herb gardens for its pungent leaves, despite the fact that it is mildly toxic.

Citrus reticulata

Citrus bergamia

Citrus maxima

maple family

FAMILY **Sapindaceae**

ORDER Sapindales

Occurring mainly in the tropics, though extending into temperate zones worldwide, this family includes woody and soft-stemmed plants. Flowers, usually with four or five petals, are often white or yellow, and can have an additional petallike appendage, sometimes fused at the base. Seeds often have a glossy skin and large scar. Important crops include lychees (*Litchi chinensis*) and rambutans (*Nephelium lappaceum*). Sugar maple (*Acer saccharum*) is the emblem of Canada. Its sweet sap (and that of other species) is harvested to make maple syrup. Popular garden trees include horse chestnut (*Aesculus hippocastanum*) and Japanese maples (*Acer palmatum* and *A. shirasawanum*), of which there are hundreds of cultivars.

Sapindus rarak
Engraving from C.L. Blume, *Rumphia*, 1847

Dimocarpus longan

Nephelium lappaceum

Acer saccharum subsp. *nigrum*

saxifrage family

FAMILY Saxifragaceae

ORDER Saxifragales

This family of mainly perennials is widespread across the northern hemisphere. Some species of *Saxifraga* grow in rock crevices. They usually produce basal rosettes of leaves and flowers, generally with five petals, which are carried on branching stems. Fruits normally contain numerous small seeds. Although a few genera have edible leaves, it is the ornamentals that are best known. *Astilbe* and elephant's ears (*Bergenia*) are common garden plants, both with many cultivars. Many species of saxifrage (*Saxifraga*), some encrusted with lime, are grown in rock gardens. *Astilboides tabularis* prefers damp ground, and its rhizomes have been used to make wine. Some species, such as the piggyback plant (*Tolmiea menziesii*), make good houseplants.

Bergenia ciliata (syn. *Saxifraga ciliata*)
Drawing by W.H. Fitch, *Curtis's Botanical Magazine*, 1856

Saxifraga oppositifolia

Tiarella cordifolia

Astilbe thunbergii

Bergenia crassifolia

Convolvulus arvensis var. *parviflora*
Engraving from G.C. Oeder et al., *Flora Danica*, 1761–1883

bindweed family

FAMILY Convolvulaceae
ORDER Solanales

Spread globally, although absent from polar regions and the great deserts, this is a family of annual, perennial, and woody plants (only a few of which are trees) with mainly climbing or trailing stems. Climbers twine counterclockwise. Flowers have usually five petals, fused into a variably shaped corolla. Seeds are sometimes hairy. The sweet potato (*Ipomoea batatas*) is a major tropical crop, grown for its tubers but also used to produce bioplastics and alcohol. Other species also have edible parts, sometimes used medicinally. The genus also includes several ornamentals. Some members of the family, such as bindweed (*Convolvulus arvensis*), are troublesome garden weeds.

Calystegia sepium

Ipomoea batatas

Cuscuta sp.

nightshade family

FAMILY Solanaceae

ORDER Solanales

Found globally, but absent from polar regions and the great deserts, this family is comprised of soft-stemmed and woody plants. Stems are often hairy and prickly or thorny. Flowers are cup-, funnel-, or tube-shaped, followed by seed-bearing berries or capsules, sometimes, as in *Physalis*, enclosed by persistent sepals. Most important is the potato (*Solanum tuberosum*), with edible tubers, followed by the tomato (*S. lycopersicum*). Other food plants include eggplants (*S. melongena*), peppers (*Capsicum* species), and goji berries (*Lycium chinense*). Some are toxic or hallucinogenic and have been used in folk rituals. The leaves of tobacco (*Nicotiana*) are dried for smoking. The genus also includes ornamentals; *Brugmansia*, *Cestrum*, and *Schizanthus*, among others, are represented in gardens.

Solanum diploconos **(syn. *Pionandra fragrans*),** with *Mutisia clematis* (Compositae) above it. Drawing from *La Belgique horticole, journal des jardins et des vergers*, 1864

Solanum melongena

Physalis alkekengi

Datura stramonium

Solanum lycopersicum var. *cerasiforme*

grapevine family

FAMILY Vitaceae

ORDER Vitales

Forming the sole family in its order, these woody plants are found in the tropics and warm temperate regions. Climbing species can cling by means of tendrils or adhesive pads, although these are not always present. Fruits are berries, sometimes juicy, containing one to four seeds. The most important member is *Vitis*, especially the grape (*Vitis vinifera*), cultivated in many warm regions for its fruits that can be eaten fresh or dried, or fermented as wine. Breeding with other species has produced plants that fruit successfully in cooler climates. Its leaves are also edible. Other genera also have edible parts. *Cissus* and *Parthenocissus* are used in gardens, the former also often as a houseplant.

Vitis vinifera
Drawing by W. Müller, *Köhler's Medizinal-Pflanzen*, 1887

Cissus quadrangularis

Parthenocissus tricuspidata

Ampelopsis glandulosa var. *brevipedunculata*

Parthenocissus henryana

ginger family

FAMILY Zingiberaceae

ORDER Zingiberales

Found throughout the tropics, as well as in some temperate zones, these perennials are aromatic in all their parts. They have creeping, often branching, rhizomes that can appear above ground level, and alternate leaves clasping stems. Flowers have three petals. Many are valued as spices, especially in Asian cuisine. The rhizome of ginger (*Zingiber officinale*) is used in cooking; fresh, pickled, candied, dried, and ground, it can flavor a range of dishes. Turmeric (*Curcuma longa* and *C. aromatica*) is sometimes used to color food yellow as a substitute for saffron. Other spices include green cardamom (*Elettaria cardamomum*) and black cardamom (*Hornstedtia costata*). The ginger lily (*Hedychium*) is cultivated as an ornamental.

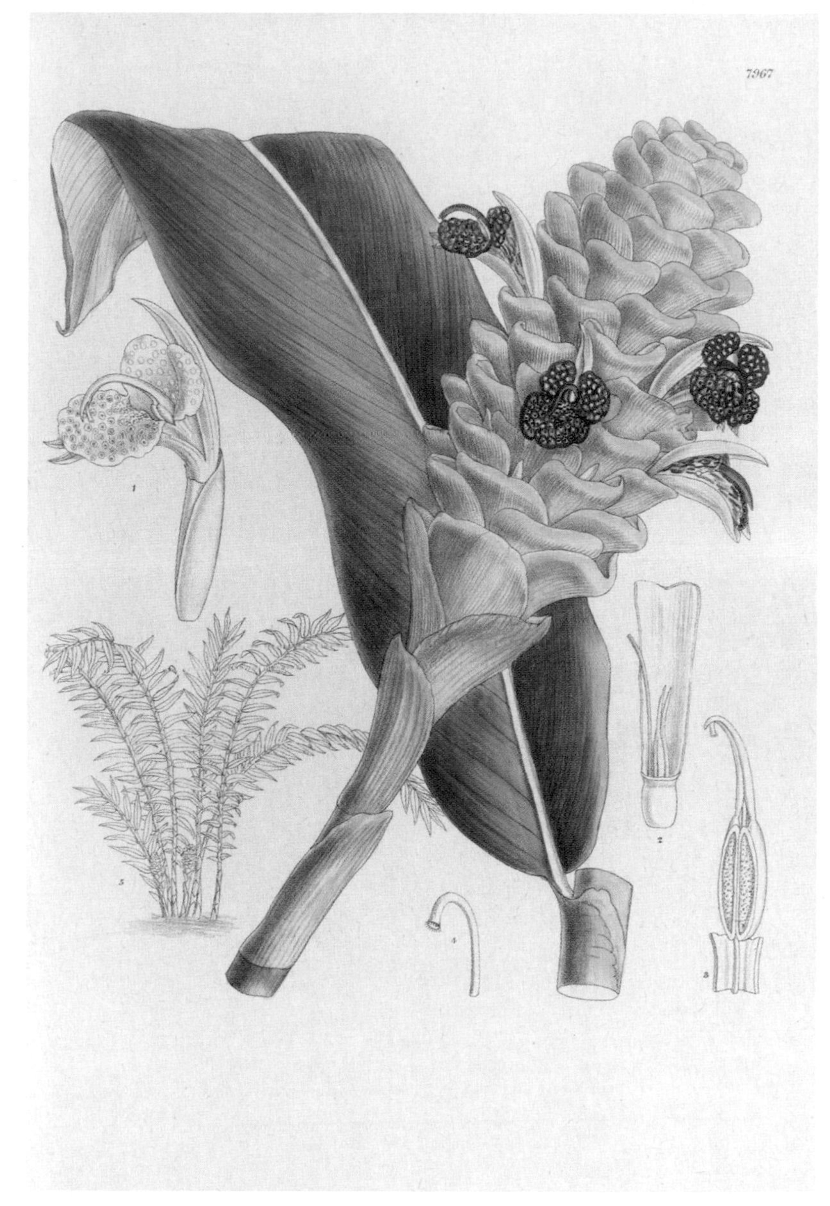

Zingiber spectabile
Drawing by M. Smith, *Curtis's Botanical Magazine*, 1904

Zingiber officinale

Hedychium yunnanense

Elettaria cardamomum

cypress family

FAMILY Cupressaceae

ORDER Cupressales

These coniferous trees and shrubs are found globally, with the exception of Antarctica. The family includes the world's largest tree species, the giant redwood (*Sequoiadendron giganteum*), and the tallest, *Sequoia sempervirens*. The largest genus, *Juniperus* (juniper), is widespread throughout the northern hemisphere; some species are extremely variable. Dwarf or prostrate forms are popular garden plants. Cones of *Juniperus communis*, which are fleshy and berrylike, are used as a foodstuff and to flavor gin. Forms of the Italian cypress (*Cupressus sempervirens*) have traditionally been planted as graveyard trees in Mediterranean areas. When growing near water, roots of the bald cypress (*Taxodium distichum*) produce kneelike growths that project above ground level.

Cupressus lusitanica (syn. *Cupressus pendula*)
Drawing by P.J. Redouté, *Traité des arbres et arbustes*, 1806

Taxodium distichum

Cunninghamia lanceolata

Juniperus communis

Cupressus sempervirens

Pinus bungeana
Drawing by M. Smith, *Curtis's Botanical Magazine*, 1909

pine family

FAMILY Pinaceae
ORDER Pinales

The order Pinales consists of this single family that comprises evergreen and deciduous trees and shrubs, all parts of which are resinous and aromatic. They are distributed mainly in the northern hemisphere and have spirally arranged, needlelike to linear leaves. Male cones are ovoid to ellipsoid or cylindrical. Seeds are fine and dispersed by wind. This order provides most of the world's softwood and is also used in the production of paper, tar, turpentine, and household cleaning products. Some species of *Pinus* have edible seeds—pine nuts—used to make pesto. The biblical Temple of Solomon was made from cedar of Lebanon (*Cedrus libani*), and today species of *Picea* and *Abies* are grown for use as Christmas trees.

Abies alba

Pinus sylvestris

Larix decidua

tree-fern family

FAMILY Cyatheaceae

ORDER Cyatheales

The order Cyatheales was formerly divided into eight families that are now treated as subfamilies of the single family Cyathaceae. Spread throughout the tropics, subtropics, and wet regions of the southern hemisphere, it includes some of the largest and smallest ferns. So-called tree ferns, such as *Cibotium* and *Dicksonia*, have erect rhizomes that resemble tree trunks. Species with creeping rhizomes are sometimes confused with polypods (see p.418). Some produce woolly rhizomes that have been used to stuff cushions. *Cibotium barometz* has a woolly rhizome known as the mythical "Vegetable Lamb of Tartary"—supposedly, the plant produced live sheep as its fruit.

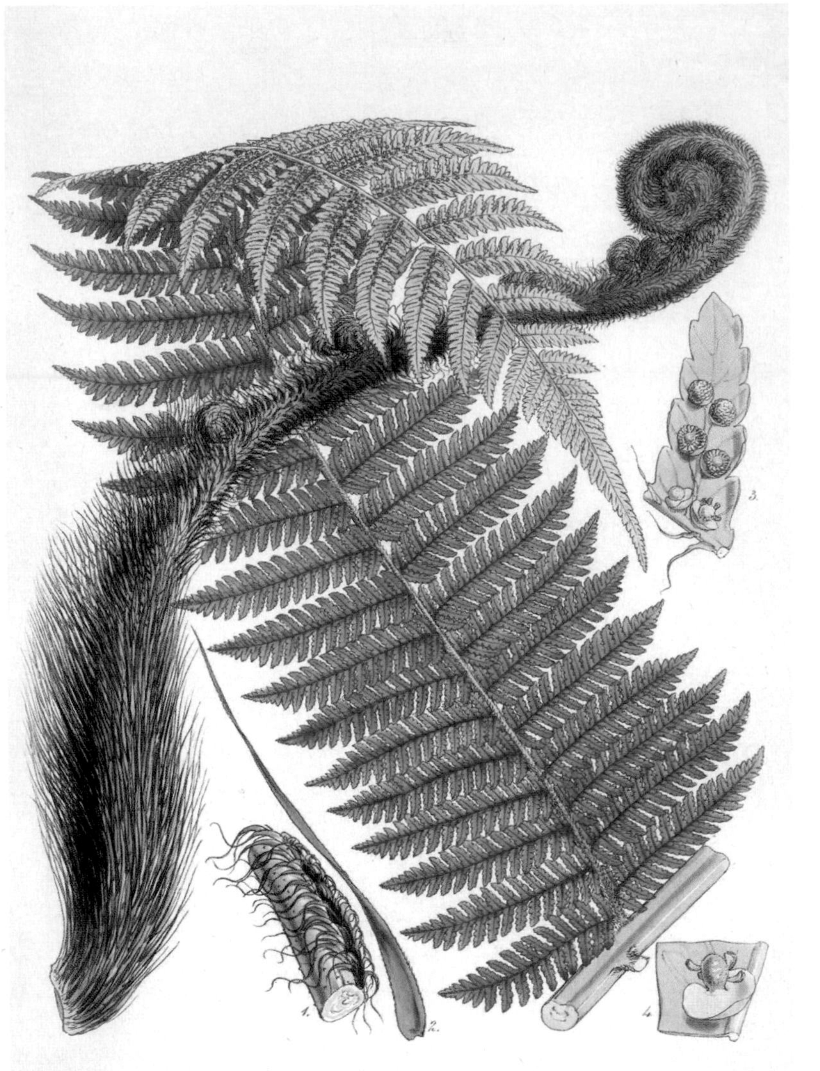

Cyathea smithii
Drawing by W.H. Fitch, *The botany of the Antarctic voyage of* H.M. *discovery ships Erebus and Terror in the years 1839-1843, under the command of Captain Sir James Clark Ross*, 1855

Cibotium barometz

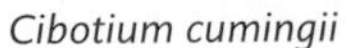

Cibotium cumingii

Balantium antarcticum (syn. *Dicksonia antarctica*)

horsetail family

FAMILY Equisetaceae

ORDER Equisetales

A single genus, *Equisetum*, represents this family. A member of the fern plant group, it is the sole survivor of a formerly diverse family that included many trees. Its 20 species, found mainly in the northern hemisphere and South America, are herbaceous plants with jointed, hollow stems and hollow underground rhizomes. The leaves are reduced to sheaths that surround the nodes. Plants hybridize freely. They contain a high concentration of silicates, which makes them potentially toxic, although some have been used as foodstuffs and medicines. Stems of *Equisetum hyemale* have been used to scour cooking pots and to polish wood. Field horsetail (*Equisetum arvense*) is an invasive garden weed in many territories.

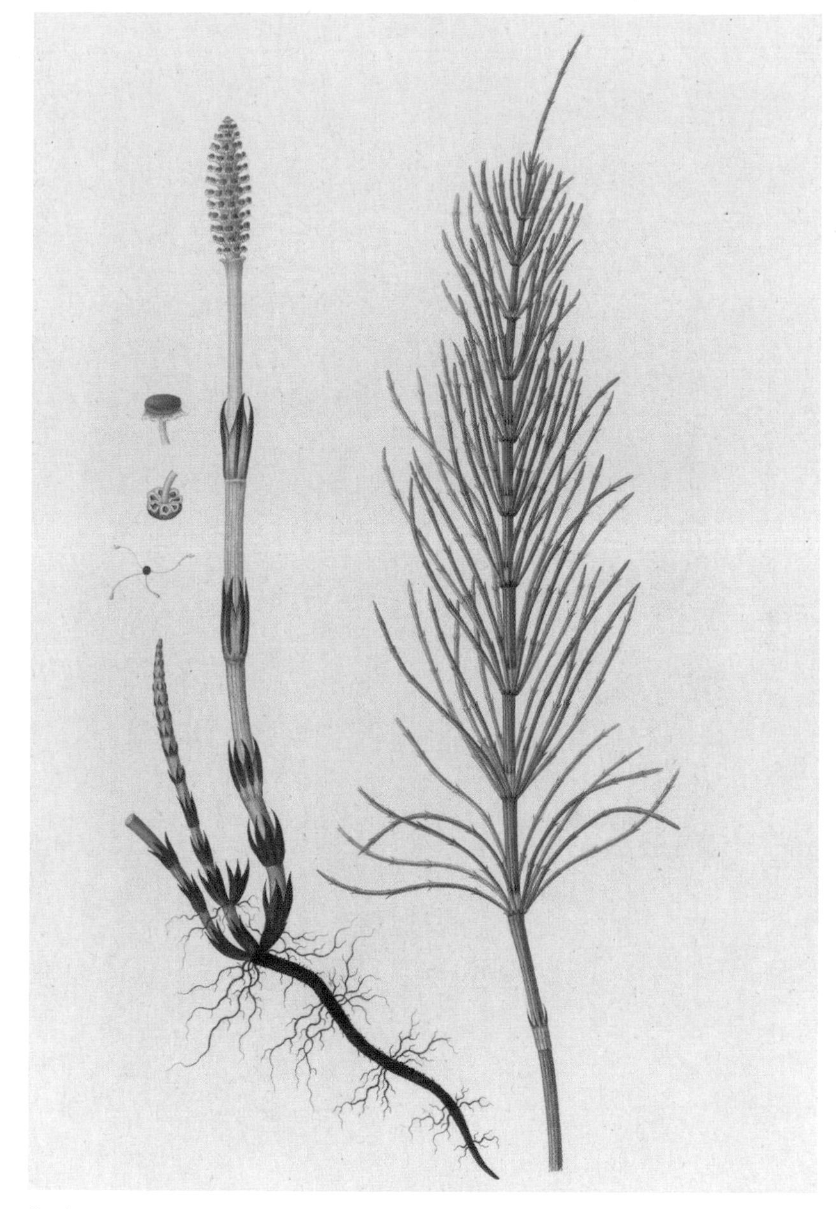

Equisetum arvense
Engraving from G.C. Oeder et al., *Flora Danica*, 1761–1883

Equisetum sylvaticum

Equisetum telmateia

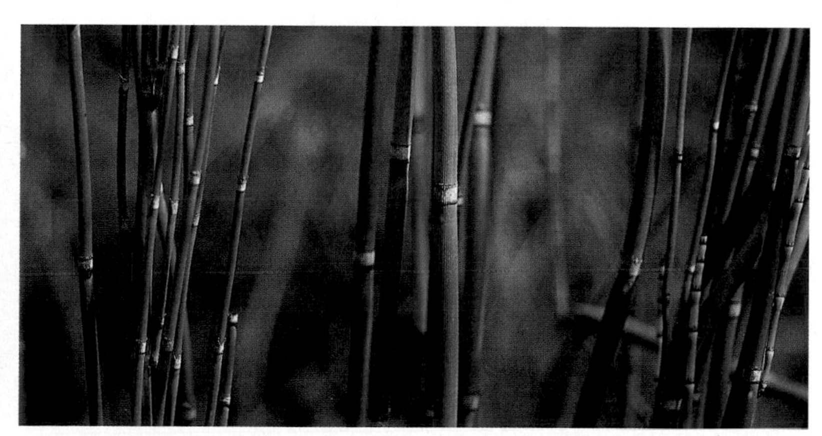

Equisetum hyemale

polypody family

FAMILY Polypodiaceae

ORDER Polypodiales

This is the largest fern family, occurring worldwide and comprising terrestrial, epiphytic, and rock-dwelling plants, as well as some climbers, occasionally with branching rhizomes. The leaves are variable and can be simple or divided. This family is most diverse in rainforests. Some are edible; the leaves of several species are cooked as a vegetable in parts of Asia. The rhizomes of certain *Polystichum* and *Polypodium* species were eaten by Native Americans. Many species are grown as ornamentals, both in gardens and as houseplants; the Boston fern (*Nephrolepis exaltata* 'Bostoniensis') is particularly popular. *Rumohra adiantiformis* is grown commercially to provide material for the cut-flower industry.

Polypodium vulgare
Drawing by T. Meehan, *The native flowers and ferns of the United States in their botanical, horticultural, and popular aspects*, 1878–1879

Phymatosorus grossus

Phlebodium aureum

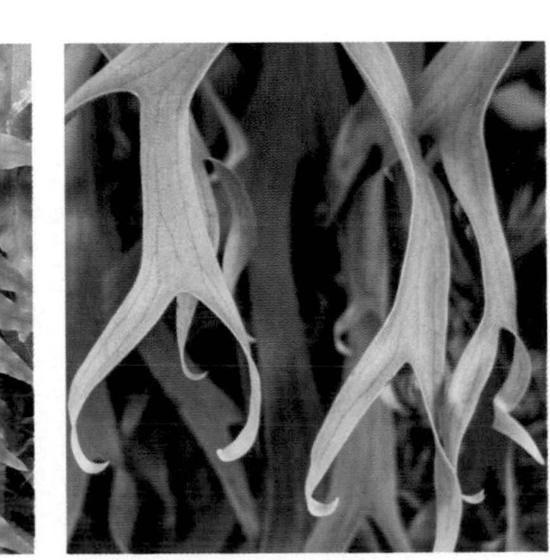

Platycerium hillii

Polypodium vulgare

clubmoss family

FAMILY **Lycopodiaceae**

ORDER Lycopodiales

The order Lycopodiales comprises a single family of perennials with branching stems found worldwide apart from arid regions. Their treelike ancestors once dominated the vegetation but those seen today have no close affinity with any other group of plants. Spores of the stag's-horn clubmoss (*Lycopodium* species) are rich in oils and highly flammable; they are often used for special effects in the film industry. The spores were also used in cosmetics and as a coating in latex products such as condoms and surgical gloves but are no longer, as they can cause an allergic reaction. Other species have been woven into mats and fish traps.

Lycopodium clavatum
Drawing by O.W. Thomé, *Flora von Deutschland, Österreich und der Schweiz*, 1885

Huperzia sp.

Lycopodium annotinum

Diphasiastrum digitatum

families of the plant kingdom

These pages list the 639 vascular and nonvascular plant families currently identified by scientists. In each of the major groups—flowering plants, conifers and cycads, ferns, lycopods, hornworts, mosses, and liverworts—the families are arranged alphabetically within their orders to reveal their closest relatives.

FLOWERING PLANTS (ANGIOSPERMS)

Acorales
ACORACEAE Sweet-flag family

Alismatales
ALISMATACEAE Water-plantain family
APONOGETONACEAE Waterblommetjie family
ARACEAE Calla-lily family *see p.347*
BUTOMACEAE Flowering-rush family
CYMODOCEACEAE Turtle-grass family
HYDROCHARITACEAE Frogbit family
JUNCAGINACEAE Arrowgrass family
MAUNDIACEAE Maund's-arrowgrass family
POSIDONIACEAE Tapeweed family
POTAMOGETONACEAE Pondweed family
RUPPIACEAE Tasselweed family
SCHEUCHZERIACEAE Rannoch-rush family
TOFIELDIACEAE False-asphodel family
ZOSTERACEAE Eelgrass family

Amborellales
AMBORELLACEAE Amborella family

Apiales
APIACEAE Carrot family *see p.348*
ARALIACEAE Ivy family
GRISELINIACEAE Kapuka family
MYODOCARPACEAE Mousefruit family
PENNANTIACEAE Kaikomako family
PITTOSPORACEAE Cheesewood family
TORRICELLIACEAE Ivy-palm family

Aquifoliales
AQUIFOLIACEAE Holly family *see p.349*
CARDIOPTERIDACEAE Citronella family
HELWINGIACEAE Flowering-rafts family
PHYLLONOMACEAE Flowering-leaf family
STEMONURACEAE Buff-beech family

Arecales
ARECACEAE Palm family *see p.350*
DASYPOGONACEAE Saviour-grass family

Asparagales
AMARYLLIDACEAE Onion family *see p.351*
ASPARAGACEAE Hyacinth family *see p.352*
ASPHODELACEAE Daylily family *see p.353*
ASTELIACEAE Pineapple-grass family
BLANDFORDIACEAE Christmas-bells family
BORYACEAE Pincushion-lily family
DORYANTHACEAE Gymea-lily family
HYPOXIDACEAE Stargrass family
IRIDACEAE Iris family *see p.354*
IXIOLIRIACEAE Tartar-lily family
LANARIACEAE Lambtails family
ORCHIDACEAE Orchid family *see p.355*
TECOPHILAEACEAE Chilean-crocus family
XERONEMATACEAE Poor-Knights-lily family

Asterales
ALSEUOSMIACEAE Toropapa family
ARGOPHYLLACEAE Silverleaf family
ASTERACEAE Daisy family *see p.356*
CALYCERACEAE Balsamleaf family
CAMPANULACEAE Bellflower family *see p.357*
GOODENIACEAE Fanflower family
MENYANTHACEAE Bogbean family
PENTAPHRAGMATACEAE Scorpion's-tail family
PHELLINACEAE Corkfruit family
ROUSSEACEAE Putaweta family
STYLIDIACEAE Triggerplant family

Austrobaileyales
AUSTROBAILEYACEAE Austrobaileya family
SCHISANDRACEAE Star-anise family
TRIMENIACEAE Bittervine family

Berberidopsidales
AEXTOXICACEAE Olivillo family
BERBERIDOPSIDACEAE Coral-vine family

Boraginales
BORAGINACEAE Forget-me-not family *see p.358*

Brassicales
AKANIACEAE Turnipwood family
BATACEAE Turtleweed family
BRASSICACEAE Cabbage family *see p.359*
CAPPARACEAE Caper family
CARICACEAE Papaya family
CLEOMACEAE Spiderflower family
EMBLINGIACEAE Slippercreeper family
GYROSTEMONACEAE Buttoncreeper family
KOEBERLINIACEAE Allthorn family
LIMNANTHACEAE Meadowfoam family
MORINGACEAE Horseradish-tree family
PENTADIPLANDRACEAE Oubli family
RESEDACEAE Mignonette family
SALVADORACEAE Toothbrush-tree family
SETCHELLANTHACEAE Azulita family
TOVARIACEAE Stinkbush family
TROPAEOLACEAE Nasturtium family

Bruniales
BRUNIACEAE Buttonbush family
COLUMELLIACEAE Andean-holly family

Buxales
BUXACEAE Box family

Canellales
CANELLACEAE Canella-bark family
WINTERACEAE Winter's bark family

Caryophyllales
ACHATOCARPACEAE Snake-eyes family
AIZOACEAE Dewplant family
AMARANTHACEAE Spinach family *see p.360*
ANACAMPSEROTACEAE Love-plant family
ANCISTROCLADACEAE Kardal family
ASTEROPEIACEAE Manoko family
BARBEUIACEAE Liane-barbeu family
BASELLACEAE Malabar-spinach family
CACTACEAE Cactus family *see p.361*
CARYOPHYLLACEAE Carnation family *see p.362*

DIDIEREACEAE Octopus-tree family
DIONCOPHYLLACEAE Hookleaf-vine family
DROSERACEAE Sundew family
DROSOPHYLLACEAE Dew-pine family
FRANKENIACEAE Sea-heath family
GISEKIACEAE Stork's-henna family
HALOPHYTACEAE Verdolaga family
KEWACEAE Suring family
LIMEACEAE Lizard-foot family
LOPHIOCARPACEAE Sandaarbossie family
MACARTHURIACEAE Macarthuria family
MICROTEACEAE Jumby-pepper family
MOLLUGINACEAE Carpetweed family
MONTIACEAE Blinks family
NEPENTHACEAE Asian-pitcherplant family
NYCTAGINACEAE Four-o'clocks family
PETIVERIACEAE Henwood family
PHYSENACEAE Balloonfruit family
PHYTOLACCACEAE Pokeweed family
PLUMBAGINACEAE Thrift family *see p.363*
POLYGONACEAE Knotweed family *see p.364*
PORTULACACEAE Purslane family
RHABDODENDRACEAE Clubfruit-tree family
SARCOBATACEAE Greasewood family
SIMMONDSIACEAE Jojoba family
STEGNOSPERMATACEAE Cuban-tangle family
TALINACEAE Fameflower family
TAMARICACEAE Salt-cedar family

Celastrales
CELASTRACEAE Spindle family
LEPIDOBOTRYACEAE Snail-cedar family

Ceratophyllales
CERATOPHYLLACEAE Hornwort family

Chloranthales
CHLORANTHACEAE Pearl-orchid family

Commelinales
COMMELINACEAE Spiderwort family
HAEMODORACEAE Kangaroo-paw family
HANGUANACEAE Susum family
PHILYDRACEAE Frogsmouth family
PONTEDERIACEAE Water-hyacinth family

Cornales
CORNACEAE Dogwood family *see p.365*
CURTISIACEAE Assegai family
GRUBBIACEAE Koolhout family
HYDRANGEACEAE Hortensia family
HYDROSTACHYACEAE Waterspike family
LOASACEAE Blazingstar family
NYSSACEAE Tupelo-tree family

Crossosomatales
APHLOIACEAE Mountain-peach family
CROSSOSOMATACEAE Rockflower family
GEISSOLOMATACEAE Cape-cups family
GUAMATELACEAE Guatemalan-bramble family
STACHYURACEAE Spiketail family
STAPHYLEACEAE Bladdernut family
STRASBURGERIACEAE Tawari family

Cucurbitales
ANISOPHYLLEACEAE Leechwood family
APODANTHACEAE Stemsucker family
BEGONIACEAE Begonia family *see p.366*
CORIARIACEAE Tanner-bush family
CORYNOCARPACEAE Cribwood family
CUCURBITACEAE Cucumber family *see p.367*
DATISCACEAE Durango-root family
TETRAMELACEAE False-hemp-tree family

Dilleniales
DILLENIACEAE Guineaflower family

Dioscoreales
BURMANNIACEAE Bluethreads family
DIOSCOREACEAE Yam family
NARTHECIACEAE Bog asphodel family

Dipsacales
ADOXACEAE Elder family *see p.368*
CAPRIFOLIACEAE Honeysuckle family

Ericales
ACTINIDIACEAE Kiwifruit family
BALSAMINACEAE Balsam family
CLETHRACEAE Lily-of-the-valley-tree family
CYRILLACEAE Leatherwood family
DIAPENSIACEAE Pincushion-plant family
EBENACEAE Persimmon family
ERICACEAE Heather family *see p.369*
FOUQUIERIACEAE Ocotillo family
LECYTHIDACEAE Cannonball-tree family
MARCGRAVIACEAE Shingle-vine family
MITRASTEMONACEAE Nippledaisy family
PENTAPHYLACACEAE Saintedwood family
POLEMONIACEAE Jacob's-ladder family *see p.370*
PRIMULACEAE Primrose family *see p.371*
RORIDULACEAE Flycatcher-bush family
SAPOTACEAE Sapodilla family
SARRACENIACEAE American-pitcherplant family
SLADENIACEAE Ribfruit family
STYRACACEAE Storax family
SYMPLOCACEAE Sweetleaf family
TETRAMERISTACEAE Punah family
THEACEAE Tea family *see p.372*

Escalloniales
ESCALLONIACEAE Currybush family

Fabales
FABACEAE Pea family *see p.373*
POLYGALACEAE Milkwort family
QUILLAJACEAE Soapbark-tree family
SURIANACEAE Bay-cedar family

Fagales
BETULACEAE Birch family *see p.374*
CASUARINACEAE She-oak family
FAGACEAE Beech family *see p.375*
JUGLANDACEAE Walnut family *see p.376*
MYRICACEAE Bayberry family
NOTHOFAGACEAE Roble family
TICODENDRACEAE Tico-tree family

Garryales
EUCOMMIACEAE Chinese-rubbertree family
GARRYACEAE Tasselbush family

Gentianales
APOCYNACEAE Periwinkle family *see p.377*
GELSEMIACEAE Yellow-jessamine family
GENTIANACEAE Gentian family
LOGANIACEAE Indian-pink family
RUBIACEAE Coffee family *see p.378*

Geraniales
FRANCOACEAE Bridal-wreath family
GERANIACEAE Crane's-bill family *see p.379*

Gunnerales
GUNNERACEAE Giant-rhubarb family
MYROTHAMNACEAE Resurrection-shrub family

Huerteales
DIPENTODONTACEAE Shichi family
GERRARDINACEAE Brown-ironwood family
PETENAEACEAE Petén-linden family
TAPISCIACEAE Silverpheasant-tree family

Icacinales
ICACINACEAE False-yam family
ONCOTHECACEAE Kanak-laurel family

Lamiales
ACANTHACEAE Bear's-breeches family *see p.380*
BIGNONIACEAE Trumpetvine family *see p.381*
BYBLIDACEAE Rainbowplant family
CARLEMANNIACEAE Fragrant-princess family

GESNERIACEAE Gloxinia family *see p.382*
LAMIACEAE Mint family *see p.383*
LENTIBULARIACEAE Bladderwort family
LINDERNIACEAE Wishbone-flower family
MARTYNIACEAE Unicornplant family
MAZACEAE Cupflower family
OLEACEAE Olive family *see p.384*
OROBANCHACEAE Broomrape family
PAULOWNIACEAE Empress-tree family
PEDALIACEAE Sesame family
PHRYMACEAE Lopseed family
PLANTAGINACEAE Speedwell family *see p.385*
PLOCOSPERMATACEAE Staghorn-shrub family
SCHLEGELIACEAE Higuerito family
SCROPHULARIACEAE Figwort family *see p.386*
STILBACEAE Candlesticks family
TETRACHONDRACEAE Rustweed family
THOMANDERSIACEAE West-African-bitterbush family
VERBENACEAE Vervain family

Laurales

ATHEROSPERMATACEAE Southern-sassafras family
CALYCANTHACEAE Spicebush family
GOMORTEGACEAE Keule family
HERNANDIACEAE Lantern-tree family
LAURACEAE Bay-laurel family *see p.387*
MONIMIACEAE Boldo family
SIPARUNACEAE Fevertree family

Liliales

ALSTROEMERIACEAE Inca-lily family
CAMPYNEMATACEAE Green-mountainlily family
COLCHICACEAE Naked-ladies family
CORSIACEAE Ghost-flower family
LILIACEAE Lily family *see p.388*
MELANTHIACEAE Wake robin family
PETERMANNIACEAE Petermann's-vine family
PHILESIACEAE Chilean-bellflower family
RIPOGONACEAE Supplejack family
SMILACACEAE Catbrier family

Magnoliales

ANNONACEAE Soursop family
DEGENERIACEAE Masiratu family
EUPOMATIACEAE Bolwarra family
HIMANTANDRACEAE Pigeonberry-ash family
MAGNOLIACEAE Tuliptree family *see p.389*
MYRISTICACEAE Nutmeg family

Malpighiales

ACHARIACEAE Chaulmoogra family
BALANOPACEAE Pimplebark family
BONNETIACEAE Cascarilla family
CALOPHYLLACEAE Takamaka family
CARYOCARACEAE Souari-tree family
CENTROPLACACEAE Biku-biku family
CHRYSOBALANACEAE Cocoplum family
CLUSIACEAE Mangosteen family
CTENOLOPHONACEAE Litoh family
DICHAPETALACEAE Ratbane family
ELATINACEAE Waterwort family
ERYTHROXYLACEAE Coca family
EUPHORBIACEAE Spurge family *see p.390*
EUPHRONIACEAE Euphronia family
GOUPIACEAE Kopi family
HUMIRIACEAE Umiri family
HYPERICACEAE St. John's-wort family
IRVINGIACEAE Ogbono-nut family
IXONANTHACEAE Twentymen-tree family
LACISTEMATACEAE Cemp-wood family
LINACEAE Flax family
LOPHOPYXIDACEAE Koteb family
MALPIGHIACEAE Acerola family
OCHNACEAE Mickey-Mouse-plant family
PANDACEAE Kana-nut family
PASSIFLORACEAE Passionfruit family
PERACEAE Lightning-bush family
PHYLLANTHACEAE Leafflower family
PICRODENDRACEAE Hollyspurge family
PODOSTEMACEAE Riverweed family
PUTRANJIVACEAE Childlife-tree family
RAFFLESIACEAE Corpse-flower family
RHIZOPHORACEAE Mangrove family
SALICACEAE Willow family *see p.391*
TRIGONIACEAE Triangle-vine family
VIOLACEAE Violet family *see p.392*

Malvales

BIXACEAE Annatto family
CISTACEAE Rock-rose family
CYTINACEAE Rockrose-rape family
DIPTEROCARPACEAE Maranti family
MALVACEAE Mallow family *see p.393*
MUNTINGIACEAE Bajelly-tree family
NEURADACEAE Pietsnot family
SARCOLAENACEAE Tunic-bells family
SPHAEROSEPALACEAE Lombiry family
THYMELAEACEAE Mezereon family

Metteniusales

METTENIUSACEAE Urupagua family

Myrtales

ALZATEACEAE Wantsum family
COMBRETACEAE Bushwillow family
CRYPTERONIACEAE Bekoi family
LYTHRACEAE Pomegranate family
MELASTOMATACEAE Senduduk family
MYRTACEAE Myrtle family *see p.394*
ONAGRACEAE Fuchsia family *see p.395*
PENAEACEAE Cape-fellwort family
VOCHYSIACEAE Quaruba family

Nymphaeales

CABOMBACEAE Fanwort family
HYDATELLACEAE Watertufts family
NYMPHAEACEAE Water lily family *see p.396*

Oxalidales

BRUNELLIACEAE Palo-bobo family
CEPHALOTACEAE Albany-pitcherplant family
CONNARACEAE Zebrawood family
CUNONIACEAE Butterspoon-tree family
ELAEOCARPACEAE Fairy-petticoats family
HUACEAE Cameroon-garlic family
OXALIDACEAE Wood-sorrel family *see p.397*

Pandanales

CYCLANTHACEAE Panama hat family
PANDANACEAE Screwpine family
STEMONACEAE Baibu family
TRIURIDACEAE Threetails family
VELLOZIACEAE Baboon-tail family

Paracryphiales

PARACRYPHIACEAE Possumwood family

Petrosaviales

PETROSAVIACEAE Oze-so family

Picramniales

PICRAMNIACEAE Bitterbush family

Piperales

ARISTOLOCHIACEAE Birthwort family
PIPERACEAE Pepper family
SAURURACEAE Lizard's-tail family

Poales

BROMELIACEAE Pineapple family *see p.398*
CYPERACEAE Sedge family *see p.399*
ECDEIOCOLEACEAE Kwongan-rush family
ERIOCAULACEAE Pipewort family
FLAGELLARIACEAE Whip-vine family
JOINVILLEACEAE Ohe family
JUNCACEAE Rush family
MAYACACEAE Bog-moss family
POACEAE Grass family *see p.400*

RAPATEACEAE Tow-tow family
RESTIONACEAE Fynbos family
THURNIACEAE Palmiet family
TYPHACEAE Bulrush family
XYRIDACEAE Yellow-eyed-grass family

Proteales
NELUMBONACEAE Sacred-lotus family
PLATANACEAE Plane-tree family
PROTEACEAE Sugarbush family
SABIACEAE Pao-hua family

Ranunculales
BERBERIDACEAE Barberry family
CIRCAEASTERACEAE Witch's-star family
EUPTELEACEAE Asian-elm family
LARDIZABALACEAE Zabala-fruit family
MENISPERMACEAE Moonseed family
PAPAVERACEAE Poppy family *see p.401*
RANUNCULACEAE Buttercup family *see p.402*

Rosales
BARBEYACEAE Elm-olive family
CANNABACEAE Hemp family
DIRACHMACEAE Rachman family
ELAEAGNACEAE Oleaster family
MORACEAE Mulberry family *see p.403*
RHAMNACEAE Buckthorn family
ROSACEAE Rose family *see p.404*
ULMACEAE Elm family
URTICACEAE Nettle family *see p.405*

Santalales
BALANOPHORACEAE Snake-mushroom family
LORANTHACEAE Showy-mistletoe family
MISODENDRACEAE Feathery-mistletoe family
OLACACEAE Tallow-wood family
OPILIACEAE Bally-coma family
SANTALACEAE Sandalwood family
SCHOEPFIACEAE Whitewood family

Sapindales
ANACARDIACEAE Cashew family *see p.406*
BIEBERSTEINIACEAE Khardug family
BURSERACEAE Frankincense-and-myrrh family
KIRKIACEAE White-seringa family
MELIACEAE Neem family
NITRARIACEAE Nitrebush family
RUTACEAE Citrus family *see p.407*
SAPINDACEAE Maple family *see p.408*
SIMAROUBACEAE Tree-of-heaven family

Saxifragales
ALTINGIACEAE Sweetgum family
APHANOPETALACEAE Gum-vine family
CERCIDIPHYLLACEAE Caramel-tree family
CRASSULACEAE Stonecrop family
CYNOMORIACEAE Tarthuth family
DAPHNIPHYLLACEAE Laurel-leaf family
GROSSULARIACEAE Gooseberry family
HALORAGACEAE Water-milfoil family
HAMAMELIDACEAE Witch-hazel family
ITEACEAE Sweetspire family
PAEONIACEAE Peony family
PENTHORACEAE Ditch-stonecrop family
PERIDISCACEAE Ringflower family
SAXIFRAGACEAE Saxifrage family *see p.409*
TETRACARPAEACEAE Delicate-laurel family

Solanales
CONVOLVULACEAE Bindweed family *see p.410*
HYDROLEACEAE Fiddleleaf family
MONTINIACEAE Wild-clovebush family
SOLANACEAE Nightshade family *see p.411*
SPHENOCLEACEAE Gooseweed family

Trochodendrales
TROCHODENDRACEAE Wheel-tree family

Vahliales
VAHLIACEAE Flindersbush family

Vitales
VITACEAE Grapevine family *see p.412*

Zingiberales
CANNACEAE Canna-lily family
COSTACEAE Spiral-ginger family
HELICONIACEAE Parrot-flower family
LOWIACEAE Orchid-lily family
MARANTACEAE Prayer-plant family
MUSACEAE Banana family
STRELITZIACEAE Traveller's-palm family
ZINGIBERACEAE Zingiberaceae family *see p.413*

Zygophyllales
KRAMERIACEAE Ratany family
ZYGOPHYLLACEAE Twinleaf family

CONIFERS AND CYCADS (GYMNOSPERMS)

Araucariales
ARAUCARIACEAE Kauri-tree family
PODOCARPACEAE Yewpine family

Cupressales
CUPRESSACEAE Cypress family *see p.414*
SCIADOPITYACEAE Umbrella-pine family
TAXACEAE Yew family

Cycadales
CYCADACEAE Sago family
ZAMIACEAE Coontie family

Ephedrales
EPHEDRACEAE Jointfir family

Ginkgoales
GINKGOACEAE Maidenhair-tree family

Gnetales
GNETACEAE Emping family

Pinales
PINACEAE Pine family *see p.415*

Welwitschiales
WELWITSCHIACEAE Tumbo family

FERNS (PTERIDOPHYTES)

Cyatheales
CYATHEACEAE Tree-fern family *see p.416*

Equisetales
EQUISETACEAE Horsetail family *see p.417*

Gleicheniales
DIPTERIDACEAE Double-fern family
GLEICHENIACEAE Forking-fern family
MATONIACEAE Umbrella-fern family

Hymenophyllales
HYMENOPHYLLACEAE Filmy-fern family

Marattiales
MARATTIACEAE King-fern family

Ophioglossales
OPHIOGLOSSACEAE Adder's-tongue family

Osmundales
OSMUNDACEAE Royal-fern family

Polypodiales
ASPLENIACEAE Spleenwort family
CYSTODIACEAE Rowan-fern family
DENNSTAEDTIACEAE Bracken family
LINDSAEACEAE Lace-fern family
LONCHITIDACEAE Velvet-fern family
POLYPODIACEAE Polypody family *see p.418*
PTERIDACEAE Ribbon-fern family
SACCOLOMATACEAE Pouch-fern family

Psilotales
PSILOTACEAE Whisk-fern family

Salviniales
MARSILEACEAE Pillwort family
SALVINIACEAE Water-fern family

Schizaeales
SCHIZAEACEAE Fan-fern family

LYCOPODS

Isoëtales
ISOËTACEAE Quillwort family

Lycopodiales
LYCOPODIACEAE Clubmoss family *see p.419*

Selaginellales
SELAGINELLACEAE Spikemoss family

HORNWORTS

Anthocerotales
ANTHOCEROTACEAE

Dendrocerotales
DENDROCEROTACEAE

Leiosporocerotales
LEIOSPOROCEROTACEAE

Notothyladales
NOTOTHYLADACEAE

Phymatocerotales
PHYMATOCEROTACEAE

MOSSES (BRYOPHYTES)

Andreaeales
ANDREAEACEAE

Andreaeobryales
ANDREAEOBRYACEAE

Archidiales
ARCHIDIACEAE

Bryales
AULACOMNIACEAE
BARTRAMIACEAE
BRYACEAE
CATOSCOPIACEAE
HYPNODENDRACEAE
LEMBOPHYLLACEAE
MEESIACEAE
MITTENIACEAE
MNIACEAE
PSEUDODITRICHACEAE
RACOPILACEAE
RHIZOGONIACEAE
SPIRIDENTACEAE
TIMMIACEAE

Buxbaumiales
BUXBAUMIACEAE

Dicranales
BRUCHIACEAE
BRYOXIPHIACEAE
DICNEMONACEAE
DICRANACEAE
DITRICHACEAE
EUSTICHIACEAE
PHYLLODREPANIACEAE
PLEUROPHASCACEAE
RHABDOWEISIACEAE
SORAPILLACEAE
VIRIDIVELLERACEAE

Fissidentales
FISSIDENTACEAE

Funariales
DISCELIACEAE
EPHEMERACEAE
FUNARIACEAE
GIGASPERMACEAE
OEDIPODIACEAE
SPLACHNACEAE
SPLACHNOBRYACEAE

Grimmiales
GRIMMIACEAE
PTYCHOMITRIACEAE

Hookeriales
DALTONIACEAE
HOOKERIACEAE

Hypnales
AMBLYSTEGIACEAE
BRACHYTHECIACEAE
ENTODONTACEAE
FABRONIACEAE
HYLOCOMIACEAE
HYPNACEAE
LESKEACEAE
MYRINIACEAE
MYURIACEAE
ORTHORRHYNCHIACEAE
PHYLLOGONIACEAE
PLAGIOTHECIACEAE
PLEUROZIOPSACEAE
PTERIGYNANDRACEAE
SEMATOPHYLLACEAE
STEREOPHYLLACEAE
THAMNOBRYACEAE
THELIACEAE
THUIDIACEAE

Hypnobryales
ANOMODONTACEAE
ECHINODIACEAE

Isobryales
CYRTOPODACEAE
FONTINALACEAE
HYDROPOGONACEAE
LEPYRODONTACEAE
PRIONODONTACEAE
PTYCHOMNIACEAE
REGMATODONTACEAE
RUTENBERGIACEAE
TRACHYPODACEAE
WARDIACEAE

Leucodontales
CLIMACIACEAE
CRYPHAEACEAE
HEDWIGIACEAE
LEPTODONTACEAE
LEUCODONTACEAE
METEORIACEAE
NECKERACEAE
PTEROBRYACEAE

Orthotrichales
ERPODIACEAE
HELICOPHYLLACEAE
MICROTHECIELLACEAE
ORTHOTRICHACEAE
RHACHITHECIACEAE

Polytrichales
POLYTRICHACEAE

Pottiales
BRYOBARTRAMIACEAE
CALYMPERACEAE
CINCLIDOTACEAE
ENCALYPTACEAE
POTTIACEAE
SERPOTORTELLACEAE

Schistostegiales
SCHISTOSTEGACEAE

Seligerales
SELIGERIACEAE

Sphagnales
SPHAGNACEAE

Tetraphidae
CALOMNIACEAE

Tetraphidales
TETRAPHIDACEAE

LIVERWORTS

Blasiales
BLASIACEAE

Calobryales
HAPLOMITRIACEAE

Fossombroniales
ALLISONIACEAE
CALYCULARIACEAE
FOSSOMBRONIACEAE
MAKINOACEAE
PETALOPHYLLACEAE

Jungermanniales
ACROBOLBACEAE
ADELANTHACEAE
ANASTROPHYLLACEAE
ANTHELIACEAE
ARNELLIACEAE
BALANTIOPSIDACEAE
BLEPHARIDOPHYLLACEAE
BLEPHAROSTOMATACEAE
BREVIANTHACEAE
CALYPOGEIACEAE
CEPHALOZIACEAE
CEPHALOZIELLACEAE
ENDOGEMMATACEAE
GEOCALYCACEAE
GROLLEACEAE
GYMNOMITRIACEAE
GYROTHYRACEAE
HARPANTHACEAE
HERBERTACEAE
HYGROBIELLACEAE
JACKIELLACEAE
JUNGERMANNIACEAE
LEPICOLEACEAE
LEPIDOZIACEAE
LOPHOCOLEACEAE
LOPHOZIACEAE
MASTIGOPHORACEAE
MYLIACEAE
NOTOSCYPHACEAE
PLAGIOCHILACEAE
PSEUDOLEPICOLEACEAE
SACCOGYNACEAE
SCAPANIACEAE
SCHISTOCHILACEAE
SOLENOSTOMATACEAE
SOUTHBYACEAE
STEPHANIELLACEAE
TRICHOCOLEACEAE
TRICHOTEMNOMATACEAE

Lunulariales
LUNULARIACEAE

Marchantiales
AYTONIACEAE
CLEVEACEAE
CONOCEPHALACEAE
CORSINIACEAE
CYATHODIACEAE
DUMORTIERACEAE
MARCHANTIACEAE
MONOCLEACEAE
MONOSOLENIACEAE
OXYMITRACEAE
RICCIACEAE
TARGIONIACEAE
WIESNERELLACEAE

Metzgeriales
ANEURACEAE
METZGERIACEAE

Neohodgsoniales
NEOHODGSONIACEAE

Pallaviciniales
HYMENOPHYTACEAE
MOERCKIACEAE
PALLAVICINIACEAE
PHYLLOTHALLIACEAE
SANDEOTHALLACEAE

Pelliales
NOTEROCLADACEAE
PELLIACEAE

Pleuroziales
PLEUROZIACEAE

Porellales
FRULLANIACEAE
GOEBELIELLACEAE
JUBULACEAE
LEJEUNEACEAE
LEPIDOLAENACEAE
PORELLACEAE
RADULACEAE

Ptilidiales
HERZOGIANTHACEAE
NEOTRICHOCOLEACEAE
PTILIDIACEAE

Sphaerocarpales
MONOCARPACEAE
RIELLACEAE
SPHAEROCARPACEAE

Treubiales
TREUBIACEAE

glossary

ABAXIAL The side of an organ that faces away from the stem or supporting structure; typically used to describe the lower surface of a leaf.

ACCESSORY FRUIT A fruit consisting of the ovary together with another plant part, such as the swollen end of a flower stalk. Examples include apples and rose hips. Also known as false fruit.

ACHENE A dry, one-seeded fruit that does not open and contains a single seed.

ADVENTITIOUS Arising from places where growths do not normally occur: adventitious roots may arise from stems, for example.

AERIAL ROOT A root that grows from the stem of a plant that is located above the ground.

AGGREGATE FRUIT A compound fruit that develops from several ovaries. The ovaries are all from the carpels of a single flower and the separate fruits are joined together. Examples are raspberries and blackberries.

ALGAE A group of simple, flowerless, mainly aquatic, plantlike organisms that contain the green pigment chlorophyll but do not have true stems, roots, leaves, and vascular tissue. An example is seaweed.

ANGIOSPERM A flowering plant that bears ovules, later seeds, enclosed in ovaries. Angiosperms are divided into two main groups for classification: monocots and eudicots.

ANNUAL A plant that completes its entire life cycle—germination, flowering, seeding, and dying—in one growing season.

ANNULUS (PL. ANNULI) The ring of thick-walled cells involved in opening a fern's sporangium and releasing the spores.

ANTHER The part of a flower's stamen that produces pollen; it is usually borne on a filament.

ANTHOCYANIN Plant pigment molecules that are responsible for red, blue, and purple colors in leaves and flowers.

APEX The tip or growing point of a leaf, shoot, or root.

APOMIXIS The process of asexual reproduction.

ARIL A berrylike, fleshy, hairy, or spongy layer around some seeds.

AWN A bristle that grows from the spikelets of certain grasses, including cultivated cereals.

AXIL The upper angle between a stem and a leaf, where an axillary bud develops.

AXILLARY BUD A bud that develops in the axil of a leaf.

BALAUSTA A fruit with a tough skin, or pericarp, and many cells, each containing a seed. A typical example is pomegranate.

BARK The tough covering on woody roots, trunks, and branches.

BERRY A fruit with soft, juicy flesh surrounding one or more seeds that have developed from a single ovary.

BIENNIAL A plant that flowers and dies in the second growing season after germination.

BIPINNATE A compound leaf whose leaflets are divided into yet smaller leaflets, such as mimosa leaves.

BISEXUAL *See* Perfect flower.

BLADE The whole part of a leaf except for its stalk (petiole). The shape of a leaf blade and its edges—or margins—are important characteristics of a plant.

BRACT A leaf that has modified into an attractive or protective structure—usually to protect buds—around the base of a flower or flower cluster. Some bracts are large, brightly colored, and resemble flower petals to attract beneficial insects, while others look like leaves, although they may be smaller and shaped differently from the other leaves on the plant.

BROAD-LEAVED Describes trees and shrubs that have broad, flat, usually deciduous leaves, in contrast to the narrow, needlelike leaves of conifers.

BROMELIAD A rosette-forming plant in the pineapple family (Bromeliaceae) that may also be an epiphyte. It has spirally arranged and sometimes toothed leaves, and stems that can become woody.

BUD An immature organ or shoot enclosing an embryonic branch, leaf, inflorescence, or flower.

BULB A modified underground bud that acts as a storage organ. It consists of one or more buds and layers of swollen, colorless, fleshy scale leaves, packed with stored food, on a shortened, disklike stem.

BULBLET A small bulblike organ, often borne in a leaf axil, occasionally on a stem or flower head.

BUNDLE SHEATH A cylinder of cells surrounding a vascular bundle inside a plant leaf.

BURR A prickly or spiny dry fruit.

CALYX (PL. CALYCES) The outer part of a flower, formed from a ring of sepals, that is sometimes showy and brightly colored but usually small and green. The calyx forms a cover that encloses the petals while in bud.

CAMBIUM A layer of tissue capable of producing the new cells that increase the girth of stems and roots.

CAPITULUM (PL. CAPITULA) A group of flowers (inflorescence) on a stem that together look like a single flower head. An example is a sunflower.

CAPSULE A dry fruit containing many seeds that has developed from an ovary formed from two or more carpels. It splits open when ripe to release its seeds.

CAROTENOIDS The plant pigment molecules responsible for yellow and orange hues.

CARPEL The female reproductive part of a flower that consists of an ovary, stigma, and style.

CARYOPSIS (PL. CARYOPSES) A dry, single-seeded fruit that is indehiscent. Grasses typically have rows or clusters of edible caryopses, or grains.

CATKIN A long, thin cluster of small flowers with inconspicuous or no petals that hangs from a tree such as a hazel or birch.

CAULIFLOROUS A term used to describe flowers and fruits that develop directly on a tree's trunk or branches rather than at the ends of its twigs.

CHASMOGAMOUS Having open, blooming flowers with exposed reproductive organs for cross-pollination.

CHLOROPHYLL The green pigment inside plant cells that allows leaves and sometimes stems to absorb light and carry out photosynthesis.

CHLOROPLAST The particles inside plant cells that contain chlorophyll, where starch is formed during photosynthesis.

CIRCINATE Coiled inward, as in fern fiddleheads.

CLADOPHYLL A modified stem that both resembles and performs the function of a leaf.

CLASS In taxonomy, the rank below kingdom and above order—for example, monocots and eudicots.

CLEISTOGAMOUS A flower that pollinates itself without the bloom opening. Opposite: Chasmogamous.

COLEOPTILE A sheath that protects the shoot emerging from a monocot seed as it grows through the soil.

COMPOUND LEAF A leaf composed of two or more similar parts (leaflets).

CONE The densely clustered bracts of conifers and some flowering plants that often develop into a woody, seed-bearing structure, such as a pine cone.

CONIFERS Mostly evergreen trees or shrubs, which usually have needlelike leaves and naked seeds that develop on scales inside cones.

CORM A bulblike underground swollen stem or stem base, often surrounded by a papery tunic.

COROLLA A ring of petals on a flower head.

CORTEX The region of tissue between the epidermis or bark and the vascular cylinder.

CORYMB A broad, flat-topped, or domed inflorescence of stalked flowers or flower heads arising at different levels on alternate sides of an axis.

COTYLEDON A seed leaf that acts as a food store or unfurls shortly after germination to fuel a seed's growth.

CREMOCARP A small, dry fruit that forms in two flattened halves, each containing one seed.

CROSS-FERTILIZATION The fertilization of the ovules of a flower as a result of cross-pollination.

CROSS-POLLINATION The transfer of pollen from the anthers of a flower on one plant to the stigma of a bloom on another plant. *See also* Self-pollination.

CUCURBIT A plant from the cucumber or gourd family (Cucurbitales) which includes melon, pumpkin, and squash.

CULM The jointed, usually hollow, flowering stem of a grass or bamboo.

CULTIVAR (CV.) A contraction of "cultivated variety," used to describe a plant that usually only exists in human cultivation.

CUPULE A cup-shaped structure made of bracts joined together.

CUTICLE A protective, waxy, water-repellent coating of the outer cells of the epidermis of some plants.

CYME A branched inflorescence, flat or round-topped, with each axis ending in a flower, the oldest at the center and the youngest arising in succession from the axils of bracteoles (secondary bracts).

DECIDUOUS Used to describe plants that shed their leaves at the end of the growing season and then renew them at the beginning of the next. Semi-deciduous plants lose only some of their leaves at the end of the growing season.

DEHISCENT FRUIT A dry fruit that splits or bursts to release its seeds.

DICHASIUM *See* Inflorescence.

DICOT, DICOTYLEDON A term—now considered obsolete or incorrect in evolutionary history—that was used to describe a flowering plant with two seed leaves, or cotyledons. *See also* Eudicot, Monocot.

DIOECIOUS A plant that bears unisexual flowers, with male and female blooms occurring on separate plants. *See also* Monoecious.

DIVISION The propagation of a plant by splitting it into two or more parts, each with its own section of root system and one or more shoots or dormant buds.

DOMATIUM (PL. DOMATIUM) A structure produced by a plant that is inhabited by animals, usually cavities in the roots, stems, or leaf veins, and often associated with ants.

DRIP TIP The edge of a leaf or leaflet that helps to direct rainwater runoff.

DRUPE A fleshy fruit containing a seed with a hard coat (endocarp).

EMERGENT Coming out of, or emerging from.

ENDOCARP The innermost layer of the pericarp of a fruit.

EPIDERMIS The outer protective layer of cells of a plant.

EPIPHYTE A plant that grows on the surface of another plant without being parasitic or stealing nutrients from its host; it obtains moisture and nutrients from the atmosphere without rooting into the soil.

EUDICOT, EUDICOTYLEDON A flowering plant that has two seed leaves, or cotyledons, including many plants formerly described as "dicots." Most eudicots have large leaves with branching veins and the floral parts, such as petals and sepals, arranged in groups of multiples of four or five. *See also* Monocot.

EVERGREEN Describes plants that retain their foliage for more than one growing season; semi-evergreen plants retain only a small portion of their leaves for more than one season.

EXOCARP The outer layer of the pericarp of a fruit. The exocarp is often thin and hard, or like a skin.

EXODERMIS A specialized layer in a root beneath the epidermis or velamen.

FALSE FRUIT *See* Accessory fruit.

FAMILY In plant classification, a group of related genera; the family Rosaceae, for example, includes the genera *Rosa*, *Sorbus*, *Rubus*, *Prunus*, and *Pyracantha*.

FERN A flowerless, spore-producing plant consisting of roots, stems, and leaflike fronds. *See also* Frond.

FIDDLEHEAD The coiled young leaf of a fern.

FILAMENT The stalk that bears the anther in a flower.

FLORET A small flower, usually one of many florets that make up a composite flower such as a daisy.

FLOWER The reproductive organ of a great many plant genera. Each flower consists of an axis which bears four types of reproductive organs: sepals, petals, stamens, and carpels.

FOLLICLE A dry fruit, similar to a pod, that develops from a single-chambered ovary with one seam that splits open to release the seeds. Most follicles are aggregate fruits.

FROND 1. The leaflike organ of a fern. Some ferns produce both barren fronds and fertile fronds, which bear spores. 2. Large, usually compound leaves such as those of palms.

FRUIT The fertilized, ripe ovary of a plant, which contains one or more seeds, such as berries, hips, capsules, or nuts. The term is also used to describe edible fruits.

FRUITLET A small fruit that comprises part of an aggregate fruit such as blackberry.

FUNGUS A single-celled or multicelled organism that is a member of the separate kingdom Fungi: molds, yeasts, and mushrooms, for example.

FUNICLE The tiny stalk inside a pod from which a seed hangs.

GENUS (PL. GENERA) A category in plant classification ranked between family and species.

GERMINATION The physical and chemical changes that take place when a seed starts to grow and develop into a plant.

GLABROUS Smooth and hairless.

GLUME A scaly, protective bract, usually one of a pair, at the base of a grass or sedge spikelet.

GYMNOSPERM A plant with seeds that develop without an ovary to enclose and protect them while they mature. Most gymnosperms are conifers, whose seeds form on scales and mature within cones.

HARDY Describes plants that can withstand freezing temperatures in winter.

HAUSTORIA Specialized parasitic plant roots that protrude into the tissues of a host plant.

HEMIPARASITE A parasite that has green leaves and can photosynthesize. An example is mistletoe (*Viscum album*).

HERBACEOUS A non-woody plant in which the upper parts die down to a rootstock at the end of the growing season. The term is mainly used to describe perennial plants, although botanically it also applies to annuals and biennials.

HERMAPHRODITE Plant species with flowers in which the male stamens and female pistil are present together in single bisexual or perfect flowers.

HESPERIDIUM The fruit of a citrus plant with a thick, leathery rind, such as a lemon or an orange.

HETEROPHYLLOUS Plants with leaves of different shapes or forms on the same plant that may be adapted to perform in particular conditions such as sun or shade.

HOLOPARASITE A parasitic plant with no leaves that is totally dependent on its host for food and water.

HYBRID The offspring of genetically different parent plants. Hybrids between species of the same genus are described as interspecific hybrids; those between different—but usually closely related—genera are known as intergeneric hybrids.

HYSTERANTHOUS A plant whose flowers open before its leaves, such as forsythia or witch hazel.

IMPERFECT FLOWER Flowers that contain either only male or only female reproductive organs. Also known as unisexual.

INDEHISCENT Describes a fruit that does not split open to release its seeds, such as a hazelnut.

INDUSIUM A thin flap of tissue that covers a fern's sorus.

INFLORESCENCE A group of flowers borne on a single axis (stem), such as a raceme, panicle, or cyme.

INTERNODE The portion of stem between two nodes.

INVOLUCRE A ring of leaflike bracts below a flower head.

KEEL 1. A prominent longitudinal ridge, usually on the underside of a leaf, that resembles the keel of a boat. 2. The two lower, fused petals of a pealike flower.

LABELLUM A lip, particularly the prominent third petal of iris or orchid flowers. *See also* Lip.

LAMINA A broad, flat structure; for example, the blade of a leaf.

LATERAL A side growth that arises from a shoot or root.

LEAF Typically a thin, flat lamina (blade) growing out of a stem that is supported by a network of veins. Its main function is to collect the energy from sunlight that the plant needs in order to photosynthesize.

LEAFLET One of the subdivisions of a compound leaf.

LEGUME A dehiscent pod that splits along two sides to disperse ripe seed.

LEMMA The outermost of two bracts enclosing a grass flower. *See also* Palea.

LENTICEL A hole in the stem that allows gases to pass between the plant's cells and the air around them.

LIGNIN A hard substance in all vascular plants that enables them to grow upright and remain standing.

LINEAR Very narrow leaves that have parallel sides.

LIP A prominent lower lobe on a flower, formed by one or more fused petals or sepals. *See also* Labellum.

LIVERWORT A simple, flowerless plant that lacks true roots. It has leaflike stems or lobed leaves, reproduces by shedding spores, and is usually found in damp habitats.

LOBED Describes a part of a plant, such as a leaf, that has curved or rounded parts.

LOCULE A compartment or chamber of an ovary or anther.

MARGIN The outer edge of a leaf.

MERISTEM Plant tissue that is able to divide to produce new cells. Both shoot and root tips can contain meristematic tissue and may be used for micropropagation.

MESOCARP The middle layer of the pericarp. In many fruits the mesocarp is the fleshy part of the fruit. In some pericarps the mesocarp is missing.

MESOPHYLL The soft, inner tissue (parenchyma) of a leaf, between the upper and lower layers of the epidermis, that contains chloroplasts for photosynthesis.

MIDRIB The primary, usually central, vein of a leaf.

MONOCARPIC A plant that flowers and fruits only once before it dies; such plants may take a number of years to reach flowering size.

MONOCOT, MONOCOTYLEDON A flowering plant that has only one cotyledon, or seed leaf, in the seed; it is also characterized by narrow, parallel-veined leaves. Monocot examples include lilies, irises, and grasses. *See also* Eudicot.

MONOECIOUS A plant with separate male and female flowers that are borne on the same plant. *See also* Dioecious.

MOSS A small, green, flowerless plant that has no true roots and grows in damp habitats. It reproduces by shedding spores.

MUCILAGE A gummy secretion present in various parts of plants, especially leaves.

MULTIPLE FRUIT A fruit that develops from several close-set flowers that fuse together to form a single fruit, such as pineapple.

MYCORRHIZA A mutually beneficial (symbiotic) relationship between a fungus and the roots of a plant.

NECTAR A sweet, sugary substance secreted by a nectary, to attract insects and other pollinators to a flower.

NECTARY A gland that secretes nectar. Nectaries are most frequently located in the flower of a plant, but are sometimes found on the leaves or stems.

NODE A point of a stem from which one or more leaves, shoots, branches, or flowers arise.

NODULE 1. A small knob on a root that contains nitrogen-fixing bacteria. 2. Small swellings on a leaf (on the petiole, midrib, lamina, or margin) that contain bacteria.

NUT A one-seeded, indehiscent fruit with a tough or woody coat, for example, an acorn. Less specifically, all fruits and seeds with woody or leathery coats.

NUTRIENTS Minerals (mineral ions) used to develop proteins and other compounds required for plant growth.

OFFSET A small plant that develops from a shoot growing out of an axillary bud on the parent plant.

ORDER In taxonomy, the rank below class and above family.

OVARY The lower part of the carpel of a flower, containing one or more ovules; it may develop into a fruit after fertilization.

OVULE The part of an ovary that develops into the seed after pollination and fertilization.

OVULIFEROUS SCALES The scales of a female cone, which bear ovules that become seeds once they have been fertilized.

PALEA (PL. PALEAE) The innermost of the two bracts enclosing a grass flower. *See also* Lemma.

PALMATE Has lobed leaflets that arise from a single point.

PANICLE A branched raceme.

PAPPUS An appendage or tuft of appendages that crowns the ovary or fruit in various seed plants and functions in wind dispersal of the fruit.

PARENCHYMA Soft plant tissue consisting of cells with thin walls.

PEDICEL The stalk bearing a single flower in an inflorescence.

PEDUNCLE The main stalk of an inflorescence, which holds a group of pedicels.

PENDENT Hanging downward.

PEPO A many-seeded, hard-skinned berry that forms the fruit of cucurbits such as pumpkins, watermelons, and cucumbers.

PERENNIAL A plant that lives for more than two years.

PERFECT FLOWER Flowers that contain both male and female reproductive organs. Also known as bisexual or hermaphrodite.

PERFOLIATE Stalkless leaves or bracts that encircle the stem, so that the stem appears to pass through the leaf blade.

PERIANTH The collective term for the calyx and corolla, particularly when they are very similar in form, as in many bulb flowers.

PERICARP The wall of a fruit that develops from the maturing ovary wall. In fleshy fruits the pericarp often has three distinct layers: exocarp, mesocarp, and endocarp. The pericarp of dry fruits is papery or feathery, but on fleshy fruits it is succulent and soft.

PETAL A modified leaf, usually brightly colored and sometimes scented, that attracts pollinators. A ring of petals on a flower is known as a corolla.

PETIOLE The stalk of a leaf.

PHLOEM The vascular tissue in plants that conducts sap containing nutrients (produced by photosynthesis) from the leaves to other parts of a plant.

PHOTOSYNTHESIS The process by which the energy in sunlight is trapped by green plants and used to carry out a chain of chemical reactions to create nutrients from carbon dioxide and water. A by-product is oxygen.

PHYLLOTAXIS The arrangement of leaves on a stem or branch.

PHYTOTELMA (PL. PHYTOTELMATA) A water-filled cavity in a plant that serves as a habitat.

PIN FLOWER A flower with a long style and relatively short stamens. Opposite: thrum flower.

PINNATE The arrangement of leaflets on opposite sides of the central stalk of a compound leaf.

PIONEER SPECIES A species that colonizes a new environment, for example after volcanic eruptions or fire, and starts a plant succession.

PISTIL *See* Carpel.

PLANTLET A young plant that develops on the leaf of a parent plant.

PLUMULE The first shoot that emerges from a seed when it germinates.

PNEUMATOPHORE An erect aerial root that protrudes upward through swampy soil with the ability to exchange gases or "breathe." Often found in mangroves.

POD A flattened, dry fruit that develops from a single ovary with one chamber.

POLLEN The small grains, formed in the anther of seed-bearing plants, which contain the male reproductive cells of the flower.

POLLINATION The transfer of pollen from an anther to the stigma of a flower.

POLLINATOR 1. The agent or means by which pollination is carried out; for example, via insects, birds, wind. 2. A plant required to ensure that seeds set on another self- or partially self-sterile plant.

POME The fleshy part of an apple or related fruit, consisting of an enlarged receptacle and the ovary and seeds.

PRICKLE A sharp outgrowth from the epidermis or cortex of a plant, which can be detached without tearing the part of the plant from which it is growing.

PROPAGATE To increase plants by seed or by vegetative means.

PROTANDROUS Hermaphroditic plants that are functionally male before becoming functionally female. Opposite: Protogynous.

PROTOGYNOUS Hermaphroditic plants that are functionally female before becoming functionally male. Opposite: Protandrous.

PSEUDOBULB A thickened, bulblike stem arising from a (sometimes very short) rhizome.

PTERIDOPHYTE A fern or fern-ally, such as horsetail or club moss, that reproduces using alternating generations, with the main generation producing spores.

RACEME A cluster of several or many separate flower heads borne singly on short stalks along a central stem, with the youngest flowers at the tip.

RACHIS The main axis of a compound leaf or inflorescence.

RADICLE The root of a plant embryo. The radicle is normally the first organ to appear when a seed germinates.

RECEPTACLE The enlarged or elongated tip of the stem from which all parts of a simple flower arise.

RECURVED Arched backward.

REFLEXED Bent completely backward.

REGMA (PL. REGMATA) A type of dry fruit consisting of three or more fused carpels that break apart explosively when mature.

REPLUM A partition in certain fruits, such as some legumes.

RESIN A thick, sticky substance formed of organic compounds that a tree produces to heal wounds in its bark that have been inflicted by pests or caused by physical damage.

RHIZOME A creeping underground stem that acts as a storage organ and produces shoots at its apex and along its length.

RHIZOSPHERE A root system and the immediate surrounding substrate.

ROOT The part of a plant, normally underground, that anchors it in the soil and through which water and nutrients are absorbed.

ROOT CAP A hood-shaped cap at the root tip that continually produces new cells to protect the root from abrasion as it grows in the soil.

ROOT HAIR A threadlike growth that develops behind the root cap. Root hairs extend the surface area of a root and increase the amount of water and nutrients it can absorb.

ROSETTE A cluster of leaves radiating from approximately the same point, often at ground level at the base of a very short stem.

RUNNER A horizontally spreading, usually slender, stem that runs aboveground and roots at the nodes to form new plants. *See also* Stolon.

SAMARA A dry, indehiscent, one-seeded fruit that has "wings" to ensure dispersal by the wind. Examples are ash, maple, and sycamore.

SAP The juice of a plant contained in the cells and vascular tissue.

SCALE A reduced leaf, usually membranous, that covers and protects buds, bulbs, and catkins.

SCHIZOCARP A papery, dry fruit, which breaks up into enclosed, single-seeded units that disperse separately when the seeds are ripe.

SCLEREID A plant cell with lignified, pitted walls.

SEAM (OR SUTURE) The edge of a pod where it breaks open.

SEED The ripened, fertilized ovule, containing a dormant embryo capable of developing into an adult plant.

SEEDLING A young plant that has developed from a seed.

SELF-INCOMPATIBLE Describes a plant that is unable to produce viable seed by fertilizing itself and needs a different pollinator in order for fertilization to take place. Also known as "self-sterile."

SELF-POLLINATION The transfer of pollen from the anthers to the stigma of the same flower, or alternatively to another flower on the same plant. *See also* Cross-pollination.

SELF-STERILE *See* Self-incompatible.

SEPAL The outer whorl of the perianth of a flower, usually small and green, but sometimes colored and petallike.

SEPTUM A partition, such as a separating wall, in a fruit.

SHOOT A developed bud or young stem.

SIMPLE FRUIT A fruit that forms from a single ovary. Examples include berries, drupes, and nuts.

SIMPLE LEAF A leaf that is formed in one piece.

SORUS (PL. SORI) 1. A cluster of sporangia on the underside of a fern leaf. 2. A spore-producing structure of some lichens and fungi.

SPADIX A fleshy flower spike that bears numerous small flowers, usually sheathed by a spathe.

SPATHE A bract that surrounds a single flower or a spadix.

SPECIES (SP.) In plant classification, a class of plants whose members have the same main characteristics and are able to breed with one another.

SPIKE A long flower cluster with individual flowers borne on very short stalks or attached directly to the main stem.

SPINE A stiff, sharp-tipped, modified leaf or leaf parts such as stipules or petioles.

SPORE The minute, reproductive structure of flowerless plants, such as ferns, fungi, and mosses.

SPORANGIUM (PL. SPORANGIA) A body that produces spores on a fern.

SPUR 1. A hollow projection from a petal, which often produces nectar. 2. A short branch that bears a group of flower buds (such as those found on fruit trees).

STAMEN The male reproductive part of a flower comprised of the pollen-producing anther and usually its supporting filament or stalk.

STAMINOID A sterile structure resembling a stamen.

STANDARD The upper petal in certain flowers in the pea family.

STEM The main axis of a plant, usually aboveground, that supports structures such as branches, leaves, flowers, and fruit.

STIGMA The female part of a flower that receives pollen before fertilization. The stigma is situated at the tip of the style.

STIPULE A leafy outgrowth, often one of a pair.

STOLON A horizontally spreading or arching stem, usually aboveground, that roots at its tip to produce a new plant. Often confused with a runner.

STOMA (PL. STOMATA) A microscopic pore on the surface of aerial parts of plants (leaves and stems), allowing transpiration to take place.

STRIATE Striped.

STYLE The stalk that connects the stigma to the ovary in flowers.

SUBMERGENT A plant that lives entirely underwater.

SUBSPECIES A major division of a species, where the distinction is not complete.

SUCCULENT A drought-resistant plant with thick, fleshy leaves or stems adapted to store water. All cacti are succulents.

SUCKER A new shoot that develops from the roots or the base of a plant and rises from below ground level.

SYMBIOSIS Living together in a mutually beneficial relationship.

SYMBIOTIC Mutually beneficial.

TAPROOT The primary, downward-growing root of a plant, such as dandelion.

TENDRIL A modified leaf, branch, or stem, usually long and slender, that can attach itself to a support.

TEPAL A single segment of a perianth that cannot be distinguished as either a sepal or a petal, as in crocuses or lilies.

TERMINAL BUD A bud that forms at the apex or tip of a stem.

TESTA The hard, protective coating around a fertilized seed that prevents water from entering the seed until it is ready to germinate.

THIGMOTROPISM A plant's ability to grow, bend, and twine in response to a touch stimulus.

THORN A modified stipule or simple outgrowth from a stem that forms a sharp, pointed end.

THRUM FLOWER A flower with a short style in which only the stamens are visible in the throat of the corolla. Opposite: pin flower.

THYRSE A compound inflorescence with numerous flowering stalks that branch in pairs from the main stem.

TRANSPIRATION The loss of water by evaporation from plant leaves and stems.

TRICHOME Any type of outgrowth from the surface tissue of a plant, such as a hair, scale, or prickle.

TRIFOLIOLATE Describes a compound leaf that has three leaflets growing from the same point.

TUBER A swollen, usually underground, organ derived from a stem or a root, used for food storage.

UMBEL A flat or round-topped inflorescence in which the flower stalks grow from a single point at the top of a supporting stem.

UNISEXUAL Flower producing either pollen (male) or ovules (female).

VARIEGATED Describes irregular arrangements of pigments, usually the result of either mutation or disease, mainly in leaves.

VARIETY Another division of a species, generally differing from the type in one character only.

VASCULAR BUNDLE A strand of connecting vessels in the vein or stem of a plant leaf.

VASCULAR PLANT A plant that has food-conducting tissues (the phloem) and water-conducting tissues (xylem).

VEIN A vascular structure in a leaf, surrounded by the bundle sheath, often visible as lines on a leaf surface.

VELAMEN Water-absorbing tissue that covers the aerial roots of certain plants, including many epiphytes.

VENATION The arrangement of veins in a leaf.

VERNATION The folding of leaves in a bud.

VIVIPAROUS 1. Describes a plant that forms plantlets on leaves, flower heads, or stems. 2. Also applied loosely to plants that produce bulblets on bulbs.

WHORL An arrangement of three or more organs that all originate from the same point.

WINGED FRUIT A fruit with fine, papery structures that are shaped like wings to help carry the fruits through the air.

XYLEM The woody part of plants, consisting of supporting and water-conducting tissue.

ZYGOMORPHIC Describes a flower that can only be cut in one plane to produce two halves that are mirror images of each other.

index

Page numbers in **bold** relate to pages with the most information. Page numbers in *italics* relate to information found in diagrams.

D

E

F

Q

R

S

list of botanical art

acknowledgments

The Publisher would like to thank the directors and staff at the Royal Botanic Gardens, Kew for their enthusiastic help and support throughout the preparation of this book, in particular Richard Barley, Director of Horticulture; Tony Sweeney, Director of Wakehurst; and Kathy Willis, Director of Science. Special thanks to all at Kew Publishing, especially Gina Fullerlove, Lydia White, and Pei Chu, and to Martyn Rix for his detailed comments on the text. Thanks to the Kew Library, Art, and Archives team, particularly Craig Brough, Julia Buckley, and Lynn Parker, and also to Sam McEwen and Shirley Sherwood.

DK would also like to thank the many people who provided help and support with photoshoots in the tropical nursery and gardens at Kew and Wakehurst, and all who provided expert advice on specific details, notably Bill Baker, Sarah Bell, Mark Chase, Maarten Christenhusz, Chris Clennett, Mike Fay, Tony Hall, Ed Ikin, Lara Jewett, Nick Johnson, Tony Kirkham, Bala Kompalli, Carlos Magdalena, Keith Manger, Hugh McAllister, Kevin McGinn, Greg Redwood, Marcelo Sellaro, David Simpson, Raymond Townsend, Richard Wilford, and Martin Xanthos.

The Publisher would also like to thank Sylvia Myers and her team of volunteers at the London Wildlife Trust's Center for Wildlife Gardening (www.wildlondon.org), and Rachel Siegfried at Green and Georgeous flower farm in Oxfordshire for hosting photographic shoots. DK is also grateful to Joannah Shaw of Pink Pansy and Mark Welford of Bloomsbury Flowers for their help in sourcing plants for photoshoots, and to Dr. Ken Thompson for his help in the early stages of this book.

DK would also like to thank the following:

Additional picture research: Deepak Negi

Image retoucher: Steve Crozier

Creative Technical Support: Sonia Charbonnier, Tom Morse

Proofreader: Joanna Weeks

Indexer: Elizabeth Wise

PICTURE CREDITS

The publisher would like to thank the following for their kind permission to reproduce their photographs:

(Key: a-above; b-below/bottom; c-center; f-far; l-left; r-right; t-top)

4–5 Alamy Stock Photo: Gdns81 / Stockimo. **6 Dorling Kindersley:** Green and Gorgeous Flower Farm. **8–9 500px:** Azim Khan Ronnie. **10–11 iStockphoto.com:** Grafissimo. **12 Dorling Kindersley:** Neil Fletcher (tr); Mike Sutcliffe (tc). **14 Alamy Stock Photo:** Don Johnston PL (fcr). **Getty Images:** J&L Images / Photographer's Choice (c); Daniel Vega / age fotostock (cr). **Science Photo Library:** BJORN SVENSSON (cl). **15 Dorling Kindersley:** Gary Ombler: Center for Wildlife Gardening / London Wildlife Trust (br). **FLPA:** Arjan Troost, Buiten-beeld / Minden Pictures (c). **iStockphoto.com:** Alkalyne (cl). **16 Alamy Stock Photo:** Mark Zytynski (cb). **17 Alamy Stock Photo:** Pictorial Press Ltd. **18–19 iStockphoto.com:** ilbusca. **20 Science Photo Library:** Dr. Keith Wheeler (clb). **20–21 iStockphoto.com:** Brainmaster. **22–23 iStockphoto.com:** Pjohnson1. **23 Alamy Stock Photo:** Granger Historical Picture Archive (br). **24–25 Getty Images:** Ippei Naoi. **26 Alamy Stock Photo:** Emmanuel Lattes. **30–31 Alamy Stock Photo:** The Protected Art Archive (t). **31 Alamy Stock Photo:** ART Collection (br). **32–33 Science Photo Library:** Gustoimages. **33 Getty Images:** Universal History Archive / UIG (tr). **Science Photo Library:** Dr. Jeremy Burgess (br). **34–35 Amanita / facebook. com/Amanlta/ 500px.com/sotls. 40–41 Thomas Zeller / Filmgut. 41 123RF. com:** Mohammed Anwarul Kabir Choudhury (b). **43 © Board of Trustees of the Royal Botanic Gardens, Kew:** (br). **44 123RF.com:** Richard Griffin (bc, br). **Dreamstime.com:** Kazakovmaksim (l). **45 © Board of Trustees of the Royal Botanic Gardens, Kew. 46–47 Dreamstime.com:** Rootstocks. **47 Alamy Stock Photo:** Alfio Scisetti (tc). **50–51 Getty Images:** Michel Loup / Biosphoto. **52–53 Rosalie Scanlon Photography and Art, Cape Coral, FL., USA. 53 Alamy Stock Photo:** National Geographic Creative (br). **54 Alamy Stock Photo:** Angie Prowse (bl). **54–55 Alamy Stock Photo:** Ethan Daniels. **56–57 iStockphoto. com:** Nastasic. **58 Getty Images:** Davies and Starr (clb). **iStockphoto.com:** Ranasu (r). **59 iStockphoto.com:** Wabeno (r). **60–61 Science Photo Library:** Dr. Keith Wheeler (b); Edward Kinsman (t). **61 Science Photo Library:** Dr. Keith Wheeler (crb); Edward Kinsman (cra). **62 Alamy Stock Photo:** Interfoto / Fine Arts. **63 Alamy Stock Photo:** The Print Collector / Heritage Image Partnership Ltd (cl, clb). **64–65 Getty Images:** Doug Wilson. **66 123RF.com:** Nick Velichko (c). **Alamy Stock Photo:** blickwinkel (tc); Joe Blossom (tr). **iStockphoto.com:** Westhoff (br). **67 Dorling Kindersley:** Mark Winwood / RHS Wisley (bl). **© Mary Jo Hoffman:** (r). **68 Alamy Stock Photo:** Christina Rollo (bl). **68–69 © Aaron Reed Photography, LLC. 71 Getty Images:** Nichola Sarah (tr). **74 © Board of Trustees of the Royal Botanic Gardens, Kew:** (bl). **76 Denisse, E., Flore d'Amérique, t. 82 (1843–46):** (bc). **80–81 Getty Images:** Gretchen Krupa / FOAP. **82–83 iStockphoto.com:** Cineuno. **83 Science Photo Library:** Michael Abbey (bc). **84–85 Don Whitebread Photography. 85 Alamy Stock Photo:** Jurate Buiviene (cb). **86–87 Getty Images:** Peter Dazeley / The Image Ban. **88 FLPA:** Mark Moffett / Minden Pictures (bl). **89 © Board of Trustees of the Royal Botanic Gardens, Kew. 90–91 Alamy Stock Photo:** Juergen Ritterbach. **92 Bridgeman Images:** *Borassus flabelliformis* (Palmaira tree) illustration from *Plants of the Coromandel Coast*, 1795 (colored engraving), Roxburgh, William (fl.1795) / Private Collection / Photo © Bonhams, London, UK (tl). **© Board of Trustees of the Royal Botanic Gardens, Kew:** (crb). **93 © Board of Trustees of the Royal Botanic Gardens, Kew. 94–95 Alamy Stock Photo:** Arco Images GmbH. **95 Alamy Stock Photo:** Alex Ramsay (br). **100 123RF.com:** Curiousotter (bl). **100–101 © Colin Stouffer Photography. 103 Alamy Stock Photo:** FL Historical M (tr). **104–105 © Board of Trustees of the Royal Botanic Gardens, Kew. 106–107 Ryusuke Komori. 106 © Mary Jo Hoffman:** (bc). **108–109 iStockphoto.com:** EnginKorkmaz. **111 Alamy Stock Photo:** Avalon / Photoshot License (tr). **112 Science Photo Library:** Eye of Science (cl). **112–113 Damien Walmsley. 115 123RF.com:** Amnuay Jamsri (tl); Mark Wiens (tc). **Dreamstime.com:** Anna Kucherova (cr); Somkid Manowong (tr); Phanuwatn (cl); Poopiaw345 (bl); Yekophotostudio (br). **116–117 Alamy Stock Photo:** Granger, NYC. / Granger Historical Picture Archive. **117 Alamy Stock Photo:** Chronicle (cb). **Getty Images:** DEA / G. Cigolini / De Agostini Picture Library (tc). **118 Alamy Stock Photo:** Richard Griffin (cr); Alfio Scisetti (cl). **119 Alamy Stock Photo:** Richard Griffin. **120 © Board of Trustees of the Royal Botanic Gardens, Kew:** (cla). **124 © Pandora Sellars / From the Shirley Sherwood Collection with kind permission:** (tl). **© The Estate of Rory McEwen:** (crb). **125 © The Estate of Rory McEwen. 130–131 iStockphoto.com:** lucentius. **131 123RF.com:** gzaf (tc). **Alamy Stock Photo:** Christian Hütter / imageBROKER (tr). **iStockphoto.com:** lucentius (cla). **132 Alamy Stock Photo:** Ron Rovtar / Ron Rovtar Photography (cl). **133 Alamy Stock Photo:** Leonid Nyshko (cl); Bildagentur-online / Mc-Photo-BLW / Schroeer (tr). **Dorling Kindersley:** Gary Ombler: Center for Wildlife Gardening / London Wildlife Trust (bl). **iStockphoto.com:** ChristineCBrooks (br); joakimbkk (cr). **134 FLPA:** Ingo Arndt / Minden Pictures (br). **135 iStockphoto.com:** Enviromantic (cr). **136 Dorling Kindersley:** Batsford Garden Center and Arboretum (bc). **Dreamstime.com:** Paulpaladin (cl). **137 Dorling Kindersley:** Batsford Garden Center and Arboretum (cl). **iStockphoto.com:** ByMPhotos (c); joakimbkk (cr). **138 Alamy Stock Photo:** Val Duncan / Kenebec Images (c). **Dorling Kindersley:** Center for Wildlife Gardening / London Wildlife Trust (tr). **© Board of Trustees of the Royal Botanic Gardens, Kew:** (cl). **138–139 Dorling Kindersley:** Center for Wildlife Gardening / London Wildlife Trust. **140 Bridgeman Images:** *Rubus sylvestris* / Natural History Museum, London, UK. **141 Getty Images:** *Florilegius* / SSPL (tl, cr). **143 Getty Images:** Nigel Cattlin / Visuals Unlimited, Inc. (tl). **144 123RF.com:** Sangsak Aeiddam. **145 Science Photo Library:** Eye of Science (cra). **146 Alamy Stock Photo:** Daniel Meissner (bl). **146–147 © Anette Mossbacher Landscape & Wildlife Photographer. 148 Science Photo Library:** Eye Of Science (clb). **148–149 © Mary Jo Hoffman. 150 123RF.com:** Nadezhda Andriiakhina (bc). **Alamy Stock Photo:** allotment boy 1 (cl); Anjo Kan (cr). **iStockphoto.com:** NNehring (br). **Jenny Wilson / flickr.com/photos/ jenthelibrarian/:** (c). **151 Dreamstime. com:** Sirichai Seelanan (bl). **152–153 © Aaron Reed Photography, LLC. 153 iStockphoto. com:** xie2001 (tr). **154 Science Photo Library:** John Durham (bl). **154–155 © Mary Jo Hoffman. 156 Science Photo Library:** Power and Syred (tc). **158 123RF.com:** Grobler du Preez (bl). **158–159 Dallas Reed. 162–163 iStockphoto.com:** DNY59. **164–165 © Josef Hoflehner. 165 Benoît Henry:** (br). **172–173 Getty Images:** Ralph DeseniÃŸ. **175 Dreamstime.com:** Arkadyr (cr). **176 © Board of Trustees of the Royal Botanic Gardens, Kew:** (cla). **178 123RF. com:** ncristian (bc); Nico Smit (c). **Alamy Stock Photo:** Zoonar GmbH (cl). **Dreamstime.com:** Indigolotos (br). **179 Getty Images:** Alex Bramwell / Moment. **182–183 iStockphoto. com:** Ilbusca. **186 Getty Images:** bauhaus1000 / DigitalVision Vectors (bc). **188 123RF.com:** godrick (cl); westhimal (tc); studio306 (bl). **Alamy Stock Photo:** Lynda Schemansky / age fotostock (br); Zoonar GmbH (cr). **Dorling Kindersley:** Gary Ombler: Center for Wildlife Gardening / London Wildlife Trust (c). **189 123RF.com:** Anton Burakov (tl); Stephen Goodwin (tc, br); Boonchuay Iamsumang (tr); Oleksandr Kostiuchenko (cl); Pauliene Wessel (c); shihina (bc). **Getty Images:** joSon / Iconica (bl). **190 Alamy Stock Photo:** Paul Fearn (cla). **193 Dorling Kindersley:** Gary Ombler: Center for Wildlife Gardening / London Wildlife Trust (bl). **194–195 Bridgeman Images:** Blossoms, one of twelve leaves inscribed with a poem from an Album of Fruit and Flowers (ink and color on paper), Chen Hongshou (1768–1821) / Private Collection / Photo © Christie's Images. **195 Alamy Stock Photo:** Artokoloro Quint Lox Limited (cb). **Mary Evans Picture Library:** © Ashmolean Museum (tr). **198 Science Photo Library:** AMI Images (tl); Power and Syred (tc, tr); Steve Gschmeissner (cla, ca, cra, cl, c); Eye of Science (cr). **198–199 Alamy Stock Photo:** Susan E. Degginger. **202–203 Dorling Kindersley:** Center for Wildlife Gardening / London Wildlife Trust. **204 © Board of Trustees of the Royal Botanic Gardens, Kew. 205 Getty Images:** Katsushika Hokusai (cl). **Rijksmuseum Foundation, Amsterdam:** Gift of the Rijksmuseum Foundation (tr). **206 Alamy Stock Photo:** Wolstenholme Images. **207 Alamy Stock Photo:** Rex May (cla). **Getty Images:** Ron Evans / Photolibrary (cl). **208 © Board of Trustees of the Royal Botanic Gardens, Kew:** (br). **210–211 © Douglas Goldman, 2010. 211 Getty Images:** Karen Bleier / AFP (br). **212 Alamy Stock Photo:** Nature Picture Library (tl). **213 SuperStock:** Minden Pictures / Jan Vermeer. **214–215 Dorling Kindersley:** Gary Ombler: Center for Wildlife Gardening / London Wildlife Trust. **216 123RF.com:** Aleksandr Volkov (bl). **Getty Images:** Margaret Rowe / Photolibrary (tr). **iStockphoto.com:** narcisa (cr); winarm (c); Zeffss1 (cl). **217 Alamy Stock Photo:** Tamara Kulikova (tl); Timo Viitanen (c). **Dreamstime.com:** Kazakovmaksim (br). **Getty Images:** Frank Krahmer / Photographer's Choice RF (tc). **iStockphoto.com:** AntiMartina (bc); cjaphoto (cr). **220–221 Dreamstime.com:** Es75. **221 Getty Images:** Sonia Hunt / Photolibrary (br). **223 FLPA:** Minden Pictures / Ingo Arndt (crb). **227 FLPA:** Minden Pictures / Mark Moffett (tl). **228 Getty Images:** Jacky Parker Photography / Moment Open (tl). **230–231 Getty Images:** Chris Hellier / Corbis Documentary. **233 Alamy Stock Photo:** ST-images (tc). **234–235 Dorling Kindersley:** Gary Ombler: Center for Wildlife Gardening / London Wildlife Trust. **235 Alamy Stock Photo:** dpa picture alliance (tr). **236 Bridgeman Images:** Jimson Weed, 1936–37 (oil on linen), O'Keeffe, Georgia (1887–1986) / Indianapolis Museum of Art at Newfields, USA / Gift of Eli Lilly and Company / © Georgia O'Keeffe Museum / © DACS 2018. **237 Alamy Stock Photo:** Fine Art Images / Heritage Image Partnership Ltd / © 2018 The Andy Warhol Foundation for the Visual Arts, Inc. / Licensed by DACS, London. / © DACS 2018 (cb). **238 Alamy Stock Photo:** Neil Hardwick (cl). **© Board of Trustees of the Royal Botanic Gardens, Kew:** (bc). **242 Getty Images:** Florilegius / SSPL (cla). **243 Alamy Stock Photo:** Dave Watts (cra). **244 FLPA:** Photo Researchers (bl). **244–245 Image courtesy Ronnie Yeo. 246–247 Getty Images:** I love Photo and Apple. / Moment. **247 Alamy Stock Photo:** Fir Mamat (tr). **250 Science Photo Library:** Cordelia Molloy (bc). **252 123RF.com:** Jean-Paul Chassenet (bc). **252–253 Craig P. Burrows. 254 Alamy Stock Photo:** WILDLIFE

GmbH (tc). **257 © Board of Trustees of the Royal Botanic Gardens, Kew:** (br). **260–261 iStockphoto.com:** mazzzur (l). **261 Getty Images:** Paul Kennedy / Lonely Planet Images (tr). **263 Johanneke Kroesbergen-Kamps:** (tr). **265 Dreamstime.com:** Junko Barker (crb). **268–269 Alamy Stock Photo:** Alan Morgan Photography. **269 Alamy Stock Photo:** REDA &CO srl (br). **270 Bridgeman Images:** *Rosa centifolia*, Redouté, Pierre-Joseph (1759–1840) / Lindley Library, RHS, London, UK. **271 Alamy Stock Photo:** ART Collection (clb); The Natural History Museum (tr). **274–275 Dorling Kindersley:** Center for Wildlife Gardening / London Wildlife Trust. **275 Dorling Kindersley:** Center for Wildlife Gardening / London Wildlife Trust (tr). **280 Smithsonian American Art Museum:** Mary Vaux Walcott, White Dawnrose (*Pachyloplus marginatus*), n.d., watercolor on paper, Gift of the artist, 1970.355.724. **281 Bridgeman Images:** Roses on a Wall, 1877 (oil on canvas), Lambdin, George Cochran (1830–96) / Detroit Institute of Arts, USA / Founders Society purchase, Beatrice W. Rogers fund (cr). **284–285 iStockphoto.com:** Ilbusca. **289 123RF.com:** Valentina Razumova (br). **290–291 Dreamstime.com:** Tomas Pavlasek. **291 Dreamstime.com:** Reza Ebrahimi (crb). **292–293 Getty Images:** Mandy Disher Photography / Moment. **294 Gerald D. Carr:** (bl). **296 Alamy Stock Photo:** Picade LLC (tl). **296–297 Alamy Stock Photo:** Hercules Milas. **298–299 Dorling Kindersley:** Center for Wildlife Gardening / London Wildlife Trust. **300 123RF.com:** Vadym Kurgak (cl). **Alamy Stock Photo:** Picture Partners (tl); WILDLIFE GmbH (c/Wych Elm). **iStockphoto.com:** anna1311 (c/Achene); photomaru (tc/Hesperidium); csundahl (tc/Pepo). **301 123RF.com:** Dia Karanouh (c). **Alamy Stock Photo:** deefish (c/Legume); John Kellerman (tl); Jurij Kachkovskij (tc/Peach); Shullye Serhiy / Zoonar (tr). **Dorling Kindersley:** Center for Wildlife Gardening / London Wildlife Trust (bc). **iStockphoto.com:** anna1311 (tc/Aggregate); ziprashantzi (br). **302 123RF.com:** Saiyood Srikamon (bl). **302–303 Julie Scott Photography. 304 Getty Images:** Ed Reschke / Photolibrary (cl). **310 Bridgeman Images:** Various apples (with blossom), 1737–45 (engraved and mezzotint plate, printed in colors and finished by), Ehret, Georg Dionysius (1710–70) (after) / Private Collection / Photo © Christie's Images. **311 Alamy Stock Photo:** The Natural History Museum. **312 © Board of Trustees of the Royal Botanic Gardens, Kew:** (cla). **313 Dreamstime.com:** Artem Podobedov / Kiorio (tl). **316–317 Jeonsango / Jeon Sang O. 319 Science Photo Library:** Steve Gschmeissner (br). **322–323 Getty Images:** Don Johnston / All Canada Photos. **322 © Mary Jo Hoffman:** (bc). **326–327 Alamy Stock Photo:** Blickwinkel. **327 Alamy Stock Photo:** Krystyna Szulecka (cr); Duncan Usher (tr). **328–329 © Michael Huber. 333 iStockphoto.com:** heibaihui (br). **334 Bridgeman Images:** Papaya: *Carica papaya*, from Berthe Hoola van Nooten's *Fleurs, Fruits et Feuillages* / Royal Botanical Gardens, Kew, London, UK (tr). **© Board of Trustees of the Royal Botanic Gardens, Kew:** (cl). **335 Science & Society Picture Library:** The Board of Trustees of the Royal Botanic Gardens, Kew. **339 Getty Images:** André De Kesel / Moment Open (tc). **340 Getty Images:** Ed Reschke / The Image Bank (tl). **342–343 Getty Images:** Peter Lilja / The Image Bank. **343 Alamy Stock Photo:** Zoonar GmbH (cb). **344–345 iStockphoto.com:** Nastasic. **347 Alamy Stock Photo:** John Swithinbank / Garden World Images Ltd (bl). **© Board of Trustees of the Royal Botanic Gardens, Kew:** (tr). **348 123RF.com:** Eyewave (bl); Валентин Косилов (bc/*Anthriscus sylvestris*); Olga Ionina (bc). **© Board of Trustees of the Royal Botanic Gardens, Kew:** (tr). **349 Alamy Stock Photo:** Bluered / REDA &CO srl (bl); Manfred Ruckszio (bc). **© Board of Trustees of the Royal Botanic Gardens, Kew:** (tr). **350 123RF.com:** Skdesign (bl); Hans Wrang (br). **© Board of Trustees of the Royal Botanic Gardens, Kew:** (tl). **351 123RF.com:** Danciaba (bc/Clivia Miniata); Hancess (bl). **Alamy Stock Photo:** Rex May (bc). **© Board of Trustees of the Royal Botanic Gardens, Kew:** (tr). **352 Alamy Stock Photo:** Anna Yu (bl). **© Board of Trustees of the Royal Botanic Gardens, Kew:** (tr). **353 Dorling Kindersley:** RHS Wisley (br). **© Board of Trustees of the Royal Botanic Gardens, Kew:** (tl). **354 Dreamstime.com:** Iva Vagnerova (bl). **© Board of Trustees of the Royal Botanic Gardens, Kew:** (tr). **355 123RF.com:** Armando Frazão (bl); Erich Teister (bc/*Aranthera* 'Anne Black'). **Alamy Stock Photo:** UtCon Collection (tr). **Dreamstime.com:** Anatolii Lyzun (bc). **356 123RF.com:** Natasha Walton (bl). **© Board of Trustees of the Royal Botanic Gardens, Kew:** (tr). **357 Alamy Stock Photo:** Clement Philippe / Arterra Picture Library (bl). **Dorling Kindersley:** Crug Farm (br). **© Board of Trustees of the Royal Botanic Gardens, Kew:** (tl). **358 123RF.com:** Maxaltamor (bc/*Myosotis arvensis*); Praiwun Thungsarn (bl); Krzysztof Slusarczyk (br). **Alamy Stock Photo:** Antoni Agelet / Biosphoto / Photononstop (bc). **© Board of Trustees of the Royal Botanic Gardens, Kew:** (tl). **359 123RF.com:** Tamara Kulikova (bl); Scimmery (bc); Achim Prill (br). **© Board of Trustees of the Royal Botanic Gardens, Kew:** (tr). **360 Alamy Stock Photo:** Komkrit Tonusin (bl). **© Board of Trustees of the Royal Botanic Gardens, Kew:** (tr). **361 123RF.com:** Lindasj2 (bc); Александр Соколенко (bl). **© Board of Trustees of the Royal Botanic Gardens, Kew:** (tl). **362 123RF.com:** Susazoom (bl). **© Board of Trustees of the Royal Botanic Gardens, Kew:** (tr). **363 123RF.com:** Luca Lorenzelli (bc); Zanozaru (bl). **Dorling Kindersley:** Lindsey Stock (br). **Dreamstime.com:** Tt (bc/ Armeria maritima). **© Board of Trustees of the Royal Botanic Gardens, Kew:** (tr). **364 123RF.com:** Maksim Kazakov (bc); Jane McLoughlin (bl); Kiya (br). **© Board of Trustees of the Royal Botanic Gardens, Kew:** (tl). **365 123RF.com:** Lubov Vislyaeva (bl). **Alamy Stock Photo:** Shapencolour (br). **Dorling Kindersley:** Marle Place Gardens and Gallery, Brenchley, Kent (bc). **© Board of Trustees of the Royal Botanic Gardens, Kew:** (tr). **366 123RF.com:** Grigory bruev (bc/ Begonia Rex); Zigzagmtart (bl); Siamphotos (bc). **Dorling Kindersley:** RHS Hampton Court Flower Show 2012 (br). **© Board of Trustees of the Royal Botanic Gardens, Kew:** (tl). **367 123RF.com:** Phirakon Jaisangat (bl); Nataliia Zhekova (bc/Citrullus lanatus); Zdenek Sasek (bc). **© Board of Trustees of the Royal Botanic Gardens, Kew:** (tr). **Dreamstime.com:** Apichart Teapakdee (br). **368 123RF.com:** Robert Biedermann (bc); Westhimal (bl). **© Board of Trustees of the Royal Botanic Gardens, Kew:** (tr). **369 123RF.com:** Nikolay Kurzenko (bc); Lianem (bl). **© Board of Trustees of the Royal Botanic Gardens, Kew:** (tr). **370 123RF.com:** Marjorie A. Bull (br). **Alamy Stock Photo:** dpa picture alliance (bc); Florapix (bl). **© Board of Trustees of the Royal Botanic Gardens, Kew:** (tl). **371 Alamy Stock Photo:** Dave Marsden (br). **Dorling Kindersley:** Blooms of Bressingham (bl); RHS Wisley (bc). **© Board of Trustees of the Royal Botanic Gardens, Kew:** (tr). **372 123RF.com:** Pstedrak (bl). **Alamy Stock Photo:** Portforlio / Hd Signature Co.,Ltd (bc). **Dorling Kindersley:** RHS Wisley (br). **© Board of Trustees of the Royal Botanic Gardens, Kew:** (tr). **373 123RF.com:** Teerapat Pattanasoponpong (bc). **Alamy Stock Photo:** Wildlife GmbH (bl). **© Board of Trustees of the Royal Botanic Gardens, Kew:** (tr). **374 123RF.com:** Designpics (br). **© Board of Trustees of the Royal Botanic Gardens, Kew:** (tr). **375 123RF.com:** Bos11 (bc). **Alamy Stock Photo:** Florapix (bc/*Lithocarpus dealbatus*); Tim Gainey (bl). **© Board of Trustees of the Royal Botanic Gardens, Kew:** (tl). **376 123RF.com:** Andrey Shupilo (bl). **Alamy Stock Photo:** Blickwinkel / Hecker (br); McPhoto / Kraus (bc). **© Board of Trustees of the Royal Botanic Gardens, Kew:** (tr). **377 123RF.com:** Amnat Nualnuch (br). **Getty Images:** Dea / G. Cigolini / Veneranda Biblioteca Ambrosiana (tr). **378 123RF.com:** Arthit Buarapa (bc); Prachaya Kannika (bl). **© Board of Trustees of the Royal Botanic Gardens, Kew:** (tl). **379 123RF.com:** Maxaltamor (bl); Erich Teister (bc). **Alamy Stock Photo:** Paul Fearn (tr). **380 123RF.com:** Mansum007 (bl); Doug Schnurr (bc). **Alamy Stock Photo:** Paul Fearn (tr). **381 Dreamstime.com:** Prakobphoto (bl). **© Board of Trustees of the Royal Botanic Gardens, Kew:** (tl). **382 © Board of Trustees of the Royal Botanic Gardens, Kew:** (tr). **383 123RF.com:** Dovicsin András (bl). **Alamy Stock Photo:** Florilegius (tr). **384 Alamy Stock Photo:** Florapix (bc). **© Board of Trustees of the Royal Botanic Gardens, Kew:** (tl). **385 123RF.com:** Taras Verkhovynets (bl). **Alamy Stock Photo:** Wildlife GmbH (bc). **© Board of Trustees of the Royal Botanic Gardens, Kew:** (tr). **386 Alamy Stock Photo:** Willfried Gredler / INSADCO Photography (bl); Geoff Smith (bc). **© Board of Trustees of the Royal Botanic Gardens, Kew:** (tr). **387 123RF.com:** Nuttapong Wannavijid (br). **Alamy Stock Photo:** Florilegius (tr); Portforlio / Hd Signature Co. Ltd (bl); Nico Hermann / Westend61 GmbH (bc). **388 123RF.com:** Teresina Goia (bl). **© Board of Trustees of the Royal Botanic Gardens, Kew:** (tl). **389 Dorling Kindersley:** RHS Wisley (br). **© Board of Trustees of the Royal Botanic Gardens, Kew:** (tr). **390 123RF.com:** Kampwit (bc/Manihot Esculenta); Nicoayut (bl); Pierivb (bc); Thattep Youbanpot (br). **Bridgeman Images:** Euphorbia Peplus Petty Spurge / Private Collection / Photo © Liszt Collection (tr). **391 Alamy Stock Photo:** Chronicle (tr); Ingo Schulz / Imagebroker (br); FloralImages (bc). **Dorling Kindersley:** RHS Wisley (bl). **392 Alamy Stock Photo:** GWI / Richard McDowell (bl); Martin Siepmann / Westend61 GmbH (bc). **Dorling Kindersley:** RHS Wisley (br). **© Board of Trustees of the Royal Botanic Gardens, Kew:** (tr). **393 123RF.com:** Feathercollector (bl); Pauliene Wessel (bc). **Alamy Stock Photo:** Frank Teigler / Premium Stock Photography GmbH (bc/Tilia Cordata). **Dreamstime.com:** Digitalpress (br). **© Board of Trustees of the Royal Botanic Gardens, Kew:** (tr). **394 123RF.com:** Kariphoto (bc); Roseov (bl). **© Board of Trustees of the Royal Botanic Gardens, Kew:** (tr). **395 123RF.com:** Grigorii Pisotckii (bl); Marat Roytman (bc). **© Board of Trustees of the Royal Botanic Gardens, Kew:** (tr). **396 123RF.com:** Vladimíra Pufflerová (bl). **© Board of Trustees of the Royal Botanic Gardens, Kew:** (tr). **397 123RF.com:** Ninglu (bl); Subin pumsom (bc/Averrhoa Carambola). **Alamy Stock Photo:** Premaphotos (bc). **© Board of Trustees of the Royal Botanic Gardens, Kew:** (tl). **398 Dorling Kindersley:** RHS Tatton Park (bl). **© Board of Trustees of the Royal Botanic Gardens, Kew:** (tl). **399 123RF.com:** Vvoennyy (bl). **Alamy Stock Photo:** Fir Mamat (bc). **© Board of Trustees of the Royal Botanic Gardens, Kew:** (tr). **400 123RF.com:** Nicolette Wollentin (bl). **Dorling Kindersley:** RHS Tatton Park (br). **© Board of Trustees of the Royal Botanic Gardens, Kew:** (tr). **401 123RF.com:** Aleksandr Prokopenko (bl). **© Board of Trustees of the Royal Botanic Gardens, Kew:** (tr). **402 123RF.com:** Ihor Bondarenko (bl); Olga Kurguzova (bc). **© Board of Trustees of the Royal Botanic Gardens, Kew:** (tl). **403 123RF.com:** Lalalulustock (bc); Sinngern Sooksompong (br). **Dorling Kindersley:** Alan Buckingham (bl). **© Board of Trustees of the Royal Botanic Gardens, Kew:** (tr). **404 123RF.com:** Elena Yakimushkina (bl). **Alamy Stock Photo:** Hilary Morgan (bc). **© Board of Trustees of the Royal Botanic Gardens, Kew:** (tr). **405 123RF.com:** Grigory bruev (bc); Maksim Shebeko (bl). **Alamy Stock Photo:** Urs Hauenstein (br). **© Board of Trustees of the Royal Botanic Gardens, Kew:** (tr). **406 123RF.com:** Josef Muellek (bl); Hans Wrang (bc). **Alamy Stock Photo:** Michele Falzone (bc/Pistacia Vera). **iStockphoto.com:** S1llu (br). **© Board of Trustees of the Royal Botanic Gardens, Kew:** (tr). **407 123RF.com:** Balipadma (bl); Nucharee Sornsuwit (bc). **© Board of Trustees of the Royal Botanic Gardens, Kew:** (tl). **408 123RF.com:** Joemat (bc); Le Do (bl). **Dorling Kindersley:** RHS Wisley (br). **© Board of Trustees of the Royal Botanic Gardens, Kew:** (tr). **409 Alamy Stock Photo:** FLPA (bl). **Dorling Kindersley:** Crug Farm (bc); RHS Wisley (br). **© Board of Trustees of the Royal Botanic Gardens, Kew:** (tr). **410 123RF.com:** Sheryl Caston (bc); Mironovm (br). **© Board of Trustees of the Royal Botanic Gardens, Kew:** (tl). **411 123RF.com:** Steffstarr (bc/*Datura stramonium*); Lubov Vislyaeva (bl); Jiri Vaclavek (bc). **© Board of Trustees of the Royal Botanic Gardens, Kew:** (tr). **412 123RF.com:** Bundit Singhakul (bl). **© Board of Trustees of the Royal Botanic Gardens, Kew:** (tr). **413 Alamy Stock Photo:** Manfred Bail / Imagebroker (bl). **Dorling Kindersley:** RHS Wisley (bc, br). **© Board of Trustees of the Royal Botanic Gardens, Kew:** (tr). **414 123RF.com:** Bos11 (bl); Maurizio Polese (bc); Simona Pavan (br). **© Board of Trustees of the Royal Botanic Gardens, Kew:** (tr). **415 Dreamstime.com:** Ruud Morijn / Rmorijn (br). **© Board of Trustees of the Royal Botanic Gardens, Kew:** (tl). **416 123RF.com:** Mauro Rodrigues (bl). **Dreamstime.com:** Steven Jones (bc). **Getty Images:** Ruskpp / iStock / Getty Images Plus (tl). **417 123RF.com:** Umberto Leporini (bc); Evgeniy Muhortov (bl). **© Board of Trustees of the Royal Botanic Gardens, Kew:** (tr). **418 123RF.com:** Jamras Lamyai (bl); Wichean Pornpongswat (bc/Platycerium Hillii); Komkrit Preechachanwate (br). **Alamy Stock Photo:** Blickwinkel / Pinkannjoh (bc). **© Board of Trustees of the Royal Botanic Gardens, Kew:** (tr). **419 123RF.com:** Mansum007 (bl); Иван Ульяновский (bc). **Alamy Stock Photo:** Don Johnston_PL (br). **© Board of Trustees of the Royal Botanic Gardens, Kew:** (tr).

All other images © Dorling Kindersley
For further information see:
www.dkimages.com